ENVIRONMENTAL RESEARCH ADVANCES

RIPARIAN ZONES

CHARACTERISTICS, MANAGEMENT PRACTICES AND ECOLOGICAL IMPACTS

ENVIRONMENTAL RESEARCH ADVANCES

Additional books in this series can be found on Nova's website under the Series tab.

Additional e-books in this series can be found on Nova's website under the e-book tab.

ENVIRONMENTAL RESEARCH ADVANCES

RIPARIAN ZONES

CHARACTERISTICS, MANAGEMENT PRACTICES AND ECOLOGICAL IMPACTS

OLEG S. POKROVSKY
EDITOR
CNRS, Toulouse, France
BIO-GEO-CLIM Laboratory, Tomsk State University, Tomsk, Russia
Institute of Ecological Problems of the North, RAN, Arkhangelsk, Russia

nova publishers
New York

Library of Congress Cataloging-in-Publication Data

Names: Pokrovsky, Oleg S., editor.
Title: Riparian zones : characteristics, management practices, and ecological impacts / editor, Oleg S. Pokrovsky (Research Director at the CNRS, Geoscience and Environment Toulouse, France).
Description: Hauppauge, New York : Nova Science Publishers, 2016. | Series: Environmental research advances | Includes index.
Identifiers: LCCN 2016000066 (print) | LCCN 2016001465 (ebook) | ISBN 9781634846134 (hardcover) | ISBN 9781634846363 ()
Subjects: LCSH: Riparian ecology.
Classification: LCC QH541.5.R52 R584 2016 (print) | LCC QH541.5.R52 (ebook) | DDC 577.68--dc23
LC record available at http://lccn.loc.gov/2016000066

Published by Nova Science Publishers, Inc. † New York

CONTENTS

ACKNOWLEDGMENT

"Support from the mega-grant BIO-GEO-CLIM No 14.B25.31.0001 of Russian Ministry of Science and Education and Tomsk State University is acknowledged."

INTRODUCTION

O. S. Pokrovsky
GET UMR 5563 CNRS, University of Toulouse, France
and BIO-GEO-CLIM Laboratory, Tomsk State University, Russia

Riparian ecosystems occur in semi-terristrial areas adjacent to water bodies and influenced by freshwaters (Naiman et al., 2005). Specifically, Riparian Wetlands are defined as land areas adjacent to perennial, intermittent, and ephemeral streams, lakes or rivers. According to Reddy and DeLaune (2008) riparian areas receive water from groundwater discharge, overland and shallow subsurface flow from adjacent uplands, and flow from adjacent surface water body. As a result, these areas have high water tables and periodic flooding. These areas support a wide range of wetland vegetation including emergent macrophytes, grasses and trees (Reddy and DeLaune, 2008).

Riparian zones and wetlands are among the most vulnerable natural ecosystems to both climate change and human impact and they likely to represent important hot spots for climate change adaptation (Capon et al., 2013). The riparian ecosystems, located at the interface between water and land, are extremely dynamic environments in terms of structure, function and diversity and strength of abiotic-biotic feedbacks (Naiman and Décamps, 1997; Corenblit et al., 2007, 2015).

Despite significant rise of interest to the riparian zones over the past decade, these ecosystems are quite old from geological time scale since: according to Corenblit et al. (2015), the riparian vegetation existed on Earth surface since the middle Ordovician (450 Ma) and has been a significant controlling factor on river geomorphology since the Late Silurian (420 Ma). Nowadays, the riparian wetlands are strongly affected by both global climate change (Tockner and Stanford, 2002, Nilsson et al., 2012; Catford et al., 2013) and human activity (Naiman and Decamps, 1997; Richardson et al., 2007; Strayer and Findlay, 2010), being at the ecological threshold, a point at which ecological process and parameters change abruptly in response to relatively small changes in a driving force (Larsen and Alp, 2015).

For these reasons, we observed a steady and even abrupt increase of scientific publications linked to riparian problematics. A Web of Science search with "riparian" as topic yielded more than 16,000 papers published over 1950-2015 but 7,000 of them are published over past 5 years and 10,500 were produced during preceding 60 years. This rise of scientific interest is however, strongly biased geographically. Among all scientific papers dealing with riparian zones over past 65 years, 1629 are based on the data collected on rivers of USA, 630 of Brazil, 466 of China, 412 of Africa, 365 of Europe and only 24 of Russia. A more specific search of "riparian" term in the titles of 4,773 scientific publications (1950-2015) yields even higher geographical bias with 522, 166, 155, 136, 98, 85 papers studying all aspects of riparian zone in USA, Brazil, China, Canada, Africa and Europe, respectively, and only 1(!) in Russia. This book is intended to partially filling this gap by presenting 9 chapters describing the studies of riparian and flood plain zone of Russia.

The first chapter (R. Gatti) presents very general but comprehensive view of riparian zones and wetlands from ecological, botanical and geographical perspective. Ecotones and connectivity are considered as interrelated structural and functional attributes of lotic ecosystems. The concept of the ecotones is developed for alluvial rivers, the floodplain forms a complex gradient (coarse resolution ecotone) between the river channel and the uplands. These ecotones within the larger floodplains influence processes (e.g., nitrogen fixation) and structure species richness patterns. The author describes large variety of ecotones occurs within floodplains.

Flooding and channel migration maintain a diversity of lotic, semi-lotic and lentic water bodies on the floodplain and create a diverse mosaic of riparian vegetation across the riverine landscape. Floodplain water bodies include the main channel and side arms, dead arms connected to the main channel at their downstream ends, abandoned braids, abandoned meander bends, alluvial spring brooks, entering tributaries, swamps and marshes. Therefore, each type of water body exhibits a different pattern of hydrarch succession and each is characterized by a distinctive biotic community in both surface waters and ground waters. The threats to global lotic biodiversity can be grouped under seven interacting categories: overexploitation; water pollution; flow modification; destruction or degradation of habitat; invasion by exotic species; hydropower; and climate change.

Overexploitation primarily affects vertebrates, mainly fishes, reptiles and some amphibians, whereas the other threat categories have consequences for all freshwater biodiversity from microbes to megafauna. Floodplain ecosystems tend to be strongly impacted by the impoundment of river channels and other modifications that alter and generally reduce the natural hydrological variability of parent rivers. These floodplains lie along rivers with very low longitudinal slopes. Major river alterations for purposes such as navigation and hydroelectric generation have recently been proposed for each of these river systems, and such projects could affect extensive areas of floodplains through alteration of the natural flood regime.

The examples provided in the first chapter demonstrate the feasibility of using remote sensing indicators to measure natural variability in wetland extent. Coordinated field campaigns and remote sensing research can yield statistically rigorous relationships between remote sensing indicators and biophysical characteristics of wetland ecosystems. For rivers below the detection limits of remote sensing, watershed modelling within GIS software utilizes remotely sensed land cover information to derive hydrologic and suspended loading information that can be used to infer habitat quality parameters. Larger water bodies may be directly mapped, both in terms of aerial extent and water elevation.

Consistent with these recommendations, the second chapter describes a GIS approach for riparian zones. In Chapter 2, S. A. Abood presents an efficient, state-of-the-art GIS approach for mapping of riparian areas based on Riparian Buffer Delineation Model (RBDM) v3.0 and open source geospatial data for mapping variable width riparian areas and assessing riparian condition through a simple time serious analysis. The author considers riparian areas as the 3-D space for interaction that includes terrestrial and aquatic ecosystems that extend down into the groundwater, up above the canopy, outward across the floodplain, up the near-slopes that drain to the water, laterally into the terrestrial ecosystem, and along the water course at a variable width" this definition identifies riparian ecotones by its ecological functions and denies the idea of recognizing riparian ecotones by a single riparian characteristic such as vegetation, soil, or hydrologic regimes also the functional definition suggests that a riparian ecotone boundary does not occur at a fixed distance but at variable width along rivers and streams. It recognizes the dynamic and transitional nature of riparian areas by considering hydrologic, geomorphic and vegetative information. Results suggest that, by incorporating functional variable width riparian mapping into watershed management planning improves protection and restoration of valuable riparian functionality and biodiversity.

In Chapter 3, V. Londe presents his view of the impact of New Brazilian Forest Code on riparian forests. He describes the first law to protect flora of 1934 which was improved in 1965. In this chapter the reader will briefly know the evolution of the Brazilian Forest Code, the lack of application of this law and its consequences, and understand some modifications that occurred in the Code in 2012, verifying its ecological impacts over undamaged and degraded riparian forests. The reader will learn how vulnerable are the riparian forests of Brezil, in face of legislation that is aimed at satisfying the industry and short-term needs of local population rather than the environmental needs. The New Forest Code has only three years, but studies already shown that it contributes little to biodiversity conservation and provision of ecosystem services in the Atlantic Forest, and put in risk terrestrial and aquatic biodiversity and ecological functions of riparian forests as was alerted by scientists before its deployment. The reader will remain convinced that the minds have to be changed in Brazil and worldwide because we will live in a place where nature is not seen only as an area to deforest, build or plant.

Chapter 4 is devoted to Amazon floodplains, playing an important role in the organic carbon balance of the Amazon basin. Bonnet et al. focus their study on one of the largest floodplain systems of the Amazon, "Lago Grande de Curuai," which is representative of the central and lower reaches of the Amazon basin. A ten-year study of the "Várzea do Lago Grande de Curuai" floodplain provided comprehensive understanding of how hydrological and geochemical processes occurring in floodplains alter the dissolved and particulate flux of the Amazon River. In terms of internal hydrological functioning, they demonstrate that floodplain waters are influenced by direct rainfall, local runoff and seepage as well as flooding from the river with the relative importance of different inputs varying seasonally. Regarding the sediment and particulate carbon budget of the Curuai floodplain, the annual volume of sediment trapped in the floodplain is of the same order of magnitude as the mean annual sediment fluxes outflowing from the floodplain into the Amazon River, and the floodplains act as an important source of particulate and dissolved organic carbon. The reader will see, how the water passing through the floodplains undergoes important biogeochemical transformations involving biotic processes, sorption and redox reactions. Coupling the hydrological model with the database of elemental concentrations, the authors discuss the conservativity versus non-conservativity of certain elements and present the elemental mass balance between the floodplain and the main stream.

In the 5th chapter, Jiménez-Ruiz and Santin-Montanya present a study of integrated management approach to invasive weeds in riparian zones. The main causes of riparian zones degradation are the agriculture on the floodplain, large hydraulic pipes, diversions and dams, channels that have destroyed and altered their morphology, biological and hydrological function. The authors combined mechanical, physical, chemical and biological techniques in a management programme of riparian zones, taking account the ecological factors of plant communities and social context. In particular, they suggest a methodology to address the management and control of one of the most invasive riparian weeds in the world, the grass *Arundo donax* L. (Giant reed), from an integrated perspective, improving the ecological status of the riparian zones and producing the diminishing impact on the biodiversity. Their main findings should encourage further studies on the integrated management of invasive weeds in riparian zones, and environmental conditions that may influence field efficiency.

The next three chapters of the book describe biogeochemistry of the riparian zones of small rivers in the European Russia boreal region.

In the 6th chapter, Efimova and co-workers the interrelations between the floodplain and the streambed as well as factors of formation the floodplain and streambed complexes on the example of a small river of the Volga River basin, central Russia. These floodplain lakes, being a part of Kerzhentz River riparian zone, act as important regulator of DOC, nutrients and metals. The lakes are constrasting in terms of physical parameters, location on the flood plain, and chemical composition. One of the most important results of this study is that accumulative floodplain and streambed complexes are more sensitive to natural and man-caused impact and respond to any environmental changes more quickly than rock-defended complexes. Meander zones are the most sensitive to this impact. However, at present water chemistry of the floodplain lakes within the floodplain and streambed complex may be classified as apristine, and the lakes may serve as some kind of a registration mark to be used in connection with evaluation of environmental status of the water bodies within the territory surrounding the conservation area.

The 7th chapter by Ilina et al. addresses colloidal speciation and size fractionation of dissolved organic matter and trace elements in small subarctic watershed and its riparian zone. The authors show that Significant enrichement by organo-mineral colloids of the stream water occurs within the riparian zone of the stream where anoxic underground Fe^{2+}-rich waters interacting with basic rocks of the basement meet surface, well oxygenetaed waters rich in dissolved organic matter from the upper reaches of the river, located within the bog zone rich in vegetation leachates. It follows from the results of this study that autochthonous processes of organic matter fractionation, such as 1) transformation of initially allochthonous soil-derived colloids via photo- and bio-degradation or 2) new organic ligand production by plankton and peryphyton, cannot appreciably affect the distribution of trace elements among various size fractions of colloids and particles along the landscape gradient from soil water to terminal lake. The main features of colloidal chemical composition and size fractionation are therefore acquired during Fe-DOC interaction within the riparian zone.

The last chapter of the European Russian boreal zone (Avessalomova et al., chapter No 8) is devoted to barrier functions of the floodplain and their influence on the runoff formation. The landscape structure of the flood plain and riparian zone typical for the small rivers in East-European taiga (the Severnaya Dvina River baasin) are described in light of geomorphology, hydrology, vegetation and hydrochemistry. In this chapter, Avessalomova et al. analyze factors of the floodplain development, identify geochemical barrier zones and discuss the role of nutrients and dissolved river water components. In particular, the authors identify three main flowpaths of dissolved matter in the floodplain landscapes subjected to agricultural land use within the cathcments. The authors reveal geochemical mechanisms that are able to protect the floodplain landscapes and streamwater from pollutants washed out from the fields. Finally, they conclude that floodplain landscapes affect migration pathways by means of barrier zones. Barriers are responsible for regulation of geochemical connections between small catchments and the integrating stream. On their pathway toward floodplains the chemical elements can be intercepted by the deluvial trains. Within the floodplain interception functions are inherent for: (1) meadow communities with high productivity and high biogeochemical activity, (2) hydromorphous units providing deposition in soil histic horizons, (3) ecotone units at the contact of contrasting pH and/or Eh conditions of water migration.

Afterwards, we move to western Siberia, and the next 5 chapters are devoted to hydrological and biogeochemical functions of the riparian zones of two largest Siberian rivers, the Irtush River and the Ob River basin. In Chapter 9, Beisembayeva and co-authors investigate the water balance of the Irtush River and the impact of hydropower plant (HPP) construction on the water and vegetation dynamics in the riparian zone and the floodplain. The originality of this approach is that it shows how the hydrological balance of the large river controls the riparian zone and the floodplain vegetation. Note that, given significant number of lateral side-channels adjacent to the main water channel, the riparian zone per se extends much larger in the lateral direction from the main flox, compared to “classic” riparian zone profile. Through a historical sketch of HPP construction on the Irtush River, we learn the evolution of seasonal water fluxes and the degree of the impact of river flux regulation on the frequency and intensity of flooding events as well as on the dynamics of riparian and floodplain vegetation. Thus, the Irtysh flow regulation caused a change in the hydrological regime of the river, lead to a change in conditions of floodplain inundation. The result of these processes was the steppe formation and salinization of the floodplain. The authors also provide important recommendations for floodplain and riparian ecosystems maintenance.

In Chapter 10, S. Vorobyev and co-authors describe biogeochemistry of alluvial soils of the Ob River floodplain and its riparian zone. Soil studies of the riparian zone are rare, and this chapter contributes to our knowledge of the impact of riparian soils on river water flux formation. the floodplain area of the Ob from the confluence of the Biya and the Katun rivers to the Gulf of Ob is more than 60,000 km². This is an enormous flooded area of the riparian zone, which both accumulates weighed and dissolved in water substances from the whole catchment and modifies them significantly. The floodplain of the Ob in Western Siberia acts a gigantic geochemical barrier, which regulates the inflow of the substances from the catchment into the World Ocean. Soils have great importance in this barrier. However, there was little effort to study the floodplain as the geochemical barrier and alluvial soils as the main component of this barrier. The chapter 9 aims at filling this gap via providing a thorough pedological and geochemical analysis of soil cover of the flood plain and the processes of the river flux formation that occur within this territory.

Chapter 11, also by S. Vorobyev et al. deals with biogeochemistry of organic carbon, major and trace elements in the flooded and riparian zone of the Ob River. Up to present, there were very little studies of the continuum main river – its flooding zone – small inlet rivers – flood plain lakes. This chapter is filling this gap via presenting complete hydro-chemical study of the Ob River middle course, first order small tributaries, persisting and temporary flood lakes and upland lakes and the large flood zone in May and July allowed establishing first-order factors controlling dissolved organic carbon and related metal sources and sinks in this environmentally important ecosystem. Considering two contrasting seasons, spring flood and summer baseflow, allowed distinguishing the elements controlled by the groundwater influx (DIC, Na, Mg, Ca, SO_4, Sr, Mo, Sb, U) and those controlled by surface winter runoff via plant litter and topsoil leaching, notably during spring flood (Si, K, Rb, Mn, Zn, Cu). The main carrier of many insoluble trivalent and tetravalent elements and some divalent metals (Cd, Pb, Cu) is organo-ferric colloids stabilized by dissolved organic matter. The main river stream dissolved chemical composition can be approximated, within ± 30-40%, by that of the flood zone in May. Vorobyev et al. hypothesized that the main autochtonous processes controlling DOC and related TE transformation between different water bodies are phytoplankton uptake and microbial heterotrophic and photo-degradation of organo-mineral colloids, which in turn, strongly depend on water residence time. Finally, the authors conclude that these autochthonous processes will mostly affect the removal of colloidal TE from the water column of flood lakes, large and small rivers and upland lakes of the Ob' River middle course watershed under the on-going climate warming.

In the following chapter (12, Khromykh and Khromykh), the spatial structure and dynamics of the Tom River (a tributary of the Ob River) floodplain landscapes is analyzed based on GIS, digital elevation model and remote sensing. Using asuite of remote sensing, topographical and historical data, the authors show how the riparian and flood zone landscapes were evolved under serious authropogenic modification during XX-XXI centuries.

The authors conclude that, at present almost all riparian landscapes of the Tomsk River in the vicinity of Tomsk are exposed to anthropogenic modifications. The landscape is subjected to considerable desiccation due to lowering of the groundwater level, which occurred due to the overlapping of various anthropogenic factors.

Chapter 13 by D. Vorobiev et al. describes bethnic invertebrate community of the floodplain zone of the Vasyugan River, a tributary of the Ob River). The main accent here is given on occurring transformations of the biota under oil field contamination. The authors describe the distribution of groups of macrozoobenthos in different types of water bodies and habitats in the floodplain of the Vasyugan river basin system. A correlation between the abiotic factors of aquatic environment (water and temperature regimes) and quantitative indicators of benthic communities in the floodplain-river system of the lower portion of the Vasyugan River was detected. Using analysis of variance the effect of oil pollution, they revealed the quantitative characteristics of certain groups of benthic communities in the Vasyugan River floodplain-river system. Both the number and the biomass of benthic invertebrates decreased significantly in flood plain lakes due to oil field activity. Specifically, in place of mollusks, previously leading in the Vasyugan floodplain, the oligochaetes became dominant, capable of withstanding significant oil pollution. Depressed benthic communities of Vasyugan floodplain waters show a significant accumulation of oil components in sediments that suppress the development of benthic organisms. At the same time, the effect of human impact on river benthic community was much lower.

In the last, concluding chapter 14, V. Khromykh generalizes the knowledge of riparian zone and floodplain geomorphology and hydrology. In this chapter, the general characteristics of floodplain landscapes are considered. The author presents the main role of the hydrodynamic factors, erosion-accumulation activity of the river and the flood cycles. The dynamics of the groundwater level, and the role of soils in the floodplain landscape evolution is also shown. The author concludes that dynamic changes in the landscapes of the floodplain are best indicated by the vegetation cover.

REFERENCES

Capon, S. J., Chambers, L. E., Mac Nally, R., Naiman, R. J., Davies, P. et al., 2013. Riparian Ecosystems in the 21st century: Hotspots for climate change adaptation? *Ecosystems* 16, 359-381.

Catford, J. A., Naiman, R. J., Chambers, L. E., Roberts, J., Douglas, M., Davies, P., 2013. Predicting novel riparian ecosystems in a changing climate. *Ecosystems* 16, 382-400.

Corenblit, D., Tabacchi, E., Steiger, J., Gurnell, A. M., 2007. Reciprocal interactions and adjustments between fluvial landforms and vegetation dynamics in river corridors: a review of complementary approaches. *Earth-Science Rev.* 84, 56-86.

Corenblit, D., Davies, N. S., Steiger, J., Gibling, M. R., Bornette, G., 2015. Considering river structure and stability in the light of evolution: feedbacks between riparian vegetation and hydrogeomorphology. *Earth Surface Processes and landforms* 40, 189-207.

Larsen, S., Alp, M., 2015. Ecological thresholds and riparian wtlands: an overview for environmental managers. *Limnology* 16, 1-9.

Naiman, R. J., Décamps, H., 1997. The ecology of interfaces, riparian zones. *Annual Rev. Ecol. Syst.,* 28, 621-658.

Naiman, R. J., Décamps, H., McClain, M. E., 2005. *Riparia: ecology, conservation and management of streamside communities*. New York: Academic Press.

Nilsson, C., Jansson, R., Kuglerova, L., Lind, L., Ström, L., 2012. Boreal riparian vegetation under climate change. *Ecosystems* 16(3), 401-410.

Reddy, K. R., DeLaune, R. D., 2008. *Biogeochemistry of Wetlands. Science and Applications*. CRC Press, Taylor and Francis group, Boca Raton, FL, 774 pp.

Richardson, D. M., Holmes, P. M., Esler, K. J., Galatowitsch, S. M., Stromberg, J. C. et al., 2007. Riparian vegetation: degradation, alien plant invasions, and restoration prospects. *Divers Distrib.* 13, 126-139.

Strayer, D. L., Findlay, S. E. G., 2010. Ecology of freshwater shore zones. *Aquat Sci.* 72, 127-163.

Tockner, K., Stanford, J. A., 2002. Riverine flood plains: present state and future trends. *Environ. Conserv.* 29, 308-330.

In: Riparian Zones
Editor: Oleg S. Pokrovsky

ISBN: 978-1-63484-613-4

Chapter 1

BIOLOGICAL DIVERSITY AND CURRENT THREATS OF LOTIC ECOSYSTEMS

Roberto Cazzolla Gatti
Biological Institute and BIO-GEO-CLIM Centre of Excellence,
Tomsk State University, Russia

ABSTRACT

Although 100,000 out of approximately 1.75 million species described by scientists live in fresh water, knowledge of the total diversity of fresh waters is woefully incomplete, particularly among invertebrates and microbes, and especially in tropical latitudes that support most of the world's species. If trends in human demands for water remain unaltered, and species losses continue at current rates, the probability to conserve much of the remaining biodiversity in fresh water will be very low. Lotic ecosystems are increasingly impacted by multiple stressors that lead to a loss of sensitive species and an overall reduction in diversity. Global environmental change (such as climate) occurring at the global scale, together with nitrogen deposition and runoff, are superimposed upon all of the threats.

In this chapter, I conduct an extensive review of published studies that have qualitatively and quantitatively examined the species richness and the current major threats of a special component of freshwater systems: the lotic ecosystems.

Keywords: biodiversity, rivers, lotic systems, land use, climate change, riparian zones

1. INTRODUCTION

Fresh water consists of only 0.01% of the world's water and approximately 0.8% of the Earth's surface (Gleick, 1996), yet this tiny fraction of global water supports at least 100,000 species (Figure 1) out of approximately 1.8 million, which is almost the 6% of all described species (Dugeon et al. 2006). Inland waters and freshwater biodiversity constitute a valuable natural resource, in economic, cultural, aesthetic, scientific and educational terms. Although their conservation and management are critical, fresh waters are experiencing declines in biodiversity far greater than those in the most affected terrestrial ecosystems (Strayer and Dudgeon 2010). If trends in human demands for water remain unaltered and species losses continue at current rates the probability to conserve much of the remaining biodiversity in fresh water will be very low (Figure 2).

Over 10,000 fish species live in fresh water (Lundberg et al., 2000), approximately 40% of global fish diversity and 1/4 of global vertebrate species. When amphibians, aquatic reptiles (crocodiles, turtles, etc.) and mammals (otters, river dolphins, platypus, etc.) are added to this freshwater-species total, it becomes clear that as much as 1/3 of all vertebrates are confined to fresh water.

Althought 100,000 out of approximately 1.75 million of the species described by scientists live in fresh water (Hawksworth and Kalin-Arroyo, 1995), an additional 50,000 to 100,000 species may live in ground water (Gibert and Deharveng, 2002).

Knowledge of the total diversity of fresh waters is woefully incomplete, particularly among invertebrates and microbes, and especially in tropical latitudes that support most of the world's species. Even vertebrates are incompletely known, including well-studied taxa such as fishes (Stiassny, 2002). Between 1976 and 1994 an average of 309 new fish species, approximately 1% of known fishes, were formally described or resurrected from synonymy each year (Stiassny, 1999), and this trend has continued (Lundberg et al., 2000). Among amphibians, almost 35% of 5778 species has been described during the last decade (AmphibiaWeb, 2005).

Figure 1. Biodiversity of lotic ecosystems. First line: *Equisetum sp.*, catfish and river shrimp; second line: *Natrix sp.*, Hippos and green frog; third line: cormorants, *Bufo bufo*, river crab; fourth line: *Libellula sp.*; Eurasian spoonbill and *Ephemera sp.* (credits: Roberto Cazzolla Gatti©).

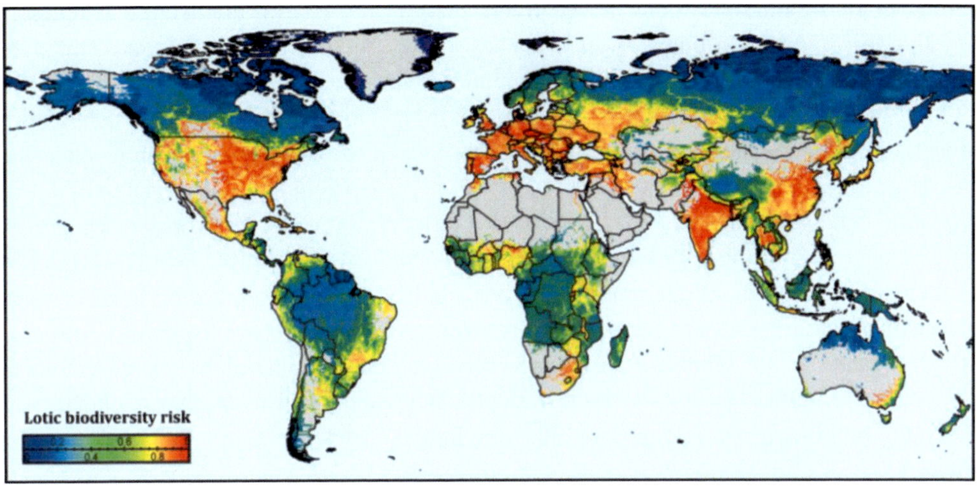

Figure 2. Biodiversity risk (from low-blue to high-red) of lotic ecosystems (adapted from Vörösmarty et al. 2010).

Adequate data on the diversity of most invertebrate groups in tropical fresh waters do not exist, but high levels of local endemism and species richness seem typical of several major groups, including mollusks, decapod crustaceans, and aquatic insects such as caddisflies and mayflies (Dudgeon, 1999, 2000; Benstead et al., 2003; Strayer et al., 2004).

Most of the prokaryote taxonomic diversity remains unexplored (Torsvik, Øvreas and Thingstad, 2002; Curtis and Sloan, 2004). It is likely that the richness of freshwater fungi and microalgae has been underestimated (Johns and Maggs, 1997; Gessner and Van Ryckegem, 2003).

In this chapter I completed an extensive review of published studies that have qualitatively and quantitatively examined the species richness and the current major threats of a special component of freshwater systems: the lotic ecosystems.

2. Diversity of Lotic Ecosystems

Among freshwater ecosystems two different types can be recognized: lentic and lotic. This latter, from the Latin *lotus*, refers to flowing waters. Lotic waters range from springs, only a few centimeters wide, to major rivers, wide kilometers in. At the opposite, lentic ecosystems include relatively still terrestrial waters such as lakes and ponds (Marsh and Fairbridge,1999).

Moreover, lotic ecosystems can be arbitrary divided in 4 categories: large rivers, temporary/small rivers, floodplain rivers and estuarine basins (Figure 3).

2.1. Large Rivers

Large river ecosystems are the zone of the Earth with the highest biological diversity and, also, of human most intense activity (Figure 4). Rivers are important habitats for a large variety of animals and plants. Fish, amphibians, birds, insects, invertebrates, and reptiles live in rivers, or find their food there. Large rivers play a vital role in connecting habitats, and their value to plants and animals extends far beyond the surface area they cover. This habitat connectivity functions both between upstream and downstream areas, and by connecting both sides of river banks (World Rivers Review, 2011). This latter evidence pushes for an approach to management that looks at the river basin as a whole, rather than an isolated water flow. Large river biodiversity is in a state of crisis, a consequence of decades of humans exploitation with large dams, water diversions and pollution. Freshwater species are even more endangered than those on land. The pressures and impacts on the world's large rivers have increased greatly in recent years, as a consequence of their exploitation to meet various human needs. Large rivers are particularly exposed to problems of multiple uses, often with conflicting aims. At the global scale, large river systems are altered by increased nutrient loads leading to eutrophication of river stretches, navigation and hydropower plants, which deteriorate ecosystem functions and further human uses, leading to other problems downstream.

Large dams harm biological diversity by flooding lands, fragmenting habitats, isolating species, interrupting the exchange of nutrients between ecosystems, and cutting off migration routes. They reduce water and sediment flows to downstream habitat, and change the nature of a river's estuary, where many of the world's fish species spawn. The impacts from dams increase the vulnerability of entire ecosystems to other threats, such as climate change (World Rivers Review, 2011).

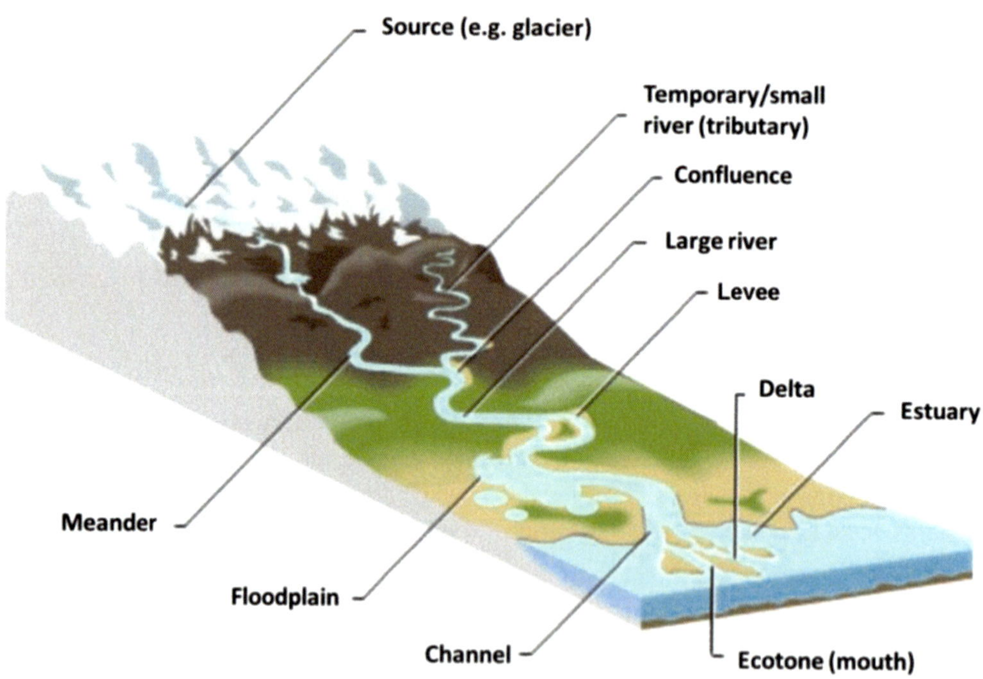

Figure 3. Diversity of the lotic systems with its different components shown.

Figure 4. An example of a large river ecosystem: the Ivindo river in Gabon (credits: Roberto Cazzolla Gatti©).

2.2. Temporary/Small Rivers

Large rivers and their riparian zones, which are considered hot spots of biodiversity (Ward et al.1999), are not the only lotic systems rich in biological diversity.

In fact, rivers that periodically cease to flow comprise a substantial proportion of the total number, length and discharge of the world's rivers (Tooth, 2000). These temporary rivers are not restricted to arid regions; they occur in most terrestrial biomes. In the next century the number and length of temporary rivers may increase in regions that experience drying trends due to climate change and to water abstraction for socio-economic uses (Larned et al., 2010). Large-scale changes in intermittence have not been considered in historical trend analyses or forecasts of future river flow patterns. However, negative trends in flow have been detected in many regions (e.g., Zhang et al., 2001; Cigizoglu, Bayazit and Onoz, 2005; Pasquini and Depetris, 2006; Milliman et al., 2008; Tockner, Uehlinger and Robinson, 2009), and interlinked climate change-runoff models predict future decreases in runoff in some mid-latitude regions (Arnell, 1999; Jones, McMahon and Bowler, 2001; Huntington, 2006; Kundzewicz et al., 2008). If these climate-driven changes will take place, increases in the occurrence and frequency of water intermittence are likely to follow.

Figure 5. A temporary river formed in a South Italy canyon during the Spring (credits: Roberto Cazzolla Gatti©).

Figure 6. A small river is an important incubator of biological diversity (credits: Roberto Cazzolla Gatti©).

Temporary rivers are important links between water stored in soils, aquifers, snowpack, glaciers, vegetation and the atmosphere (Larned, 2010).

In alpine, polar and boreal catchments, meltwater from ice and snow moves to perennial rivers and lakes through networks of temporary rivers (McKnight et al., 1999; Malard,

Tockne and Ward, 2000; Robinson and Matthaei, 2007). These networks expand during the melt season and contract during the freeze-up. In addition to their roles in the water cycle, temporary rivers provide a wide range of habitats.

2.3. Floodplain Rivers

Floodplain rivers are disturbance-dominated ecosystems (Figure 7) characterized by high levels of habitat diversity and biota adapted to exploit the spatial-temporal heterogeneity (Welcomme, 1979; Salo et al., 1986; Copp, 1989; Ward and Stanford, 1995; Petts and Amoros, 1996). The fluvial action of flooding and channel migration create a shifting mosaic of habitat patches across the riverine landscape. Ecotones, connectivity and succession play major roles in structuring the spatial-temporal heterogeneity leading to the high biodiversity that characterizes floodplain rivers (Ward, Tockner and Schiemer, 1999).

The most extensive and biologically diverse floodplain ecosystems in the world are found along lowland rivers of the humid tropics, including the Amazon, Orinoco and Paraguay rivers of South America, which largely retain natural flow regimes (Hamilton et al., 2007).

The complex floodplains of many large rivers offer a striking example of the influence of geomorphology on hydrology and, consequently, on ecosystem biodiversity. Much of this influence can be attributed to spatially variable patterns in the frequency and duration of soil saturation, and surface flooding (Winter, 2001; Hamilton, 2002). This variation is dictated by the elevation and position of fluvial landforms (i.e., fluvial geomorphology) in relation to the local water table and the annual range in river levels (Church, 2002). Thus, geomorphological patterns are fundamentally linked to biodiversity in floodplain environments (Brinson, 1993; Junk, 1997; Lewis et al., 2000; Ward et al., 2002).

Accumulation of water on floodplains can result from riverine overflow or from delayed drainage of local rainfall and runoff, and often these sources of water have distinct chemical and nutrient compositions (Hamilton et al. 2007). Riverine overflow often produces greater sediment and nutrient inputs and consequently higher biological productivity compared to areas flooded with locally derived waters (Klinge et al., 1990; Kalliola et al., 1991; Mertes, 1997), particularly in lowland rivers fed by mountainous watersheds. Riverine overflow tends to be episodic, albeit lasting for months in the largest rivers (Hamilton et al., 2002), while saturation, because of the emergence of local groundwater, may be constant and persist through the dry season. Soil saturation and surface flooding determine the species composition and relative abundance of plants and animals as well as the characteristics of soil, sediments, and detrital organic matter derived from the vegetation (i.e., theecological structure). In addition, saturation and flooding control biological productivity and rates of key ecological processes, such asdecomposition and biogeochemical transformations of elements (Hamilton et al. 2007). Biological activity, in turn, affects floodplain geomorphology and hydrology by influencing sediment accretion, soil development, and the flow paths of surface water and the movement of subsurface water (Figure 8). The spatial and temporal complexity of floodplain ecosystems makes them important components of regional biodiversity (Puhakka et al., 1992; Lewis et al., 2000). Floodplains, with permanent water bodies or long-lasting inundation, provide critical habitat for aquatic biota and are essential to maintain native riverine fisheries (Junk, 1997). Plants and animals on floodplains exhibit numerous adaptations to cope with, and benefit from, episodic or seasonal soil saturation or inundation.

Even relatively short episodes of saturation or inundation (from days to weeks) can profoundly affect the species composition of floodplain vegetation (Losos, 1995) and may be important in the life cycles of many aquatic animals (Junk, 1997). Variation in hydroperiod (the duration and temporal pattern of saturation or inundation), which can result from minor spatial variation in elevation, position, and soil composition of fluvial landforms, often produces dramatic differences in vegetation across floodplains and thereby contributes to biodiversity across the landscape (Salo et al., 1986; Hupp, 1988; Lamotte, 1990; Kalliola et al., 1991). Over long time scales, the migration of river channels produces landform gradients of varying age, soil development, and vegetation succession, and thereby enhances the biodiversity of environments of fluvial origin that are no longer subject to saturation or inundation (Puhakka et al., 1992).

Figure 7. Floodplain rivers play a fundamental role in creating temporary habitats and are key elements for local populations' supply (credits: Roberto Cazzolla Gatti©).

Figure 8. The large foodplain ecosystem created in India by the Gange shapes the landscape (credits: Roberto Cazzolla Gatti©).

2.3. Estuarine Systems

The relatively large and unpredictable variations in salinity and water level characterizing most estuarine systems tend to select particular life-forms, limiting the number of species capable of adapting to these stresses (Rey Benayas and Scheiner 1993; Kittelson and Boyd 1997). As an ecotone between fresh and marine environments (Figure 9), estuaries and estuarine wetlands are habitats for a mixture of freshwater and halophilous plant species (Holland et al. 1990; Gosz 1993). The integrity of estuarine wetlands is maintained by hydrologic and sedimentation dynamics, environmental gradients, and lateral connections between the river or the sea and the floodplain (Junk et al. 1989; Pasternack and Brush 1998; Ward 1998). Most of the large estuarine floodplain ecosystems in the world have been altered by human activities (Bravard and Petts 1993; Vitousek et al. 1997).

Biodiversity in estuarine ecosystems is generally important because of the recurrent disturbance regime caused by the variation in salinity and water level that increase environmental heterogeneity and thus diversity (Amoros and Roux 1988; Vivian-Smith 1997). At the same time, estuarine communities are known to have a relatively low taxonomic diversity (Day et al. 1989; Costanza et al. 1993) because of the high amplitude of unpredictable stresses, which select a limited set of species adapted to changing salinity, osmotic stress, and oxygen deficiency (Chabrerie et al. 2001). Because the major factors controlling biodiversity in estuaries are easily identifiable, their wetlands are interesting open-laboratories where to study how biodiversity dynamics affect ecosystem processes (Schulze and Mooney 1994).

Figure 9. The esturarine ecosystems are ecotone between fresh and marine environments and habitats for a mixture of freshwater and halophilous plant species (credits: Roberto Cazzolla Gatti©).

3. Special Features of Lotic Systems

3.1. Ecotones in the Lotic Systems

Ecotones and connectivity are interrelated structural and functional attributes of lotic ecosystems. Ecotones are transition zones between adjacent patches which, although differing from each other, exhibit high within-patch homogeneity (Ward, 1999). The ecotone concept has been a recurrent theme in ecology (Risser, 1995), with heightened interest in recent years attributable, in part, to their influence on biodiversity (Hansen and di Castri, 1992; Lachavanne and Juge, 1997). Ecotones occur over a range of scales in floodplain rivers, forming the boundaries between land and water, between surface water and ground water, and between in-stream habitat patches (Naiman and Decamps, 1990; Ward and Wiens, 1999).

In alluvial rivers, the floodplain forms a complex gradient (coarse resolution ecotone) between the river channel and the uplands, within which a variety of secondary and tertiary ecotones are embedded (Ward, 1999). These ecotones within the larger floodplains influence processes (e.g., nitrogen fixation) and structure species richness patterns (Naiman et al., 1988; Amoros et al., 1996). A variety of ecotones occur within floodplains. The littoral zone, for example, forms ecotones between the open water of floodplain lakes and the shore. These special environments also occur between different stands of floodplain vegetation (Ward, 1999). There are also vertical ecotones between surface water bodies and groundwater aquifers (Gibert et al., 1997). All of themare characterized by relatively steep gradients (e.g., thermal, chemical and organic), thereby collectively forming a high level of environmental heterogeneity across the riverine landscape.

3.2. Connectivity in the Lotic Systems

Connectivity may be defined as the ease with which organisms, matter or energy traverse the ecotones between adjacent ecological units. From a purely biological perspective, connectivity refers to gene flow between meta-populations and the extent to which ecotones alter dispersal, movement and migration (Ward, 1999).

Connectivity also refers to the extent to which nutrients, organic matter and other substances cross ecotones. Hydrological connectivity, the transfer of water between the river channel and the floodplain, and between surface and subsurface compartments, has major implications for biodiversity patterns (Welcomme, 1979; Amoros and Roux, 1988; Schiemer and Spindler, 1989; Obrdlik and Fuchs, 1991; Gibert et al., 1997; Ward, 1998). This is owing in part to the role that hydrological connectivity plays in structuring successions.

3.3. Ecological Succession in the Lotic Systems

Successional processes are responsible for much of the spatial-temporal heterogeneity of riverine floodplains (Salo et al., 1986; Amoros et al., 1987; Terborgh and Petren, 1991; Ward and Stanford, 1995; Decamps, 1996). Flooding and channel migration maintain the diversity of lotic, semi-lotic and lentic water bodies on the floodplain and create a diverse mosaic of riparian vegetation across the riverine landscape (Ward, 1999). Floodplain water bodies include the main channel and side arms (eupotamal), dead arms connected to the main channel at their downstream ends (parapotamal), abandoned braids (plesiopotamal), abandoned meander bends (palaeopotamal), alluvial spring brooks, entering tributaries, swamps and marshes (Ward, 1999). These water bodies may be arrayed along a gradient of connectivity with the main channel, from eupotamal side arms permanently connected with the main channel, to plesiopotamal habitats reconnected to the channel during the annual flood, to isolated palaeopotamal lakes that are rarely inundated by flood waters (Ward and Stanford, 1995). Therefore, each type of water body exhibits a different pattern of hydrarch succession and each is characterized by a distinctive biotic community in both surface waters (Castella et al., 1984; Copp, 1989) and ground waters (Marmonier et al., 1992).

Fluvial dynamics and channel migration also maintain a diversity of successional stages among the riparian vegetation. Salo et al. (1986) and Terborgh and Petren (1991) provide vivid descriptions of the role of natural disturbance by fluvial action in creating a mosaic of alluvial forest stands in different successional stages. Forests on the concave bends of laterally migrating rivers are undercut by erosion, and primary succession is initiated on point bars of alluvium deposited as annual increments on convex bends (Ward, 1999).

4. Threats for Lotic Ecosystems

Lotic ecosystems are increasingly impacted by multiple stressors that lead to a loss of sensitive species and an overall reduction in diversity (Meyer, 1999).

The threats to global lotic biodiversity can be grouped under seven interacting categories: overexploitation; water pollution; flow modification; destruction or degradation of habitat; invasion by exotic species; hydropower; and climate change (Arthington, 2010).

Climate change occurring at the global scale, together with nitrogen deposition, and runoff patterns (Poff, Brinson and Day, 2002, Galloway et al., 2004), are superimposed upon all of the previous threat categories.

4.1. Local Threats for Lotic Biodiversity

Overexploitation primarily affects vertebrates, mainly fishes, reptiles and some amphibians, whereas the other threat categories have consequences for all freshwater biodiversity from microbes to megafauna.

Pollution problems are pandemic, and although some industrialized countries have made considerable progress in reducing water pollution from domestic and industrial point sources, threats from excessive nutrient enrichment (Smith, 2003) and other chemicals, such as endocrine disrupters, are growing (Colburn, Dumanoski and Myers, 1996).

Table 1. Groups mostly affected by threats to lotic biodiversity. The intensity of impacts is represented by + or –

Threats to lotic biodiversity	Groups mostly affected	Intensity
Overexploitation	Plants, fishes, reptiles and some amphibians	++
Water pollution	Plants, fishes, amphibians, birds, mammals, invertebrates	+++
Flow modification	Fishes and invertebrates	++
Destruction or degradation of habitats	All groups of flora and fauna (microorganism also)	++++
Invasion by exotic species	Specialized species	+-
Hydropower	Fishes, mammals	+++

Habitat degradation is brought about by an array of interacting factors (Dudgeon et al., 2006). It may involve direct effects on the aquatic environment (such as excavation of river sand) or indirect impacts that result from changes within the drainage basin (Table 1). For example, forest clearance is usually associated with changes in surface runoff and increased river sediment loads that can lead to habitat alterations, such as shoreline erosion, smothering of littoral habitats, clogging of river bottoms, and floodplain aggradation (Dudgeon et al., 2006).

Flow modifications are ubiquitous in running waters (Dynesius and Nilsson, 1994; Vorosmarty et al., 2000; Nilsson et al., 2005). They vary in severity and type, but tend to be most aggressive in regions with highly variable flow regimes. This is because humans in these places have the greatest need for flood protection or water storage. That existing dams retain approximately 10,000 km^3 of water, the equivalent of five times the volume of all the world's rivers (Nilsson and Berggren, 2000), illustrates the global extent of human alteration of river flow. Water impoundment by dams in the Northern Hemisphere is now so great that it has caused measurable geodynamic changes in the Earth's rotation and gravitational field (Chao and Gross, 1995). Even some of the world's largest rivers now run dry for part of the year or are likely to do so as a result of large-scale water abstraction (Richter et al., 2010). Flow modifications are likely to be exacerbated by global climate change because of greater frequency of floods and droughts and, consequently, increased water-engineering responses can be anticipated (Vorosmarty et al., 2000). Impacts on river biota are likely to be severe (Dudgeon et al., 2006).

Widespread invasion and deliberate introduction of exotic species add threats to the physical and chemical impacts of humans on fresh waters, in part because exotics are most likely to successfully invade fresh waters already modified or degraded by humans (Bunn and Arthington, 2002; Koehn, 2004). There are many examples of large scale and dramatic effects of exotics on indigenous species (e.g., Nile perch, *Lates niloticus*, in Lake Victoria, the crayfish plague in Europe, salmonids in Southern Hemisphere lakes and streams; see Rahel, 2002), and impacts are projected to increase further (Sala et al., 2000).

Indirect impacts can arise from exotic terrestrial plants such as *Tamarix spp.* (Tamaricaceae), which alter the water regime of riparian soils and affect stream flows in Australia and North America (Tickner et al., 2001).

The particular vulnerability of freshwater biodiversity also reflects the fact that fresh water is a resource for humans that may be extracted, diverted, contained or contaminated in ways that compromise its value as a habitat for organisms. In some instances, impacts have been sustained over centuries and, in the case of many of the major rivers of China, they have persisted for more than 4,000 years (Dudgeon, 2000). Indeed, some authors now believe it unlikely that there remain a substantial number of water bodies that have not been irreversibly altered from their original state by human activities (Leveque and Balian, 2005).

The extent of most freshwater systems is not confined to the wetted perimeter, but includes the catchment from which water and material are drawn (Naiman and Latterell, 2005). Their position in the landscape (almost always in valley bottoms) makes lakes and rivers "receivers" of wastes, sediments and pollutants in runoff.

This is also true of seas and oceans, but inland water bodies typically lack the volume of open marine waters, limiting their capacity to dilute contaminants or mitigate other impacts (Dudgeon, 2006).

In addition, in many parts of the world fresh water is subject to severe competition among multiple human stakeholders, to the point that armed conflicts can arise when water supplies are limiting and rivers traverse political boundaries (Poff et al., 2010). There are 263 international rivers, draining 45% of the Earth's land surface, and this area supports more than 40% of the global human population (Postel and Richter, 2003). In the vast majority of disagreements over multiple uses of water, whether they are international or on a local scale, allocation of water to maintain aquatic biodiversity is largely disregarded (Poff et al., 2003). In China and India, where approximately 55% of the world's large dams are situated, hardlyany consideration has been given to the downstream allocation of water for biodiversity (Tharme, 2003; Poff et al., 2010).

The combined and interacting influences of the major threat categories have resulted in population declines and range reduction of freshwater biodiversity worldwide.

Qualitative data suggest reductions in numerous wetland and water margin vertebrates (e.g., a loss of 19 mammals, 92 birds, 72 reptiles and 44 fish species), while population trends indicate declines averaging 54% among freshwater vertebrates (mainly waterfowl), with a tendency toward higher values in tropical latitudes (Groombridge and Jenkins, 2000; Loh, 2000). Furthermore, 32% of the world's amphibian species now are threatened with extinction, a much higher proportion than threatened birds (12%) or mammals (23%), and 168 species may already be extinct (AmphibiaWeb, 2005). The well-known global decline of amphibians started during the 1950s and 1960s and has continued at the current rate of approximately 2% per year, with more pronounced decreases in tropical streams (Houlahan et al., 2000; Stuart et al., 2004). This is close to the estimate of 2.4% for declines in populations of freshwater vertebrates over the period 1970–1999 (Balmford et al., 2002). These estimates are extremely alarming. The limited data on extinction rates from few continent are believed to be indicative of a global freshwater "biodiversity crisis" (Kottelat and Whitten, 1996).

Rates of species loss from fresh waters in non-temperate latitudes are not known with any degree of certainty. They are likely to be high because species richness of many freshwater taxa (e.g., fishes, macrophytes, decapod crustaceans) increases toward the tropics. The drainage basins of many large tropical and subtropical rivers (e.g., the Ganges and Yangtze) are densely populated, with large dams, altered flow patterns and gross pollution from a variety of sources being the inevitable outcomes (Dudgeon, 2000, 2002). For larger species in these rivers, the situation is parlous: the Yangtze dolphin (*Lipotes vexillifer*) is probably the most endangered mammal on Earth (now numbering fewer than 100 individuals; Dudgeon, 2005), and the Ganges dolphin (*Platanista gangetica*) is "Endangered" (IUCN, 2003). The crocodilian fauna of the Ganges and Yangtze likewise consists entirely of threatened or highly endangered species. Many other species of water associated reptiles,a primarily tropical group, are gravely threatened (Gibbons et al., 2000), most particularly turtles, as are large freshwater fishes in most rivers (Hogan et al., 2004), and many freshwater fish stocks are over-fished to the point of population collapse (Dudgeon, 2002).

Floodplain ecosystems tend to be strongly impacted by the impoundment of river channels and other modifications that alter and generally reduce the natural hydrological variability of parent rivers. These floodplains lie along rivers with very low longitudinal slopes. Major river alterations for purposes such as navigation and hydroelectric generation have recently been proposed for each of these river systems, and such projects could affect extensive areas of floodplains through alteration of the natural flood regime (Hamilton, 2002).

Hydropower, which has substantial unexploited potential in many developing countries, can potentially mitigate greenhouse gas emissions by displacing fossil fuel production of energy, but large scale hydropower systems, in particular, can have adverse biodiversity and social effects (Secretariat of the Convention on Biological Diversity, 2009).

The environmental and social impacts of hydropower projects vary widely (Table 2), depending upon pre-dam conditions, maintenance of upstream water flows and ecosystem integrity, design and management of the dam (e.g., water-flow management) and the area, depth and length of the reservoir. Run of the river dams typically have fewer adverse environmental and social effects. Sectorial environmental assessments can assist in designing systems with minimum adverse consequences for ecological systems (Secretariat of the Convention on Biological Diversity, 2009).

Table 2. Environmental impacts of hydropower stations and operations (adapted from McCully, 1996)

Impacts due to Hydropower presence	Impacts due to Hydropower operation
Upstream change from river valley to reservoir Flooding of terrestrial habitats Conversion of terrestrial habitats Changes in fish migration patterns Changes in vegetation	Changes in hydrology Changes in downstream and total flows Changes in seasonal and short term timing and fluctuation in flows Changes in high and low flows Local extinctions of fish species Changes in vegetation
Changes in downstream morphology of banks, delta, riverbeds, estuary and coastline due to altered sediment load	Changes in downstream morphology due to altered flow patterns Reduction of connectivity
Changes in water quality, river temperature, nutrient load, turbidity, dissolved organic material, heavy metals and minerals	Changes in downstream water quality due to altered flow patterns Changes in microorganism temporary communities
Loss of biological diversity Loss of animal migrations Changes in ecological succession	Decrease of riverine, riparian and floodplain habitat diversity Elimination of floods

Well-designed installations, for example using modern technologies that cascade the water through a number of smaller dams and power plants, may reduce the adverse environmental impacts of the system (Secretariat of the Convention on Biological Diversity, 2009).

Small and micro-scale hydroelectric schemes normally have low environmental impacts, but the cumulative effects of many projects within a watercourse may have considerable impact on the biodiversity within a larger area. In general, run-of-river projects will have fewer impacts than storage dams with large reservoirs but they may also have serious effects on biodiversity.

These impacts are mainly due to the blocking of fish migration, either because of the physical barrier of the dam wall or through the dewatering of a stretch of river below the dam (Table 2).

Another important determinant of the impacts of dams is their location within the river system. Dams near the headwaters of tributaries will tend to have fewer impacts than mainstream dams that may cause perturbations throughout the whole watershed (Pringle 1997). The protection of dams from siltation may be a mayor incentive for biodiversity conservation in the form of reforestation or afforestation measures within the watershed. The World Commission on Dams has published a comprehensive list of guidelines for water and energy planning which might be helpful in that respect (World Commission on Dams 2000).

4.2. Global Threats to Lotic Biodiversity

Global biodiversity hotspots contain exceptional concentrations of endemic species in areas of escalating habitat loss. However, most hotspots are geographically constrained and consequently vulnerable to climate change (Table 3) as there is limited ability for the movement of species to less hostile conditions. Predicted changes to rainfall and temperature will undoubtedly further impact on freshwater ecosystems in these hotspots (Davies, 2010). For instance, current and predicted water temperatures may exceed thermal tolerances of aquatic fauna (Pörtner and Knust, 2007). Consequently, fauna cannot change their distribution southwards or with altitude as a response to increasing temperatures. Therefore, any mitigation responses need to be in situ to produce a suitable biophysical envelope to enhance species' resilience (Davies, 2010). This could be done through "over restoration" by increased riparian replanting at a catchment scale.

According to IPCC (2007), inland aquatic ecosystems are highly vulnerable to climate change, especially in Africa. Higher temperatures will cause water quality to deteriorate and will have negative impacts on microorganisms and benthic invertebrates. Plankton communities and their associated food-webs are likely to change in composition. Distributions of fish and other aquatic organisms are likely to shift northwards and some extinctions are likely. Changes in hydrology and abiotic processes induced by changes in precipitation, as well as other anthropogenic pressures, will have large impacts on aquatic ecosystems. Boreal peatlands will be affected most and suffer major changes in species composition. Many lakes will dry out. Increases in the variability of precipitation regimes will also have important impacts and may cause biodiversity loss in some wetlands. Seasonal migration patterns of wetland species will be disrupted (Secretariat of the Convention on Biological Diversity, 2009). The impacts of increased CO_2 will differ among wetland types, but may increase NPP in some systems and stimulate methane production in others. On the whole, ecosystem goods and services from aquatic systems are expected to deteriorate.

Climate change is expected to impact lotic ecosystems in two major ways: first, through changes in the water cycle; second, through associated changes in the terrestrial ecosystems within a given catchment (Table 3).

For inland wetlands, changes in rainfall and flooding patterns, across large areas of arid land, will adversely affect bird species that rely on a network of wetlands and lakes that are alternately, or even episodically, wet and fresh, and drier and saline (Roshier and Rumbachs, 2004).

Responses to these climate induced changes may also be affected by fragmentation of habitats or disruption or loss of migration corridors or, even, changes to other biota, such as increased exposure to predators by wading birds (Butler and Vennesland, 2000).

The lack of thermal refugia and migratory routes in streams and rivers may cause contraction of the distributions of many fish species. For example, warmer lake water temperature will reduce dissolved oxygen concentration and lower the level of the thermocline, most likely resulting in a loss of habitat for cold-water fish species. In addition, reduced summer flows and increased temperatures will cause a loss of suitable habitat for cool water fish species in riverine environments (Bunn and Arthington, 2002).

For streams, the effects of temperature-dependent changes would be least in the tropics, moderate at mid-latitudes, and pronounced in high latitudes where the largest changes in temperature are projected. Increased temperatures will alter thermal cycles of lakes and solubility of oxygen and other materials, and thus affect ecosystem structure and function (Secretariat of the Convention on Biological Diversity, 2009). Changes in rainfall frequency and intensity, combined with land-use change in watershed areas, has led to increased soil erosion and siltation in rivers. This, along with increased use of manure, chemical fertilizers, pesticides, and herbicides as well as atmospheric nitrogen deposition, affects river chemistry and has led to eutrophication, with major implications for water quality, species composition, and fisheries. The extent and the duration of the ice cover is projected to decrease in some high latitude lakes and, thus, the biodiversity may be affected by the shorter ice cover season (Christensen et al., 2003).

Climate change will have most pronounced effects on wetlands through altering the hydrological regime as most inland wetland processes are intricately dependent on the hydrology of the catchments (river basin) or coastal waters. This is expected to affect biodiversity and the phenology of wetland species (van Dam et al. 2002)

As with terrestrial ecosystems, adaptation strategies to climate change in lotic ecosystems include conservation and spatial linkages (Secretariat of the Convention on Biological Diversity, 2009).

Adaptation options to these changes should consider all components of the watershed (Sparks 1995). River biota, within reasonable limits, is naturally well adapted to rapid and unpredictable changes in environmental conditions (Puckridge et al. 1998). For rivers, it may be essential to conserve or restore ecosystem connectivity, both longitudinally along the river course and laterally between the river and its wetlands, in order to sustain ecosystem function (Ward et al. 2001). However, many of the natural aquatic corridors are already blocked through dams and embankments. This increases the vulnerability of lotic biodiversity to climate change and constrains implementing adaptive strategies. In their lower reaches, coastal rivers enter the estuarine and coastal zone where they have a major influence. These areas should be considered a contiguous part of inland water ecosystems and managed together under the ecosystem approach. The identification of the degree of vulnerability of the various components of complex inland water ecosystems, and the subsequent development of appropriate ecosystem management plans based upon this information, is a critical requirement for adaptation to climate change for inland waters (Andrade, Herrera, and Cazzolla Gatti, 2010).

Any management that favours near natural hydrological function in inland water ecosystems is likely to have major benefits for the conservation of biodiversity. In particular, modern approaches to the management of rivers recognise that for many systems change is inevitable. This has stimulated much interest in the concept of sustaining "environmental flows" as a management target for rivers (Tharme, 2003). Such approaches need to take on board climate change if they are to be adaptive. The increase in extreme weather events that climate change may bring (for freshwaters particularly the frequency and extent of droughts and floods) is likely to be more of a concern with isolated lakes and wetlands. The issue of extreme hydrological events is, however, of major significance to integrated water resources planning and management. For example, maintaining river floodplains and wetlands helps restore water balance and hence mitigate catastrophic flooding (Secretariat of the Convention on Biological Diversity, 2009).

Table 3. Impacts of climate change on lotic biodiversity and potential adaptation strategies

Impacts of climate change	Involved organisms	Potential adaptations	Extinction risk level
Water temperature increase	Fishes, plants, crustaceans, benthic species, turtles and mammals	Migrations northwards and upwards when not limited	Very High
Precipitation patterns changes	Fishes, vegetation, benthic invertebrates	Dispersal and in situ adaptations	Moderate
Water quality decline due to higher temperature	Microorganisms, benthic invertebrates, plankton communities	Dispersal and in situ adaptations	High
Fragmentation of habitats, disruption or loss of migration corridors	Fishes, reptiles and mammals	Behavioural changes and phenotypic plasticity for generalist species	High
Invasion of species	Specialized species	Avoidance of competition (shifting of niches)	Low-moderate
Extreme events	All species (benthic species less affected)	Long distance dispersal	Moderate
Community composition changes	All species	New interactions and mutualism partnerships	High

Climate change, therefore, can be regarded as providing additional incentives to manage inland waters better and both the financial and conservation benefits of doing so are considerable. Maintaining natural river form and related ecosystem processes is likely to provide significant benefits for coastal regions and populations. For instance, while maintaining their traditional way of life, lotic species also play a significant role in nutrition of the nomadic reindeer-herders who take long stops at rivers and lakes for fishing. Hunting is the oldest occupation of mankind, and many reindeer hunters migrated across vast distances of Siberia throughout historic times (Secretariat of the Convention on Biological Diversity, 2009).

Major impacts of climate change that need to be addressed in water management include increasing flood risk, increasing risk of drought, and change in timing of flow regimes. Common technical approaches to flood risk include the construction of dykes and dams. Technical solutions are also often applied to address problems of water shortage, including the construction of reservoirs and canals, facilities for water diversion and abstraction from rivers, and alterations to river beds to improve shipping capacity during low-water periods. Hard structures can have significant environmental impacts, such as destruction or alteration of wetlands, reducing connectivity between lakes, rivers and riparian zones, and changing sediment flows (Secretariat of the Convention on Biological Diversity, 2009). Restoration of upland watersheds and floodplain restoration are ecologically viable alternatives that deserve attention.

Climate change is leading to increased inland flooding in many regions through more variable rainfall events. Restoring and maintaining ecosystems in upland watersheds, including through the management of soils and vegetation, can contribute to reduce the risk of flooding and maintain regular water supplies. Run-off from mountainous areas in small islands is often the major supply of water, and in many countries, watersheds form a critical part of the national economy (Secretariat of the Convention on Biological Diversity, 2009). Often these watersheds are degraded, and their rehabilitation is one adaptation option.

Wetland ecosystems in watersheds can reduce flooding and sediment deposition whilst improving water quality downstream. A study of upland forests in a watershed in Madagascar (Kramer et al., 1997) has estimated their flood protection value at $126,700, and peat bogs in Sri Lanka that buffer floodwaters from rivers have an estimated annual value of more than $5 million (Emerton and Kekulandala, 2003). In the Morogoro region of the United Republic of Tanzania, reduced river flow and increased flooding has been attributed to deforestation in the mountains, and it has been suggested that effective management of soil, forests and water resources are needed as adaptation measures, along with improved social capacity (Chamshama and Nduwayezu, 2002). Ecuador and Argentina have integrated forests and wetlands into their living with floods strategies, and reforestation is recognised as an important option for adaptation in the watersheds of the Philippines. Vietnam includes measures such as integrated management of watersheds in its disaster reduction planning, along with forest management, and soil and water conservation. Large-scale afforestation projects in China have been carried out with the aim of reducing flooding and increasing water conservation, and countries of Central America are collaborating to protect watersheds and forests (Campbell et al., 2009).

Climate change is causing an increase in the scale of flooding and dry periods in many flood plains. In some systems dams are no longer a viable adaptation strategy, and in some cases dams have had negative environmental and socio-economic impacts. Also in these circumstances ecosystem management is an effective adaptation strategy at the river basin scale and an alternative to the development of small-scale dams. In developed countries, cost-effective flood reduction strategies, which allow re-growth of vegetation alongside rivers and establish vegetation buffers along streams, combined with the reduced development of infrastructure, are being promoted in some areas. Restoration of floodplain ecosystems can also help to reduce the levels of water pollution following extreme events (CBD, 2009).

There is new observational evidence of climatic impacts on inland waters. A recent research has shown that there has been a fourfold increase in permanently dry ponds in Yellowstone over the last 16 years and that this can be linked directly to dramatic declines in amphibian populations and diversity (McMenamin et al. 2008). Modelling work suggests that climatic warming, in combination with other environmental changes, may cause the nature of river channels to change in the Russian Arctic (Anisimov and Reneva, 2006). In the Arctic, factors such as reduced ice-cover duration on lakes, especially in northern Arctic areas, increased rates of rapid stratification and primary production, and decreased oxygenation at depth. This will possibly result in a reduction in the quality and quantity of habitat for species such as lake trout. Decreased water flow in summer is also likely to decrease habitat availability and possibly deny or shift access for migrating fish (Reist et al., 2006; Wrona et al., 2006). In monsoonal Asia, where ecological processes surrounding rivers are mediated by flow, disruptions in timing and velocity will have large environmental impacts (Dudgeon, 2011). The interaction between climate change and land cover change is likely to lead to reduced discharge from many rivers that will in turn lead to significant loss of freshwater fish species (Xenopoulos and Lodge, 2006).

There is also new observational evidence of compositional change in fish communities in France over the last 15-25 years (Daufresne and Boet, 2007): species richness, proportions of warm water species, and total abundance increased. Modelling has demonstrated negative impacts on the habitats of native fish species, including freshwater salmon (Xenopoulos and Lodge, 2006; Battin et al., 2007), especially at higher elevations and in headwater areas (Buisson, 2008). In the Arctic, there is an expected decrease of native fish as southern Arctic and sub-Arctic fish species migrate northwards. The broad whitefish (Arctic char complex), and the Arctic cisco are particularly vulnerable to displacement. Decreased water flow in summer is likely to decrease habitat availability and possibly deny or shift access for migrating fish (Reist et al. 2006; Wrona et al.; Anisimov et al. 2007; Berry 2008). Temperature is a very important determinant of distribution and survival of aquatic macro-invertebrates at high latitudes, and changes in species composition have already been shown for boreal inland waters (Heino et al. 2009).

It has been suggested that species characteristic of lentic systems may disperse more effectively than those of lotic systems (Hof et al. 2008), and therefore that lentic systems may show more rapid compositional change in response to changing climate (Heino et al. 2009).

Models show that climate change will also affect wetland species composition through its effects on river flow, especially low water flows (Xenopoulos and Lodge 2006; Harrison et al. 2008), though the interaction with socio-economic drivers of flow management is also very important.

There is considerable and growing concern about the linkages between climate change impacts on aquatic systems (including warmer water temperatures, shorter duration of ice cover, altered streamflow patterns, increased salinization, and increased demand for water storage and conveyance structures) and aquatic invasive species (Rahel and Olden 2008).

Climate change is influencing invasive establishment by eliminating adverse winter conditions and will alter the distribution and ecological impacts of existing invasive species, by enhancing their competitive and predatory effects on native species, and by increasing the virulence of some diseases (Hellmann et al., 2008; Rahel and Olden, 2008).

Predictions done for Canada indicate that water temperature may change as much as 18° C by 2100, which would mean that a number of lakes will be newly vulnerable to invasion by smallmouth bass (Sharma et al., 2007).

Other factors that will interact with climate change in determining compositional change in inland waters include acidification (Barnes and Conlan, 2007; Durance and Ormerod, 2007), eutrophication (Heino et al. 2009) and land cover change, with change in composition of terrestrial systems (Chapin et al., 2005; Heino et al., 2009) and agricultural expansion (Heino et al., 2009).

There is very little information on real or projected changes in aquatic ecosystems in the tropics, but it is clear that some major tropical wetlands are at risk from altered flows of freshwater (Gopal and Chauhan, 2006; Xenopoulos et al., 2005; Xenopoulos and Lodge, 2006).

Some others significant impacts of climate change have been projected for both carbon storage and fisheries services from inland waters. While this has been projected particularly strongly for the Arctic (Wrona et al., 2006), there is also a growing body of model-based evidence relating to other regions (Xenopoulos and Lodge, 2006). A new concern is the effect of sea level rise on carbon storage in coastal wetlands, including 150,000 km^2 of freshwater peatlands worldwide below 5 m elevation and vulnerable to sea level rise, which are likely to emit significant amounts of carbon when they are inundated (Henman and Poulter, 2008). The protection and biodiversity conservation roles of coastal wetlands are also at risk as, for example, in the case of the Sundarbans, the world's largest wetland, which is threatened by altered freshwater flows and sea level rise, which are both influenced by climate change (Gopal and Chauhan, 2006). Ecosystem services provided by peatlands are also at risk, as temperature changes are expected to reduce their function as carbon sinks (Lloyd, 2008).

4.3. Global Scenarios for Lotic Biodiversity

Scenarios and projection of recent trends suggest that a combination of climate change, water withdrawal, pollution, invasive species, and dam construction will further deteriorate the current state of freshwater biodiversity (Leadley, 2010). The particular vulnerability of freshwater species to global changes reflects the fact that both fish and freshwater are resources that have been heavily managed.

Scenarios for freshwater biodiversity are limited compared to terrestrial and marine biodiversity (Sala et al., 2000). Moreover, global scenarios tend to address water resources for people, but rarely include models of freshwater biodiversity. Those that do it, model a limited number of drivers and lack or treat only qualitatively major drivers, such as dam construction, eutrophication and invasive species.

Habitat loss and/or fragmentation are among the greatest threats to biodiversity worldwide, and this certainly holds true for riverine fish (Dudgeon et al., 2006). It is almost certain that disturbances to freshwater ecosystems, such as dams, reservoirs and diversions for irrigation and industry, will endanger or extinguish many freshwater fish species in the future, by creating physical barriers to normal movements and migration of the biota and by decreasing habitat availability.

Currently it is difficult to make precise predictions about how climate change will affect fish biodiversity, even though climate niche modelling suggests that, locally, the number of warm-water species may increase in temperate areas and some cryophilic (i.e., cold-water) species may regionally vanish (Leadley, 2010).

Narrow endemic riverine fishes can be particularly threatened by climate change. The biggest problems occur in basins which have an East-West configuration, while in basins with a North-South configuration, there will be more opportunities for migration and adaptation, as long as the rivers are not blocked by dams (Leadly, 2010). Models also project that in shallow lakes in northern latitudes there will be summer fish kills of cold-water species due to both increased water temperatures and decreased dissolved oxygen.

Other negative impacts of climate change on freshwater ecosystems are changes in snow melt timing and flow volumes. Global climate scenarios have been applied to known relationships between fish diversity and river discharge. Results predict decreased freshwater biodiversity in about 15% of the world's rivers in 2100, from a combination of reduced run-off (caused by climate change) and increased water withdrawals for human use. However, these predictions should be considered with great caution, as the approach does not provide true extinction rates, but instead a percentage of species "committed to extinction" with an unspecified time lag. These predictions also do not include other current stresses on freshwater fish, such as pollution or river fragmentation (Olden et al., 2010).

Based on the established relationship between the number of non-native fish species and human activity, we expect that river basins of developing countries will host an increasing number of non-native fish species as a direct result of economic development (Leprieur, 2008).

Furthermore impoundments and climate change may facilitate the expansion of invasive species and diseases associated with lake ecosystems (Johnson et al., 2008).

Pressure on freshwater ecosystem services (Rapport et al., 1998) and wetland degradation will increase leading to the deterioration of regulating services such as regulation of water quality and flood protection.

The combination of population growth, increasing water use and climate change will lead to an increase in human population living in river basins facing severe water stress. This will not only increase the risks of chronic water shortages in these regions, but will also cause major negative impacts on freshwater ecosystems (Leadley, 2010).

Many studies report that eutrophication of freshwater systems will increase in the developing world as fertilizer use and untreated sewage effluents continue to increase (Tilman et al., 2001). This may be further exacerbated in some regions by decreasing precipitation and increasing water stress. The transition to eutrophic conditions is in some instances difficult to reverse and can lead to loss of fish species, loss of recreational value, and in certain cases health risks for humans and livestock (Leadley, 2010).

4.4. Remote Sensing to Monitor Biodiversity of Lotic Ecosystems

Most small rivers are not directly detectable using data at, or below the, 30 m resolution threshold, considered a practical limit for national assessments. Often, analysts turn to watershed analysis within geographic information systems (GIS) software (Gardiner and Díaz-Delgado, 2007). These methods use digital elevation model (DEM) data to infer the direction that water would flow if travelling overland and downhill from any point in a landscape (O'Callaghan and Mark, 1984). Once flow direction information has been extracted, it is possible to infer the total number of cells, and hence area, above every cell in a DEM. The GIS database could include watershed and river attributes such as forested area, land use, river discharge at measured locations, resident species, important features such as dams, water withdrawal points, discharge points (Figure 10), or other information pertinent to hydrology or biodiversity of wetlands (Hutchinson, 1991). These GIS methods provide a means for tracking physical and biological conditions in rivers and watersheds.

For instance, HydroSHEDS (Hydrological data and maps based on SHuttle Elevation Derivatives at multiple Scales) represents waterbodies, waterways, watersheds, and surface hydrology on a near-global basis and at multiple resolutions (Lehner et al. 2006). The data were built from NASA's SRTM data, which describe surface elevations for Earth's land area lying between ±57 degrees latitude. HydroSHEDS data may be downloaded free of charge (http://hydrosheds.cr.usgs.gov/). The goal of developing this database was to generate key data layers to support watershed analyses, hydrologic modelling, and freshwater conservation planning at previously inaccessible quality, resolution, and extent. The seamless coverage of HydroSHEDS makes this dataset useful for continental analyses because it eliminates the need to blend multiple data sources (Gardiner and Díaz-Delgado, 2007).

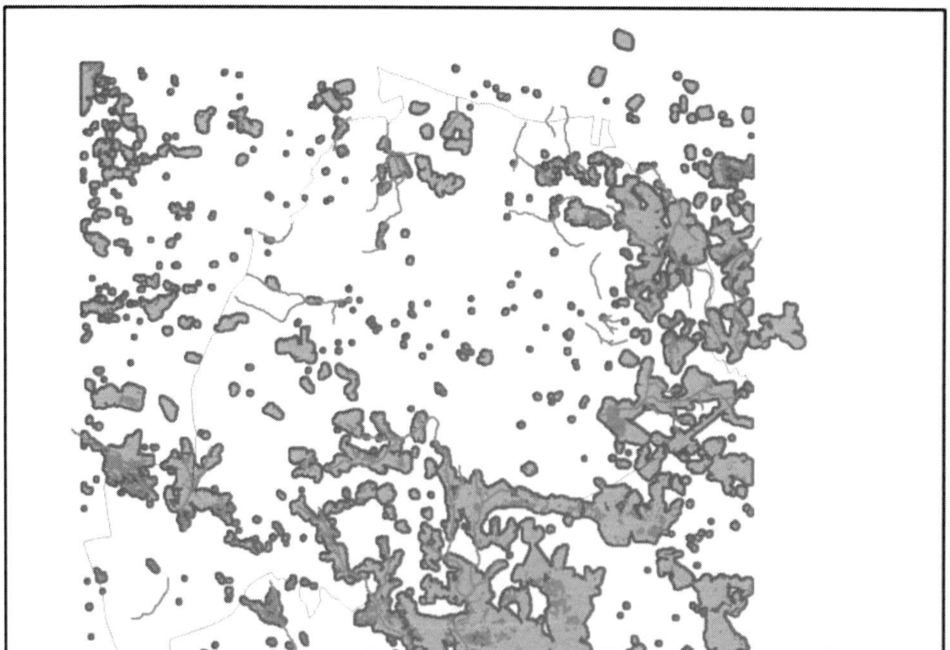

Figure 10. An elaboration derived from satellite maps of a region in South Italy that shows the hydrographic network and vegetation cover (green: woodlands; brown: grasslands). This map has been used by WWF to address conservation policies of the region (from Cazzolla Gatti, 2009).

Data acquired at high- and low-water are needed to accurately map wetland extent because these ecosystems are defined by water level fluctuation and function differently at different water levels. The acquisition time of archived data may be compared to the best-known stage information for a given area in order to choose imagery that will be of most use for characterizing high- and low water regimes of the targeted ecosystem. For upland wetlands, antecedent soil moisture and precipitation affect river discharge, wetland extent, and surface water levels and therefore influence the optimal date for acquiring imagery. The situation is more complex for seasonally inundated systems that receive water input from rivers because river flood stage and peak flow are influenced by upstream, catchment-wide hydrological factors that delay peak discharge relative to peak precipitation events (Gardiner and Díaz-Delgado, 2007). Temperate rivers reach bank full stage with a recurrence interval of between 1.5 and 2.5 years, suggesting that optimal data for analysis might appropriately come from different years in order to capture both low- and high-water events within the data record for a site. Mapping the high- and low-water stages of rivers with active floodplain systems, such as the Amazon, is important since the function of these systems is defined relative to that variation in water level and extent.

The behaviour, function, and aerial extent of wetlands changes through time, so characterizing changes in habitat quality requires a thorough understanding of natural variability. Water and nutrients are transported from headwaters to downstream river ecosystems, but there is a bidirectional interaction of rivers with active floodplains (Gardiner and Díaz-Delgado, 2007). Nutrient flux from floodplains into river food webs is an important linkage between rivers and the floodplain habitat associated with them. When researchers first mapped and compared the extent of floodplain inundation at high and low-water levels, they surmised that the mass of carbon emitted from rivers in the form of CO_2 during high water stages throughout the year was comparable in magnitude to the amount of carbon transported down river (see Melack, 2004). Thus, accurate mapping at high and low-water levels has led to significant new insights into the structure and function of aquatic systems.

Recent remote sensing efforts have demonstrated how to map floodplain forests using a variety of satellite sensors and data available at multiple resolutions. Hamilton and colleagues (2007) used remotely sensed data in combination with HydroSHEDS river network data to characterize wetlands in floodplains of the Madre de Dios River. They mapped floodplains, standing water, and vegetation associated with unique geomorphic settings in this flood-dominated ecosystem. This research employed object-oriented, contextual classification, a set of techniques that utilizes the spatial setting of landscape features to help identify and classify imagery. Image data included Landsat 7 ETM+ data, elevation profiles from NASA's SRTM, and JERS-1 L-band radar scatterometer mosaics.

To evaluate in situ biological and physical properties of river ecosystems, researchers often use "multi-metric" indicators, statistical descriptions that simultaneously describe a site's species and local habitat relative to undisturbed sites with similar landscape settings (Gardiner and Díaz-Delgado, 2007). Multi-metric indicators are derived empirically from a set of sites or through time, for example from density or relative proportion of taxa collected at a site or group of sites. The categories and point assignments used to derive multi-metric scoring systems must be calibrated to the fishes, macroinvertebrates, or microbes found in streams and rivers within a bioregion, so this work is conducted by a biologist with the requisite expertise in regional fauna and flora. Fish and invertebrate ecologists have the most experience using multi-metrics to describe and categorize ecosystem health in rivers and streams, but taxonomists and ecologists are studying how to develop indicators of stream health that focus on the microbes found at a stream sampling site. Multi-metric scores can be compared statistically to land use data derived from remote sensing and extracted on a watershed basis using GIS software (Gardiner and Díaz-Delgado, 2007). This statistical approach guides inferences about the effect of watershed practices on streams or rivers. This procedure is widely practiced, but should be conducted only through direct collaboration among experts in GIS, remote sensing, and freshwater biology.

Hyperspectral technologies have also been used to study primary productivity of inland waters (Hoogenboom et al., 1998), although these studies focus on very small areas and use data not available on a global basis. Optical data are also used to estimate suspended solids concentrations in large water bodies (Dekker et al., 2001).

Change and variability are inherent to the structure and functioning of wetlands. Just as one may assess the natural variability of water extent, exogenous inputs, and biota within wetlands, it is possible to measure long-term trends and changes to wetlands using the same or similar methods. Some changes to wetlands can be evaluated somewhat directly, for example the influence of land cover change on the timing and delivery of water and suspended constituents to rivers, or the effect of global warming on boreal wetlands (Gardiner and Díaz-Delgado, 2007).

Land cover change upstream of receiving waters alters the hydrologic, nutrient, and physical templates of those ecosystems. When forested catchments are clear-cut in temperate forest ecosystems, recovery of some parameters, such as nutrient retention and turnover, requires up to several years to re-establish pre-disturbance regimes. Other physical characteristics require decades for recovery. For example, sediment delivered to rivers and streams following a major disturbance, such as watershed-wide clear-cutting, may require infrequent, episodic torrential rain events in order to generate sufficient hydrologic power to redistribute large quantities of sediment downstream. Once vegetation recovers, the legacy of historic deforestation events can therefore have a very long-lasting impact on the habitat template of stream ecosystems (Gardiner and Díaz-Delgado, 2007).

Spatial data describing regional climate patterns, physiography, land cover, and land use lend insight into how to manage watersheds. Conservation planners prioritize their effort using the best available data describing a region of interest. Often, data describing biodiversity are absent, so planning must move forward in the absence of biological information using surrogate measures, such as climate information (Gardiner and Díaz-Delgado, 2007).

Planners and researchers from WWF-US, Michigan State University, Woods Hole Research Center, and WWF-Peru prioritized conservation recommendations for a 160,000 km2 headwater region of the Madre de Dios and Orthon rivers in Peru (Thieme et al. 2007). Each river is a tributary of the Amazon River, and the study area as a whole is within the south-western Amazonian Moist Forests ecoregion of the Global 200 priority regions identified by Olsen and Dinerstein (1998). The work used GIS-based analyses of terrain, vegetation, and existing protected areas to recommend areas for conservation attention. The study's authors hope the work will prevent problems arising from road building and other land-clearing activities that are likely to accompany oil and gas exploration in the region. Activities like these will remove vegetation and expose soil, thereby increasing sediment delivery to waterways through erosion and transport of disturbed soil. Sedimentation is among the most common processes that degrade river ecosystems (Beechie et al., 2010). GIS data describing watershed boundaries, stream channels, and watershed morphometry provided requisite data for evaluating potential discharge along stream segments, percentage of watershed area found within the Andes, and connectedness of river segments among protected areas that have already been identified. Watershed-based analyses, such as those conducted in the Madre de Dios River basin, are an essential component of evaluating the potential influence of land use decisions on wetlands (Gardiner and Díaz-Delgado, 2007).

Remote sensing analyses complement campaigns focused on the structure and function of wetland ecosystems. For example, due to changes in freeze-thaw cycles and permafrost conditions stemming from global warming, there is increasing attention and interest in greenhouse gas emissions from boreal forests (Turner et al., 2009).

Approximately 25% of the carbon that is bound within terrestrial ecosystems is likely found in high-latitude peat lands (Hess and Melack 1994). When they dry out, peat lands respire CO_2 and CH_4 into the atmosphere, so monitoring inundation in these areas is important for quantifying greenhouse gas emissions from peat lands. In boreal Siberia and eastern Canada, Gorham (1991) argued that satellite data may be used to monitor the declining area of open water as an indicator of global warming effects on peat land ecosystems, but that effort also requires contribution from biogeochemists with expertise in quantifying and evaluating outgassing. Conversely, biogeochemists recently estimated that methane emissions from seasonally melted lakes in permafrost regions of Siberia may contribute about twice as much CH_4 to the atmosphere as previously thought (Walter et al. 2006). Their estimates were based on field-collected samples from a handful of lakes, and results were extrapolated based on estimates using GIS and remote sensing. Kimball and colleagues (2006) have used microwave data to show that seasonality has changed in recent decades, with warm temperatures arriving earlier in the year and cold temperatures arriving later.

CONCLUSION

Lotic ecosystems support thousands of species, although their conservation and management are critical, and freshwater biodiversity constitutes a valuable, natural resource. These ecosystems are experiencing declines in biodiversity far greater than those of the most affected terrestrial ecosystems, and the probability to conserve much of the remaining biodiversity in fresh water seems to be very low.

These ecosystems are threatened by many anthropogenic stresses such as overexploitation, water pollution, flow modification, destruction or degradation of habitat, invasion by exotic species, hydropower, and climate change. For instance, cumulative impacts of small dams on biodiversity need to be considered, even when individual installations may have only a small.

Climate change is already affecting the ability of ecosystems to regulate water flows. The regulation of water quality and quantity is a key ecosystem service worldwide. Higher temperatures, changing insolation and cloud cover, and the degradation of ecosystem structure result in the occurrence of more and higher peak-flows on the one hand and, at the same time, impede the ability of ecosystems to regulate water flow. This has major consequences for both ecosystems, with associated species assemblages, and people in the scale of whole catchment areas. In addition to freshwater and wetlands, riverine and alluvial ecosystems, and many forest types are affected by changes in the hydrological regime.

Loss of wetlands due to over-extraction of groundwater, drainage for human uses (reclamation), reduced runoff, and increasing sea level rise, will reduce biodiversity and negatively impact the regulation services of wetlands, such as water purification and flood mitigation (Leadley, 2010).

Biodiversity can play a role in adaptation strategies to both drought and floods through the management of watershed, wetland, forest, and agricultural systems. Maintenance or restoration of forest and wetlands, for example, can reduce run-off in times of floods and also increase water retention during droughts.

Planting trees on slope fields, mini-terracing for soil and moisture conservation, and improved pasture management can also complement actions such as building of small-scale infrastructure in water resources management (CBD, 2009).

The examples provided in this chapter demonstrate also the feasibility of using remote sensing indicators to measure natural variability in wetland extent. Coordinated field campaigns and remote sensing research can yield statistically rigorous relationships between remote sensing indicators and biophysical characteristics of wetland ecosystems. For rivers below the detection limits of remote sensing, watershed modelling within GIS software utilizes remotely sensed land cover information to derive hydrologic and suspended loading information that can be used to infer habitat quality parameters. Larger water bodies may be directly mapped, both in terms of aerial extent and water elevation.

REFERENCES

Amoros, C. and Roux A. L. (1988). Interactions between water bodies within the floodplains of large rivers: Function and development of connectivity, p. 125–130. In K. F. Schreiber (eds.), Connectivity in Landscape Ecology. Munstersche Geographische Arbeiten, Munster.

Amoros, C., Gibert, J. and Greenwood, M. T. (1996). Interactions between units of the fluvial hydrosystem, in Petts, G. E. and Amoros, C. (Eds), Flu6ial Hydrosystems. Chapman and Hall, London. 184–210.

AmphibiaWeb, (2005). AmphibiaWeb species numbers. AmphibiaWeb: Information on Amphibian Biodiversity and Conservation. Berkeley, California, U.S.A. http://amphibiaweb.org/ (accessed 2 April, 2005).

Andrade Pérez, A., Herrera Fernández, B. and Cazzolla Gatti, R. (2010). Building Resilience to Climate Change: Ecosystem-based adaptation and lessons from the field (No. 9). IUCN.

Anisimov, O. and Reneva, S. (2006). Permafrost and changing climate: the Russian perspective. *AMBIO: A Journal of the Human Environment*, *35*(4), 169-175.

Arnell, N. W. (1999). Climate change and global water resources. *Global Environmental Change*, *9*, S31–S46.

Arthington, Á. H., Naiman, R. J., Mcclain, M. E. and Nilsson, C. (2010). Preserving the biodiversity and ecological services of rivers: new challenges and research opportunities. *Freshwater Biology*, *55*(1), 1-16.

Balmford, A., Bruner, A., Cooper, P., Costanza, R., Farber, S., Green, R. E. and Turner, R. K. (2002). Economic reasons for conserving wild nature. *Science*, *297*(5583), 950-953.

Barnes, D. K. and Conlan, K. E. (2007). Disturbance, colonization and development of Antarctic benthic communities. *Philosophical Transactions of the Royal Society B: Biological Sciences*, *362*(1477), 11-38.

Battin, J., Wiley, M. W., Ruckelshaus, M. H., Palmer, R. N., Korb, E., Bartz, K. K. and Imaki, H. (2007). Projected impacts of climate change on salmon habitat restoration. Proceedings of the national academy of sciences, *104*(16), 6720-6725.

Beechie, T. J., Sear, D. A., Olden, J. D., Pess, G. R., Buffington, J. M., Moir, H. and Pollock, M. M. (2010). Process-based principles for restoring river ecosystems. *BioScience*, *60*(3), 209-222.

Benstead, J. P., De Rham, P. H., Gattoliat, J. L., Gibon, F. M., Loiselle, P. V., Sartori, M., Sparks, J. S. and Stiassny, M. L. J. (2003). Conserving Madagascars freshwater biodiversity. *BioScience*, *53*, 1101–1111.

Bravard, J. P. and Petts G. E. (1993). Hydrosystemes fluviaux: Interferences avec les interventions humaines, p. 3–17. In C. Amoros and G. E. Petts (eds.), Hydrosystemes Fluviaux. Masson, Collection d'Ecologie, Paris.

Brinson, M. M. (1993). A hydrogeomorphic classification for wetlands. Technical Report WRP-DE-4. U.S. *Army Engineer Waterways Experiment Station*, *Vicksburg*, *MS*, p. 101.

Buisson, L., Thuiller, W., Lek, S., Lim, P. U. Y. and Grenouillet, G. (2008). Climate change hastens the turnover of stream fish assemblages. *Global Change Biology*, *14*(10), 2232-2248.

Bunn, S. E. and Arthington, A. H. (2002). Basic principles and ecological consequences of altered flow regimes for aquatic biodiversity. *Environmental management*, *30*(4), 492-507.

Bunn, S. E. and Arthington, A. H. (2002). Basic principles and ecological consequences of altered flow regimes for aquatic biodiversity. *Environmental management*, *30*(4), 492-507.

Butler, R. W. and Vennesland, R. G. (2000). Integrating climate change and predation risk with wading bird conservation research in North America. *Waterbirds*, 535-540.

Campbell, A., Kapos, V., Chenery, A., Kahn, S. I., Rashid, M., Scharlemann, J. P. W. and Dickson, B. (2009). The linkages between biodiversity and climate change adaptation. UNEP World Conservation Monitoring Centre.

Castella, E., Richardo-Coulet, M., Roux, C. and Richoux, P. (1984). Macroinvertebrates as "describers" of morphological and hydrological types of aquatic ecosystems abandoned by the Rhone River, *Hydrobiologia*, *119*, 219–225.

Cazzolla Gatti, R. (2010). Ambienti, flora e fauna delle Murge di sud-est, Adda Editore, Bari, Italy.

Chabrerie, O., Poudevigne, I., Bureau, F., Vinceslas-Akpa, M., Nebbache, S., Aubert, M. and Alard, D. (2001). Biodiversity and ecosystem functions in wetlands: A case study in the estuary of the Seine river, France. *Estuaries*, *24*(6), 1088-1096.

Chamshama, S. A. O. and Nduwayezu, J. B. (2002). Rehabilitation of degraded sub-humid lands in Sub-Saharan Africa: a synthesis. *Sokoine University of Agriculture*, *Morogoro*, *Tanzania*, 3-35.

Chao, B. F. and Gross, R. S. (1995). Changes in the Earth's rotational energy induced by earthquakes. *Geophysical Journal International*, *122*(3), 776-783.

Chapin, F. S., Sturm, M., Serreze, M. C., McFadden, J. P., Key, J. R., Lloyd, A. H. and Welker, J. M. (2005). Role of land-surface changes in Arctic summer warming. *science*, *310*(5748), 657-660.

Christensen, V., Guenette, S., Heymans, J. J., Walters, C. J., Watson, R., Zeller, D. and Pauly, D. (2003). Hundred-year decline of North Atlantic predatory fishes. *Fish and fisheries*, *4*(1), 1-24.

Church, M. (2002). Geomorphic thresholds in riverine landscapes. *Freshw. Biol.*, *47*, 541–557.

Cigizoglu, H. K., Bayazit, M. and Onoz, B. (2005). Trends in the maximum, mean, and low flows of Turkish rivers. *Journal of Hydrometeorology*, *6*, 280–290.

Colburn, T., Myers, J. P. and Dumanoski, P. (1996). Hormonal sabotage. Natural history (New York, NY, USA).

Copp, G. H. (1989). The habitat diversity and fish reproductive function of floodplain ecosystems, *Environ. Biol. Fish.*, *26*, 1–27.

Costanza, R., Kemp, W. M. and Boynton, W. R. (1993). Predictability, scale, and biodiversity in coastal and estuarine ecosystems: Implications for management. *Ambio*, *22*, 88–96.

Curtis, T. P. and Sloan, W. T. (2004). Prokaryotic diversity and its limits: microbial community structure in nature and implications for microbial ecology. *Current Opinion in Microbiology*, *7*, 221–226.

Daufresne, M. and Boët, P. (2007). Climate change impacts on structure and diversity of fish communities in rivers. *Global Change Biology*, *13*(12), 2467-2478.

Davies, P. M. (2010). Climate change implications for river restoration in global biodiversity hotspots. *Restoration Ecology*, *18*(3), 261-268.

Day, JR. J. W., Hall, C. A. S., Kemp, W. M. and Yanez-Arancibia, A. (1989). Estuarine Ecology. J. Willey, New York.

Decamps, H. (1996). The renewal of floodplain forests along rivers: a landscape perspective, *Verh. Int. Ver. Limnol.*, *26*, 35–59.

Dekker, A. G., Vos, R. J. and Peters, S. W. M. (2001). Comparison of remote sensing data, model results and in situ data for total suspended matter (TSM) in the southern Frisian lakes. *Science of the Total Environment*, *268*, 197-214.

Dudgeon, D. (1999). Tropical Asian Streams: Zoobenthos, Ecology and Conservation. Hong Kong University Press, Hong Kong.

Dudgeon, D. (2000). The ecology of tropical Asian rivers and streams in relation to biodiversity conservation. *Annual Review of Ecology and Systematics*, *31*, 239–263.

Dudgeon, D. (2011). Asian river fishes in the Anthropocene: threats and conservation challenges in an era of rapid environmental change. *Journal of Fish Biology*, *79*(6), 1487-1524.

Dudgeon, D., Arthington, A. H., Gessner, M. O., Kawabata, Z. I., Knowler, D. J., Lévêque, C. and Sullivan, C. A. (2006). Freshwater biodiversity: importance, threats, status and conservation challenges. *Biological reviews*, *81*(2), 163-182.

Durance, I. and Ormerod, S. J. (2007). Climate change effects on upland stream macroinvertebrates over a 25-year period. *Global change biology*, *13*(5), 942-957.

Dynesius, M. and Nilsson, C. (1994). Fragmentation and flow regulation of river systems in the northern third of the world. *SCIENCE-NEW YORK THEN WASHINGTON-*, 753-753.

Ecological Applications, *1*, 182-195.

Emerton, L. and Kekulandala, L. D. C. B. (2003). Assessment of the. by: IUCN-Sri Lanka.

Galloway, J. N., Dentener, F. J., Capone, D. G., Boyer, E. W., Howarth, R. W., Seitzinger, S. P. and Vöosmarty, C. J. (2004). Nitrogen cycles: past, present, and future. *Biogeochemistry*, *70*(2), 153-226.

Gardiner, N. and Díaz-Delgado, R. (2007). Trends in selected biomes, habitats and ecosystems: Inland waters. In Sourcebook on remote sensing and biodiversity indicators. *Secretariat of the Convention on Biological Diversity, Montreal*, (pp. 83-102).

Gessner, M. O. and Van Ryckegem, G. (2003). Water fungi as decomposers in freshwater ecosystems. In Encyclopaedia of Environmental Microbiology (ed. G. Bitton), John Wiley and Sons, New York, U.S.A. (online edition: DOI 10.1002/0471263397. env314).

Gibbons, J. W., Scott, D. E., Ryan, T. J., Buhlmann, K. A., Tuberville, T. D., Metts, B. S. and Winne, C. T. (2000). The Global Decline of Reptiles, Déjà Vu Amphibians Reptile species are declining on a global scale. Six significant threats to reptile populations are habitat loss and degradation, introduced invasive species, environmental pollution, disease, unsustainable use, and global climate change. *BioScience*, *50*(8), 653-666.

Gibert, J. and Deharveng, L. (2002). Subterranean Ecosystems: A Truncated Functional Biodiversity This article emphasizes the truncated nature of subterranean biodiversity at both the bottom (no primary producers) and the top (very few strict predators) of food webs and discusses the implications of this truncation both from functional and evolutionary perspectives. *BioScience*, *52*(6), 473-481.

Gibert, J., Mathieu, J. and Fournier, F. (Eds). (1997). Groundwater/Surface Water Ecotones: Biological and Hydrological Interactions. Cambridge University Press, Cambridge, UK.

Gleick, P. H. (1996). Water resources. In Encyclopedia of Climate and Weather (ed. S. H. Schneider), pp. 817–823. Oxford University Press, New York, USA.

Gopal, B. and Chauhan, M. (2006). Biodiversity and its conservation in the Sundarban Mangrove Ecosystem. *Aquatic Sciences*, *68*(3), 338-354.

Gorham, E. (1991). Northern peatlands: role in the carbon cycle and probable responses to climatic warming. *Ecological applications*, *1*(2), 182-195.

Gosz, J. R. (1993). Ecotone hierarchies. *Ecological Apllications*, *3*, 369–376.

Groombridge, B. and Jenkins, M. D. (2000). Global biodiversity: Earth's living resources in the 21st century. World Conservation Press.

Hamilton, S. K. (2002). Hydrological controls of ecological structure and function in the Pantanal wetland (Brazil). The Ecohydrology of South American Rivers and Wetlands. *International Association of Hydrological Sciences, Special Publication*, *6*, 133-158.

Hamilton, S. K., Kellndorfer, J., Lehner, B. and Tobler, M. (2007). Remote sensing of floodplain geomorphology as a surrogate for biodiversity in a tropical river system (Madre de Dios, Peru). *Geomorphology*, *89*(1), 23-38.

Hamilton, S. K., Sippel, S. J. and Melack, J. M. (2002). Comparison of inundation patterns in South American floodplains. *J. Geophys. Res.*, *107* (D20).

Hansen, A. J. and di Castri, F. (Eds). (1992). Landscape Boundaries. Springer, New York.

Harrison, P. J., Yin, K., Lee, J. H. W., Gan, J. and Liu, H. (2008). Physical–biological coupling in the Pearl River Estuary. *Continental Shelf Research*, *28*(12), 1405-1415.

Hawksworth, D. L., Kalin-Arroyo, M. T., Hammond, P. M., Ricklefs, R. E., Cowling, R. M., Samways, M. J. and Stace, C. A. (1995). Global Biodiversity Assessment: Ch 3 Magnitude and distribution of biodiversity. *In Global biodiversity assessment*, (pp. 107-192).

Heino, J., Virkkala, R. and Toivonen, H. (2009). Climate change and freshwater biodiversity: detected patterns, future trends and adaptations in northern regions. *Biological Reviews*, *84*(1), 39-54.

Hellmann, J. J., Byers, J. E., Bierwagen, B. G. and Dukes, J. S. (2008). Five potential consequences of climate change for invasive species. *Conservation biology*, *22*(3), 534-543.

Hess, L. L. and Melack, J. M. (1994). Mapping wetland hydrology and vegetation with synthetic aperture radar. *International Journal of Ecology and Environmental Sciences*, *20*(1-2), 74-81.

Hogan, Z. S., Moyle, P. B., May, B., Zanden, M. J. V. and Baird, I. G. (2004). The imperiled giants of the Mekong. *American Scientist*, *92*(3), 228-237.

Holland, M. M., Whigham, D. F. and Gopal, B. (1990). The charasteristics of wetland ecotones, p. 171–198. In R. J. Naiman and H. De´champs (eds.), The Ecology and Management of Aquatic-Terrestrial Ecotones. The Parthenon Publishing Group, Paris.

Hoogenboom, H. J., Dekker, A. G. and Althuis, I. A. (1998). Simulation of AVIRIS sensitivity for detecting chlorophyll over coastal and inland waters. *Remote Sensing of Environment*, *65*, 333-340.

Houlahan, J. E., Findlay, C. S., Schmidt, B. R., Meyer, A. H. and Kuzmin, S. L. (2000). Quantitative evidence for global amphibian population declines. *Nature*, *404*(6779), 752-755.

Huntington, T. G. (2006). Evidence for intensification of the global water cycle: review and synthesis. *Journal of Hydrology*, *319*, 83–95.

Hupp, C. R. (1988). Plant ecological aspects of flood geomorphology and paleoflood history. In: Baker, V. R., Kochel, R. C., Patton, R. C. (Eds.), Flood Geomorphology. Wiley, 335–356.

Hutchinson, C. F. (1991). Uses of satellite data for famine early warning in sub-Saharan Africa. *International Journal of Remote Sensing*, *12*(6), 1405–1421. *International Journal of Ecological and Environmental Science*, 20, 197-205.

IPCC (2007). Intergovernmental panel on climate change. Climate change 2007: Synthesis report.

Johns, D. M. and Maggs, C. A. (1997). Species problems in eukaryotic algae: a modern perspective. In Species: the Units of Biodiversity (eds. M. F. Claridge, H. Dawah and M. R. Wilson), pp. 82–107. Chapman and Hall, London, U.K.

Johnson, P. T., Olden, J. D. and Vander Zanden, M. J. (2008). Dam invaders: impoundments facilitate biological invasions into freshwaters. *Frontiers in Ecology and the Environment*, *6*(7), 357-363.

Jones, R. N., McMahon, T. A. and Bowler, J. M. (2001). Modelling historical lake levels and recent climate change at three closed lakes, Western Victoria, Australia (c.1840–1990). *Journal of Hydrology*, *246*, 159–180.

Junk, W. J., Bayley, P. B. and Sparks, R. E. (1989). The flood pulse concept in river floodplain systems. *Canadian Special Publication of Fisheries and Aquatic Sciences*, *106*, 110–127.

Junk, W. J. (1997). The central Amazon floodplain: ecology of a pulsing system. *Ecological Studies*, *vol.*, *126*. Springer, New York.

Kalliola, R., Puhakka, M. E., Salo, J., Tuomisto, H. and Ruokolainen, K. (1991a). The dynamics, distribution and classification of swamp vegetation in Peruvian Amazonia. *Ann. Bot. Fenn.*, *28*, 225–239.

Kimball, J. S., McDonald, K. C. and Zhao, M. (2006). Terrestrial vegetation productivity in the western arctic observed from satellite microwave and optical remote sensing. *Earth Interactions*, *10*, 22.

Kittelson, P. M. and Boyd M. J. (1997). Mechanisms of expansion for an introduced species of cordgrass, Spartina densiflora, in Humboldt Bay, California. *Estuaries*, *20*, 770–778.

Klinge, H., Junk, W. J. and Revilla, C. J. (1990). Status and distribution of forested wetlands in tropical South America. For. *Ecol. Manag.*, *33*/34, 81–101.

Koehn, J. D. (2004). Carp (Cyprinus carpio) as a powerful invader in Australian waterways. *Freshwater biology*, *49*(7), 882-894.

Kottelat, M. and Whitten, T. (1996). Freshwater fishes of Western Indonesia and Sulawesi: additions and corrections. Hong Kong: Periplus Editions.

Kramer, R. A., Richter, D. D., Pattanayak, S. and Sharma, N. P. (1997). Ecological and economic analysis of watershed protection in Eastern Madagascar. *Journal of Environmental Management*, *49*(3), 277-295.

Kundzewicz, Z. W., Mata, L. J., Arnell, N. W., Do¨ll, P., Jimenez, B., Miller, K., Oki, T., Sen Z. and Shiklomanov, I. (2008). The implications of projected climate change for freshwater resources and their management. *Hydrological Sciences Journal*, *53*, 3–10.

Lachavanne, J. B. and Juge, R. (Eds). (1997). Biodi6ersity in Land-Inland Water Ecotones. Parthenon, Pearl River, New York.

Lamotte, S. (1990). Fluvial dynamics and succession in the lower Ucayali River Basin, Peruvian Amazonia. *For. Ecol. Manag*., 33-4, 141–156.

Larned, S. T., Datry, T., Arscott, D. B. and Tockner, K. (2010). Emerging concepts in temporary-river ecology. *Freshwater Biology*, *55*(4), 717-738.

Leadley, P. (2010). Biodiversity Scenarios: Projections of 21st Century Change in Biodiversity, and Associated Ecosystem Services: a Technical Report for the Global Biodiversity Outlook 3 (No. 50). UNEP/Earthprint.

Lehner, B., Verdin, K. and Jarvis, A. (2006). HydroSHEDS Technical Documentation, Version 1.0. http://gisdata.usgs.net/HydroSHEDS/downloads/HydroSHEDS_TechDoc_v10. pdf.

Leprieur, F., Beauchard, O., Blanchet, S., Oberdorff, T. and Brosse, S. (2008). Fish invasions in the world's river systems: when natural processes are blurred by human activities. *PLoS Biol*, *6*(2), e28.

Lévêque, C. and Balian, E. V. (2005). Conservation of freshwater biodiversity: does the real world meet scientific dreams? *In Aquatic Biodiversity II*, (pp. 23-26). Springer Netherlands.

Lewis, Jr. W. M., Hamilton, S. K., Lasi, M. A., Rodríguez, M. and Saunders III, J. F. (2000). Ecological determinism on the Orinoco floodplain. *BioScience*, *50*, 681–692.

Lloyd, J. and Farquhar, G. D. (2008). Effects of rising temperatures and [CO2] on the physiology of tropical forest trees. Philosophical Transactions of the Royal Society B: *Biological Sciences*, *363*(1498), 1811-1817.

Loh, J. (2000). Living planet report 2000. *WWF--World Wide Fund for Nature*.

Losos, E. (1995). Habitat specificity of 2 palm species — experimental transplantation in Amazonian successional forests. *Ecology*, *76*, 2595–2606.

Lunderberg, G., Kottelat, M., Smith, G. R., Stiassny, M. L. J. and Gill, A. C. (2000). So many fishes, so little time: an overview of recent ichthyological discovery in continental waters. *Annals of the Missouri Botanical Gardens*, *87*, 26–62.

Malard, F., Tockne, K. and Ward, J. V. (2000). Physicochemical heterogeneity in a glacial riverscape. *Landscape Ecology*, *15*, 679–695.

Marmonier, P., Dole-Olivier, M. J. and Creuze des Chatelliers, M. (1992). Spatial distribution of interstitial assemblages in the floodplain of the Rhone River, Regul. *Rivers*, *7*, 75–82.

Marsh, G. A. and Fairbridge, R. W. (1999). Lentic and lotic ecosystems. *InEnvironmental Geology*, (pp. 381-388). Springer Netherlands.

McCully, P. (1996). Silenced Rivers. The Ecology and Politics of Large Dams. London: Zed Books.

McKnight, D. M., Niyogi, D. K., Alger, A. S., Bomblies, A., Conovitz, P. A. and Tate, C. M. (1999). Dry valley streams in Antarctica: ecosystems waiting for water. *Bioscience*, *49*, 985–995.

McMenamin, S. K., Hadly, E. A. and Wright, C. K. (2008). Climatic change and wetland desiccation cause amphibian decline in Yellowstone National Park. *Proceedings of the national Academy of Sciences*, *105*(44), 16988-16993.

Melack, J. M. (2004). Remote sensing of tropical wetlands in S. Ustin, editor. Manual of Remote Sensing, 3rd edition. Remote Sensing for Natural Resources Management and Environmental Monitoring. 3 ed., Vol. 4. John Wiley and Sons, New York. Pages 319-343.

Mertes, L. A. K. (1997). Documentation and significance of the perirheic zone on inundated floodplains. *Water Resour. Res.*, *33*, 1749–1762.

Meyer, J. L., Sale, M. J., Mulholland, P. J. and LeRoy Poff, N. (1999). Impacts of climate change on aquatic ecosystem functioning and health. *Journal of the american Water Resources association*, *35*(6), 1373-1386.

Milliman J. D., Farnsworth K. L., Jones P. D. and Smith L. C. (2008) Climatic and anthropogenic factors affecting river discharge to the global ocean, 1951–2000. *Global and Planetary Change*, *62*, 187–194.

Naiman, R. J. and Latterell, J. J. (2005). Principles for linking fish habitat to fisheries management and conservation. *Journal of Fish Biology*, *67*(sB), 166-185.

Naiman, R. J. and Decamps, H. (Eds). (1990). The Ecology and Management of Aquatic Terrestrial Ecotones. Parthenon, Pearl River, New York.

Naiman, R. J., Decamps, H., Pastor, J. and Johnston, C. A. (1988). The potential importance of boundaries to fluvial ecosystems, *J. North Am. Benthol. Soc.*, *7*, 289–306.

Nilsson, C. and Berggren, K. (2000). Alterations of Riparian Ecosystems Caused by River Regulation Dam operations have caused global-scale ecological changes in riparian ecosystems. How to protect river environments and human needs of rivers remains one of the most important questions of our time. *BioScience*, *50*(9), 783-792.

Nilsson, C., Reidy, C. A., Dynesius, M. and Revenga, C. (2005). Fragmentation and flow regulation of the world's large river systems. *Science*, *308*(5720), 405-408.

O'Callaghan, J. F. and Mark, D. M. (1984). The extraction of drainage networks from digial elevation data. *Computer Vision, Graphics and Image Processing*, *28*, 323-344.

Obrdlik, P. and Fuchs, U. (1991). Surface water connection and the macrozoobenthos of two types of floodplains on the upper Rhine, Regul. *Ri6ers*, *6*, 279–288.

Olden, J. D., Kennard, M. J., Leprieur, F., Tedesco, P. A., Winemiller, K. O. and García-Berthou, E. (2010). Conservation biogeography of freshwater fishes: recent progress and future challenges. *Diversity and Distributions*, *16*(3), 496-513.

Olsen, D. M. and Dinerstein, E. (1998). The Global 200: a representative approach to conserving the Earth's most biologically valuable ecoregions. *Conservation Biology*, *12*, 502–515.

Pasquini, A. I. and Depetris, P. J. (2006) Discharge trends and flow dynamics of South American rivers draining the southern Atlantic seaboard: an overview. *Journal of Hydrology*, *333*, 385–399.

Pasternack, G. B. and Brush G. S. (1998). Sedimentation cycles in a river-mouth tidal freshwater marsh. *Estuaries*, *21*, 407–415.

Petts, G. E. and Amoros, C. (1996). (Eds), Flu6ial Hydrosystems. Chapman and Hall, London.

Poff, N. L., Brinson, M. M. and Day, J. W. (2002). Aquatic ecosystems and global climate change. *Pew Center on Global Climate Change, Arlington, VA, 44*.

Poff, N. L., Richter, B. D., Arthington, A. H., Bunn, S. E., Naiman, R. J., Kendy, E. and Warner, A. (2010). The ecological limits of hydrologic alteration (ELOHA): a new framework for developing regional environmental flow standards. *Freshwater Biology*, *55*(1), 147-170.

Pörtner, H. O. and Knust, R. (2007). Climate change affects marine fishes through the oxygen limitation of thermal tolerance. *Science*, *315*(5808), 95-97.

Postel, S. and Richter, B. (2012). Rivers for life: managing water for people and nature. Island Press.

Pringle, C. M. (1997). Exploring how disturbance is transmitted upstream: going against the flow. *Journal of the north american Benthological society*, 425-438.

Puckridge, J. T., Sheldon, F., Walker, K. F. and Boulton, A. J. (1998). Flow variability and the ecology of large rivers. *Marine and freshwater research*, *49*(1), 55-72.

Puhakka, M. E., Kalliola, R., Rajasilta, M. and Salo, J. (1992). River types, site evolution and successional vegetation patterns in Peruvian Amazonia. *J. Biogeogr.*, *19*, 651–665.

Rahel, F. J. (2002). Homogenization of freshwater faunas. *Annual Review of Ecology and Systematics*, 291-315.

Rahel, F. J. and Olden, J. D. (2008). Assessing the effects of climate change on aquatic invasive species. *Conservation Biology*, *22*(3), 521-533.

Rapport, D. J., Costanza, R. and McMichael, A. J. (1998). Assessing ecosystem health. *Trends in Ecology and Evolution*, *13*(10), 397-402.

Reist, J. D., Wrona, F. J., Prowse, T. D., Power, M., Dempson, J. B., Beamish, R. J. and Sawatzky, C. D. (2006). General effects of climate change on Arctic fishes and fish populations. *AMBIO: A Journal of the Human Environment*, *35*(7), 370-380.

Rey Benayas, J. M. and Scheiner, S. M. (1993). Diversity pattern of wet meadows along geochemical gradients in central Spain. *Journal of Vegetation Science*, *4*,103–108.

Richter, B. D., Postel, S., Revenga, C., Scudder, T., Lehner, B., Churchill, A. and Chow, M. (2010). Lost in development's shadow: The downstream human consequences of dams. *Water Alternatives*, *3*(2), 14-42.

Risser, P. G. (1995). The status of the science of examining ecotones, *BioScience*, *45*, 318–325.

Robinson, C. T. and Matthaei, S. (2007). Hydrological heterogeneity of an alpine stream–lake network in Switzerland. *Hydrological Processes*, *21*, 3146–3154.

Roshier, D. A. and Rumbachs, R. M. (2004). Broad-scale mapping of temporary wetlands in arid Australia. *Journal of arid environments*, *56*(2), 249-263.

Sala, O. E., Chapin, F. S., Armesto, J. J., Berlow, E., Bloomfield, J., Dirzo, R. and Wall, D. H. (2000). Global biodiversity scenarios for the year 2100. *science*, *287*(5459), 1770-1774.

Salo, J., Kalliola, R., Häkkinen, I., Mäkinene, Y., Niemelä, P., Puhakka, M. E. and Coley, P. D. (1986). River dynamics and diversity of Amazon lowland forest. *Nature*, *322*, 254–258.

Schiemer, F. and Spindler, T. (1989). Endangered fish species of the Danube River in Austria, Regul. *Ri6ers*, *4*, 397–407.

Schulze, E. D. and Mooney, H. A. (1994). Biodiversity and Ecosystem Function. Springer Verlag, New York.

Secades, C., O'Connor, B., Brown, C. and Walpole, M. (2014). Earth Observation for Biodiversity Monitoring: A review of current approaches and future opportunities for tracking progress towards the Aichi Biodiversity Targets. Secretariat of the Convention on Biological Diversity, Montréal, Canada. *Technical Series No.*, *72*, 183 pages.

Secretariat of the Convention on Biological Diversity (CBD, 2009). Connecting Biodiversity and Climate Change Mitigation and Adaptation: Report of the Second Ad Hoc Technical Expert Group on Biodiversity and Climate Change. *Montreal*, *Technical Series No.*, *41*, 126 pages.

Sharma, S., Jackson, D. A., Minns, C. K. and Shuter, B. J. (2007). Will northern fish populations be in hot water because of climate change? *Global Change Biology*, *13*(10), 2052-2064.

Smith, V. H. (2003). Eutrophication of freshwater and coastal marine ecosystems a global problem. *Environmental Science and Pollution Research*, *10*(2), 126-139.

Sparks, R. E. (1995). Need for ecosystem management of large rivers and their floodplains. *BioScience*, 168-182.

Stiassny, M. L. J. (1999). The medium is the message: freshwater biodiversity in peril. In The Living Planet in Crisis: Biodiversity Science and Policy (eds. J. Cracraft and F. T. Grifo), pp. 53–71. Columbia University Press, New York, U.S.A.

Stiassny, M. L. J. (2002). Conservation of freshwater fish biodiversity: the knowledge impediment. *Verhandlungen der Gesellschaft fu¨r Ichthyologie*, *3*, 7–18.

Strayer, D. L. and Dudgeon, D. (2010). Freshwater biodiversity conservation: recent progress and future challenges. *Journal of the North American Benthological Society*, *29*(1), 344-358.

Strayer, D., Downing, J. A., Haag, W. R., King, T. L., Layer, J. B., Newton, T. J. and Nichols, S. J. (2004). Changing perspectives on pearly mussels, North America's most imperiled animals. *BioScience*, *54*, 429–439.

Stuart, S. N., Chanson, J. S., Cox, N. A., Young, B. E., Rodrigues, A. S., Fischman, D. L. and Waller, R. W. (2004). Status and trends of amphibian declines and extinctions worldwide. *Science*, *306*(5702), 1783-1786.

Terborgh, J. and Petren, K. (1991). Development of habitat structure through succession in an Amazonian floodplain forest, in Bell, S. S., McCoy, E. D., and Mushinsky, H. R. (Eds), Habitat Structure. *Chapman and Hall*, *London.*, 28–46.

Tharme, R. E. (2003). A global perspective on environmental flow assessment: emerging trends in the development and application of environmental flow methodologies for rivers. *River research and applications*, *19*(5-6), 397-441.

Thieme, M., Lehner, B., Abell, R., Hamilton, S. K., Kellndorfer, J., Powell, G. and Riveros, J. C. (2007). Freshwater conservation planning in data-poor areas: an example from a remote Amazonian basin (Madre de Dios River, Peru and Bolivia). *Biological Conservation*, *135*, 500-517.

Tickner, D. P., Angold, P. G., Gurnell, A. M. and Mountford, J. O. (2001). Riparian plant invasions: hydrogeomorphological control and ecological impacts. *Progress in Physical Geography*, *25*(1), 22-52.

Tilman, D., Fargione, J., Wolff, B., D'Antonio, C., Dobson, A., Howarth, R. and Swackhamer, D. (2001). Forecasting agriculturally driven global environmental change. *Science*, *292*(5515), 281-284.

Tockner, K., Uehlinger, U. and Robinson, C. T. (2009). Rivers of Europe. Academic Press, San Diego.

Tooth, S. (2000) Process, form and change in dryland rivers: a review of recent research. *Earth-Science Reviews*, *51*, 67–107.

Torsvik, V., Øvreas, L. and Thingstad, T. F. (2002). Prokaryotic diversity – magnitude, dynamics, and controlling factors. *Science*, *296*, 1064–1066.

Turner, J., Bindschadler, R., Convey, P., Di Prisco, G., Fahrbach, E., Gutt, J. and Summerhayes, C. (2009). Antarctic climate change and the environment.

Vitousek, P. M., Chair, A. J., Howarth, R. W., Likens, G. E., Matson, P. A., Schindler, D. W., Schlesinger, W. H. and Tilman, D. (1997). Human alteration of the global nitrogen cycle: Cause and consequences. *Issues Ecology*, *Ecological Society of America*, *1*, 1–15.

Vivian-Smith, G. (1997). Microtopographic heterogeneity and floristic diversity in experimental wetland community. *Journal of Ecology*, *85*, 71–82.

Vörösmarty, C. J., Green, P., Salisbury, J. and Lammers, R. B. (2000). Global water resources: vulnerability from climate change and population growth.science, 289(5477), 284-288.

Vörösmarty, C. J., McIntyre, P. B., Gessner, M. O., Dudgeon, D., Prusevich, A., Green, P., et al. (2010). Global threats to human water security and river biodiversity. *Nature*, *467*(7315), 555-561.

Walter, K. M., Zimov, S. A., Chanton, J. P., Verbyla, D. and Chapin III, F. S. (2006). Methane bubbling from Siberian thaw lakes as a positive feedback to climate warming. *Nature 443*(7), 71-75.

Ward, J. V., Tockner, K. and Schiemer, F. (1999). Biodiversity of floodplain river ecosystems: Ecotones and connectivity. *Regulated Rivers: Research Management*, *15*, 125 39.

Ward, J. V. (1998). Riverine landscapes: Biodiversity patterns, disturbance regimes, and aquatic conservation. *Biological Conservation*, *83*, 269–278.

Ward, J. V. and Stanford, J. A. (1983). The intermediate-disturbance hypothesis: an explanation for biotic diversity patterns in lotic ecosystems, in Fontaine, T. D. and Bartell, S. M. (Eds), Dynamics of Lotic Ecosystems. *Ann Arbor Science Publishers*, *Ann Arbor*, *MI*. 347–356.

Ward, J. V. and Stanford, J. A. (1995). Ecological connectivity in alluvial river ecosystems and its disruption by flow regulation, Regul. *Rivers*, *11*, 105–119.

Ward, J. V. and Wiens, J. A. (1999). 'Ecotones of riverine ecosystems: role and typology, spatio-temporal dynamics, and river regulation,' in Zalewski, M., Thorpe, J. E., and Schiemer, F. (Eds), Fish and Land/lnland Water Ecotones—The Need for Integration of Fisheries Science, Limnology and Landscape Ecology. Parthenon, Pearl River, New York.

Ward, J. V., Tockner, K., Arscott, D. B. and Claret, C. (2002). Riverine landscape diversity. *Freshw. Biol.*, *47*, 517–539.

Welcomme, R. L. (1979). Fisheries Ecology of Floodplain Ri6ers. Longman, London.

Winter, T. C. (2001). The concept of hydrologic landscapes. *J. Am. Water Resour. Assoc.*, *37*, 335–349.

World Commission on Dams. (2000). Dams and Development: A New Framework for Decision-making: the Report of the World Commission on Dams. Earthscan.

Wrona, F. J., Prowse, T. D., Reist, J. D., Hobbie, J. E., Lévesque, L. M. and Vincent, W. F. (2006). Climate change effects on aquatic biota, ecosystem structure and function. AMBIO: *A Journal of the Human Environment*, *35*(7), 359-369.

Xenopoulos, M. A. and Lodge, D. M. (2006). Going with the flow: using species-discharge relationships to forecast losses in fish biodiversity. *Ecology*, *87*(8), 1907-1914.

Zhang, X., Harvey, K. D., Hogg, W. D. and Yuzyk, T. R. (2001) Trends in Canadian streamflow. *Water Resources Research*, *37*, 987–998.

In: Riparian Zones
Editor: Oleg S. Pokrovsky

ISBN: 978-1-63484-613-4

Chapter 2

The Science of Mapping Riparian Areas Utilizing GIS and Open Source Geospatial Data

S. A. Abood*

USDA Forest Service-Watershed, Fish, Wildlife,
Air and Rare Plants (WFWARP) Program, Washington D.C., US

Abstract

Variable width riparian areas delineation is an important step in riparian areas assessment and monitoring. An accurate riparian areas base map can aid projects managers in their riparian inventory and restoration efforts. Riparian areas delineation techniques are different with different riparian definitions and data to achieve the mapping goal. Previous approaches have primarily utilized fixed width buffers. However, these methodologies only take the watercourse into consideration and ignore critical geomorphology, associated vegetation and soil characteristics. Other approaches utilize remote sensing technologies such as aerial photography interpretation and/or satellite imagery to delineate and classify riparian vegetation. Such techniques require expert knowledge, high spatial resolution imagery, and become expensive when mapping riparian areas at a landscape scale. In this chapter we present details on using the Riparian Buffer Delineation Model (RBDM) v3.0 and open source geospatial data for mapping variable width riparian areas and assessing riparian land use/cover classes distribution over a period of time. The model is an add-on toolbox for ArcMap 10.x and uses elevation, soils, wetlands, and land use/cover inputs. It recognizes the dynamic and transitional nature of riparian areas by considering hydrologic, geomorphic and vegetative information. Results suggest that, by incorporating functional variable width riparian mapping into watershed management planning improves protection and restoration of valuable riparian functionality and biodiversity.

Keywords: riparian, ecotone, RBDM, variable width buffer, open source geospatial data

* Corresponding author: Email: sinanayadabood@fs.fed.us.

INTRODUCTION

Riparian buffers are important ecotones with dynamic boundaries in transitional location between two different ecosystems, aquatic and terrestrial. Riparian ecotones have a well-defined vegetation and soil characteristics (Mitsch and Gosselink, 1993). Cowardin et al. 1979 suggests that riparian ecotones are "*Land inclusive of hydrophytes and/or with soil that is saturated by ground water for at least part of the growing season within the rooting depth of potential native vegetation.*" This definition includes wetlands and adjacent lands that have a moderate or well-balanced supply of moisture (Mitsch and Gosselink, 1993).

Riparian ecotones bear out many physical ecological and biological functions beside many economic and social values. Functions such as sediments trapping, stream banks stabilizing, moderate sun light and stream temperatures, acting as filters improving water quality, provides nutrients and woody debris and also create wildlife habitat and natural corridors (Klemas et al., 2014). Riparian vegetation is an important source of particulate and dissolved organic matter for adjacent aquatic ecosystems and helps regulate the nutrient, pesticide, and sediment transport between agricultural lands and aquatic ecosystems (Naiman et al., 1997).

The ecological term "ecotone" serves as the zone of interactions between a stream ecosystem and a terrestrial ecosystem which combines geomorphology and functional characteristics of a riparian area. The term "ecotone" implies that riparian area boundary is not at a fixed distance from stream ecosystem but has a variable width (Ilhardt et al., 2000). Applying the term riparian ecotone would minimize confusion among different scientific fields and management agencies and cancels the single riparian characteristic mapping approach such as riparian vegetation or hydric soils and it also denies that riparian ecotones occurs at fixed width buffer along the stream channel floodplain (Verry et al., 2004).

There are three properties allowing to distinguish riparian ecotones from adjacent ecosystems (Mitsch and Gosselink, 1993):

- Riparian ecotones generally have a liner form as a consequence of their proximity to rivers and streams;
- Energy and material from the surrounding landscape pass through riparian ecotones in much greater amounts than those of any other wetland ecosystem; and
- Riparian ecosystems are functionally connected to upstream and downstream ecosystems and are laterally connected to upslope (upland) and down slope (aquatic) ecosystems.

Riparian ecotones mapping approaches ranged from a simple fixed width buffer around selected streams and rivers to more complicated remote sensing and geographic information systems (GIS) methodologies using aerial photos and high resolution satellite images. The most common is fixed width riparian buffer. This methodology proved to be incompetent to map riparian ecotones as it takes into consideration only the stream channel course and fails to incorporate the surrounding geomorphology and associated vegetation. Fixed width buffer failed to map natural occurring riparian floodplains since fixed buffer have no functional relationship to the naturally varying stream watercourse (Palik et al., 2000). Skally and Sagor (2001) showed that riparian ecotones 2.5 times farther form the stream than the proposed fixed width buffer while evaluating riparian buffers guidelines from Minnesota Forest Resources Council.

Developing all-encompassing definition which is the main foundation in developing any sound mapping approach for riparian ecotones delineation is challenging due to their high variability. However, there are two factors that all riparian ecotones are dependent on: the watercourse and its associated floodplain.

In this chapter we present the functional riparian ecotones definition developed by Ilhardt et al., 2000 to map variable width riparian ecotones utilizing RBDMv3.0 "*Riparian areas are the three-dimensional space of interaction that includes terrestrial and aquatic ecosystems that extend down into the groundwater, up above the canopy, outward across the floodplain, up the near-slopes that drain to the water, laterally into the terrestrial ecosystem, and along the water course at a variable width*" this definition identifies riparian ecotones by its ecological functions and denies the idea of recognizing riparian ecotones by a single riparian characteristic such as vegetation, soil, or hydrologic regimes. The functional definition suggests that a riparian ecotone boundary does not occur at a fixed distance but at variable width along rivers and streams (Ilhardt et al., 2000).

RBDMv3.0 expands the functionality of the GIS riparian areas delineation model developed by Abood et al. (2011) and Abood et al. (2012) to map riparian ecotones adequately and efficiently by hydrologically defining riparian ecotone to occur at the 50-year floodplain. The updated model discussed here incorporates optional data such as National Wetland Inventory data (NWI) and Soil Survey Geographic (SSURGO) Database to improve the delineation process utilizing available ecological attributes. Also incorporating the National Land Cover Database (NLCD) or Cropland Data Layer (CDL) to provide a better understanding of riparian ecotones land use/cover and assess any change on the ground. Throughout this chapter the terms riparian ecotone and riparian area will be used interchangeably.

STUDY AREA

Hiawatha national forest located in the eastern part of the Upper Peninsula Michigan, divided into two sections eastern and western. Upper peninsula, Michigan formed from two major landforms, Grand Marais sandy end moraine and outwash and Seney sand lake plain, both of lacustrine origin. The Grand Marais landform is composed of sandy ridges of end moraine. The moraine contains droughty sand dunes and beach ridge deposits, as well as poorly and very poorly drained glacial lacustrine deposits (Albert, 1995). The Seney sand lake plain contains broad, poorly drained embayments with beach ridges and swales, sand spits, transverse sand dunes and sand bars. Along the northern margins of the embayment deltaic deposits occur where glacial streams carried massive amounts of sand into shallow waters (Albert, 1995). The western section consists of 63 HUC-12 watersheds with a total area of 534,806.4 ha. The eastern section consists of 43 HUC-12 watersheds with a total area of 321,967 ha. Hiawatha national forest is a north woods forest with a wide range of tree species such as jack pine, white pine, maple, basswood, and hemlock. The national forest has vast coastal areas with three great lakes, Lake Superior at the northern boarders and Lake Michigan and Huron at the southern boarders. There are 418 inland lakes and 450 miles of rivers and streams (Mohlenbrock 2006). For this chapter our study area is a HUC-12 watershed located within the western section of Hiawatha national forest (Figure 1).

MODEL DATA INPUTS

The RBDMv3.0 utilizes open source geospatial data freely available for downloads from federal and states agencies with known data collection/production methodologies, and accuracies. Table 1 lists the model geospatial raw data sources and description used to delineate variable width riparian areas. The National Hydrography Dataset (NHD) serves as the source for the hydrologic input in the model. The NHD is comprised of water-related entities such as natural river courses, lakes, watersheds, ditches, industrial discharges, drinking water supplies, etc., Each entity has an assigned address that establishes its location and connections to other entities in the drainage network (USGS, 2010). National Wetlands Inventory (NWI) provides national wetlands information coverage such as wetlands locations, wetlands distribution and wetlands vegetative attributes. The NWI is produced and maintained by the U.S. Fish and Wildlife Service (FWS) (FWS, 2015). The Soil Survey Geographic (SSURGO) database provides soil map units information in tabular and spatial formats. The required soil characteristics are soil formation condition, soil wetness, soil infiltration and frequently flooded areas found in hydric soil rating, drainage class, hydrological soil group, and flood frequency soil mapping attributes. The (SSURGO) database is maintained by the Natural Resource Conservation Service (NRCS) (NRCS, 2015).

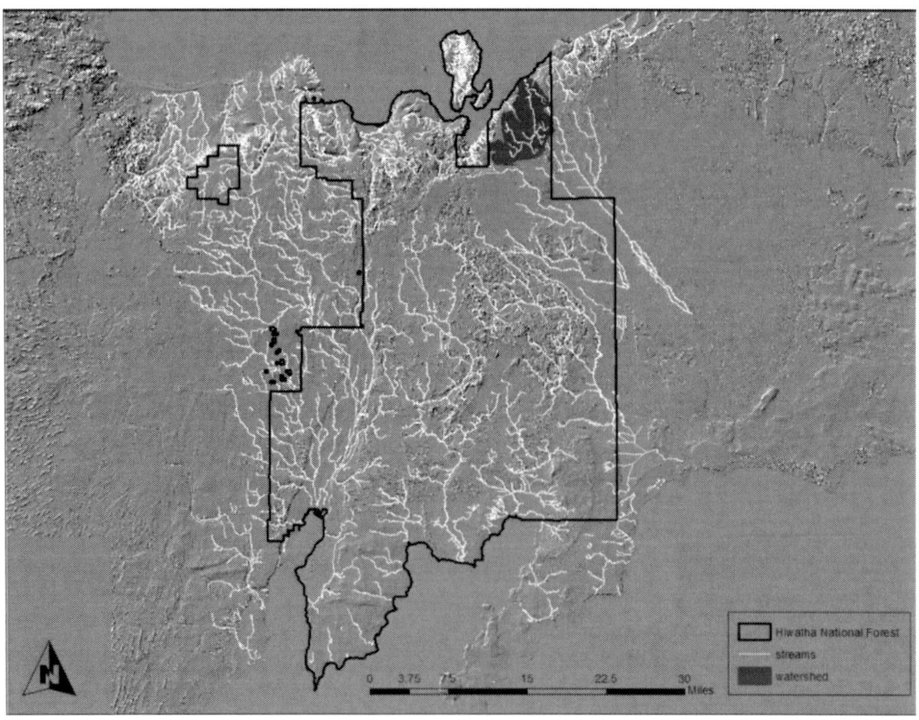

Figure 1. The project study area, western part Hiawatha National Forest.

Table 1. RBDMv3.0 data inputs

Input Data*	Sources	Description
Streams, Lakes, and Watersheds	USGS National Hydrology Dataset (NHD) http://nhd.usgs.gov/	Vector data format
Wetlands	National Wetlands Inventory (NWI) http://www.fws.gov/wetlands/	Vector data format
Soil	Natural Resources Conservation Service (NRCS) http://soildatamart.nrcs.usda.gov/or http://websoilsurvey.nrcs.usda.gov/app/HomePage.htm	Vector data and tabular data formats
Elevation	National elevation Dataset http://ned.usgs.gov/ GIS Data Depot http://data.geocomm.com/	Raster data format
Land Cover	National Land Cover Database http://www.mrlc.gov/ Corp land Data Layer http://www.nass.usda.gov/research/Cropland/SARS1a.htm	Classified raster data
50 year flood height	Calculated using Masson (2007) approach	numeric

Digital Elevation Model (DEM) is a raster based elevation information representation of the earth surface maintained by the USGS National Elevation Dataset (NED) (USGS, 2015). There are different (DEM) spatial resolutions available on the national elevation map with different range of coverage across the United States. 30m (1 arc-second) DEM is the lower resolution dataset that is available throughout the United States, part of Alaska, most of Canada and Mexico. 10m (1/3 arc-second) DEM available over the lower 48 states with a partial coverage in Alaska. 5m DEM derived from RADAR data available over Alaska only. 3m (1/3 arc-second) DEM available over approximately one third of the country. Lastly 1m LiDAR derived DEM available over limited geographic locations with a Universal Transverse Mercator (UTM) projection. In this chapter a 10m DEM is used in the riparian delineation process.

The (RBDM) model utilizes two sets of classified land cover rasters interchangeably. National Land Cover Dataset (NLCD) (MRLC, 2011) and Cropland Data layer (CDL). The (NLCD) is maintained through the Multi-Resolution Land Characteristic Consortium (MRLC) which represents a joint partnership among many federal agencies such as the U.S. Geological Survey (USGS), the National Oceanic and Atmospheric Administration (NOAA), the U.S. Environmental Protection Agency (EPA), the U.S. Department of Agriculture (USDA), the Natural Resources Conservation Service (NRCS), the U.S. Forest Service (USFS), the National Park Service (NPS), the U.S. Fish and Wildlife Service (USFWS), the Bureau of Land Management (BLM), the National Aeronautics and Space Administration (NASA), and the Office of Surface Mining (OSM). The NLCD was developed to serve two main objectives. Continues land use/cover layer for the conterminous United States, Alaska, Hawaii, and Puerto Rico at 30m spatial resolution and to provide a standard land use/cover classification system that can be updated and maintained frequently with applications across different scientific and commercial fields (Homer et al., 2004). The classification system consists of 16 land use/cover classes with average accuracy of 83.9% (Figure 2) (Homer et al., 2004 and 2007). The MRLC maintains the NLCD land use/cover layer on a 5 years cycle starting from 2001 (NLCD, 2011).

CDL is maintained by the National Agricultural Statistics Service (NASS) (NASS, 2014). The developed data is continuous classified land use/cover raster with emphasis on crops types, and crops spatial distribution at 56m (before 2008)) and 30m spatial resolutions (Johnson et al., 2010). The classified raster accuracy ranges between 90% in intensive agricultural areas such as the US Corn Belt to 80% classification accuracy in less widely planted crops like potatoes, barley, sunflowers, and canola with an average of 78% classification accuracy in all crops lands. Non-agricultural classes' classification accuracy is the same as of NLCD (Johnson et al., 2010).

The last input parameter in Table 1 is the 50 year flood height. This calculated parameter is the hydrological descriptor of variable width riparian areas which intersects with the channel first terrace or upward sloping and supports the same geomorphology and microclimate as the stream channel (Ilhardt et al., 2000). Two stream measurements syncs with the 50 year flood height, entrenchment ratio and belt width ratio. Entrenchment ratio is the stream channel width at first terrace divided by stream channel bankfull and belt ratio is the stream channel belt width detectible on aerial photos or images. Together both measurements provide characteristics and contrast distinction between vegetation on the floodplain and vegetation on upward slopes (Ilhardt et al., 2000).

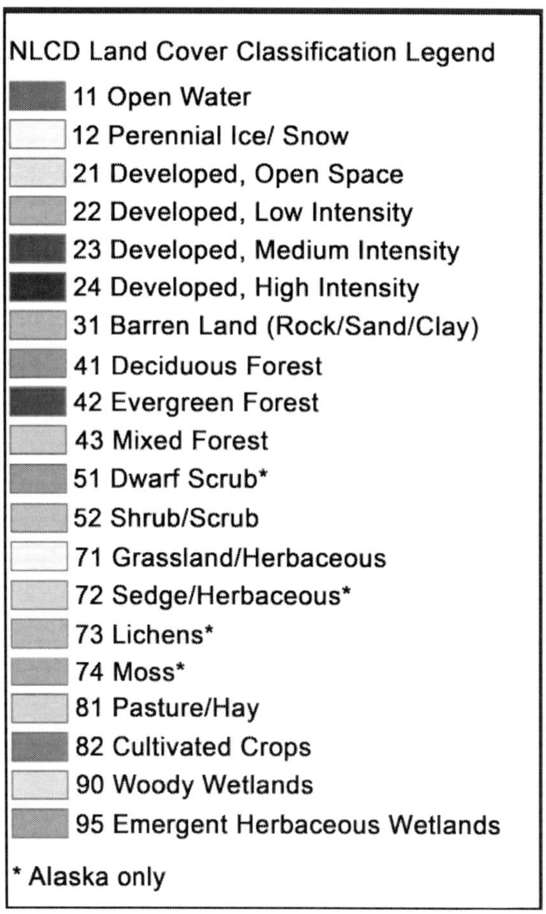

Figure 2. 2001 Land use/cover classes (MRLC, 2011).

Data Preparation

Input data preparation is a critical step before running the RBDMv3.0 model since each data input maintained by different federal agency or data sources. The three main preparation steps are: calculating the 50 year flood height value, correcting streams location, and preparing SSURGO data layers (Abood et al., 2011a and b).

Mason 2007 presented a simple methodology to calculate the 50 year flood height value at a given stream location based on (Bedient and Huber 2002) recommendations. This methodology utilizes the USGS National Water Information System (NWIS) real time water data (USGS, 2007) at each gauging station location within or in close proximity of the study area. The water data includes annual average stream flows and the periodic channel measurements includes channel discharge, channel width, and channel cross sectional area. Average annual flow data is organized and ranked from fastest to slowest where the fastest flow ranked as 1 (Figure 3). Recurrence interval developed by regressing flow against ranking to estimate the 50 year flow. Regressing cross sectional area versus channel discharge and channel width versus channel cross sectional area would be the next step to calculate the 50 year flood height. An R –squared value above 0.85 were observed in all calculations per each gauge station and the calculated minimum 50 year flood height is 1.40m.

NHD database compiles water attributes in two different formats. Spatially such as streams flowlines, watersheds boundaries, and waterbodies and important tabular attributes such as stream type/code, stream level, and stream reachcode (USGS, 2011). These are required inputs of the RBDMv3.0 model. Each stream feature within the NHD database inherits a 40 feet positional inaccuracy. Figure 4 explains the inherited position inaccuracies within the NHD database that causes stream fail to capture it valley bottom. Correcting streams location before starting the delineation process is a critical step (Abood et al., 2012a and b and Abood et al., 2015).

The RBDMv3.0 model utilizes four soils data layers representing four major riparian soils characteristics, hydric rating, Drainage class, hydrologic soil group, and flood frequency (Abood et al., 2012a and b and Abood et al., 2015). SSURGO data is formatted in spatial and tabular formats. Spatial data contains the soil map units and tabular data contains the soil attributes organized in many categories such as land classification, soil chemical properties, soil physical properties, etc., both spatial and tabular data are applied in the Soil Data Viewer developed by the NRCS to produce four ArcGIS feature classes with riparian soil attributes (Abood et al., 2012a and b and Abood et al., 2015).

All the model data input must be compiled in an ArcGIS for Desktop File Geodatabase (FGDB) format and the user must have access to the spatial analyst extension.

Riparian Model

RBDMv3.0 is a developed ArcGIS for Desktop toolbox (ESRI, 1999-2015) utilizing the management power and spatial analysis capabilities of ArcGIS for Desktop software and Python 2.7 coding language. The model is based on a procedure originally presented by Aunan et al., 2005 and further developed by Mason, 2007 and Abood et al., 2011a and b, Abood et al., 2012, and Abood et al., 2015).

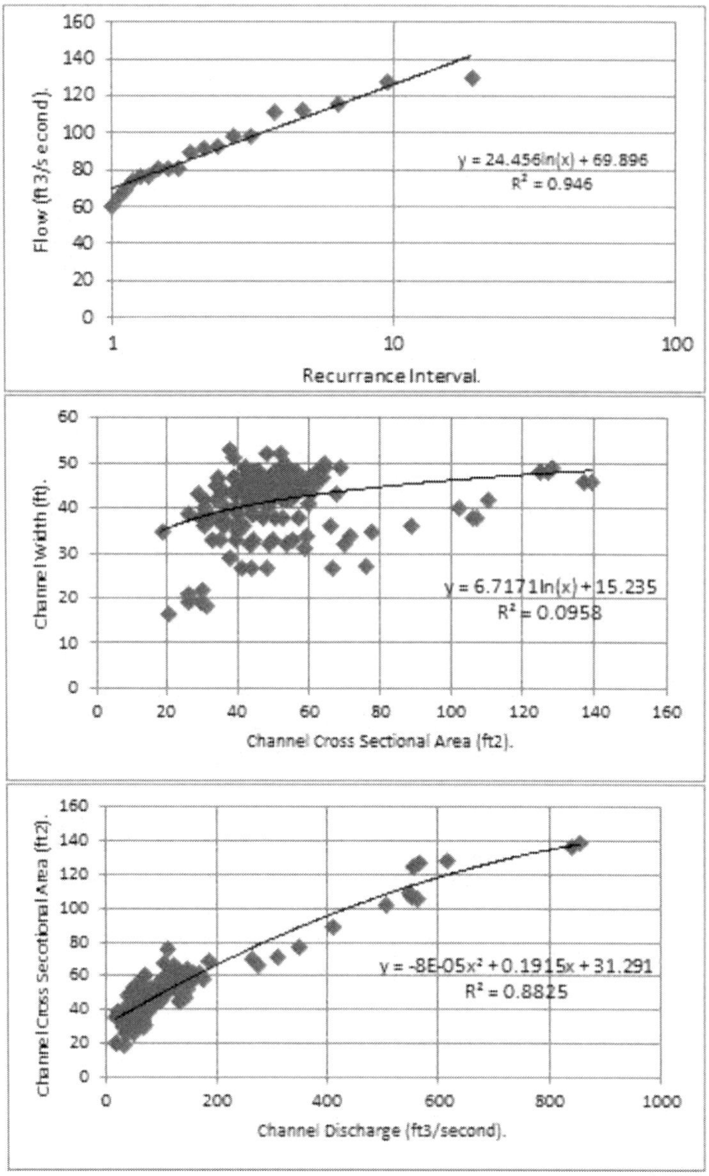

Figure 3. 50 year flood height hydrological estimation.

RBDMv3.0 consists of a simple graphical interface (Figure 5) with 19 inputs for different riparian delineation scenarios. 8 inputs are required to run the model and the rest are optional inputs (Abood et al., 2015). All geospatial layers in Table 1 are set to ESRI FGDB (ESRI, 1999-2015) for all data manipulation and spatial analysis. The new model utilizes new optional inputs such as NWI, SSURGO, and land cover data to map and classify variable width riparian areas within and beyond the occurring 50 year flood height to fully enclose all riparian areas functional, hydrological, and ecological characteristics based on recommendations by Palik et al., 2004 and Verry et al., 2004.

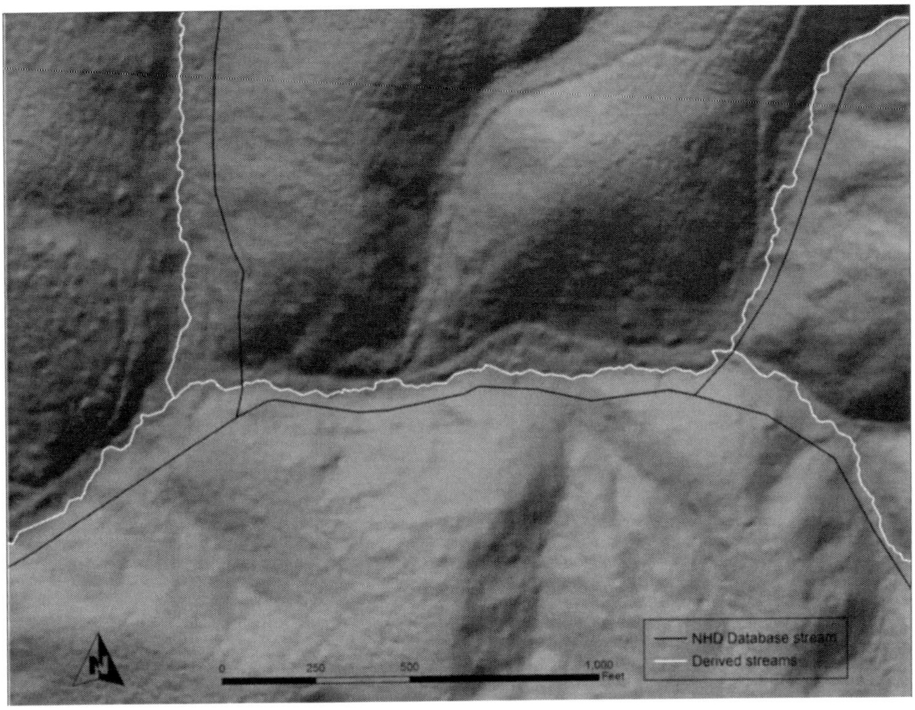

Figure 4. NHD positional inaccuracies. From Abood et al., 2011b.

The current riparian model introduces many new improvements over the last model versions RBDMv2.0, RBDMv2.2 (Abood et al., 2011a and b and 2012). A new and improved transects sampling technique. Transects are produced in 360° around streams segments in 11 directions forming transects points cloud to ensure realistic riparian areas delineation by capturing all elevation variations and changes in stream channel direction. A flexibility to choose transects vector sampling distance (Figure 5). In other words how far form a stream segment the model looks for riparian areas. This option controls the sampling distance and the size of transects points cloud. Setting the transects distance to 500 meter means the riparian algorithm would produce transects points cloud up to 500 meter away on both sides of the selected stream. The sampling distance has been limited to 3000m (9842.5 feet) to optimize the model processing time (Abood et al., 2011a and b). Lastly soil module has been updated with a new soil attribute "Flood Frequency" to map riparian soils that extends beyond the occurring 50 year flood height (Palik et al. 2004).

Table 2 lists all queries inputs utilized to run the RBDMv3.0 to delineate riparian areas in the understudy watershed. There are three sets of queries. Streams query controls the type of streams included in the riparian delineation process such as perennial, intermittent, or ephemeral. Wetlands query controls wetlands type extending within and in close proximity of the occurring 50 year flood height. The last set is soil query. It is designed to identify riparian soils according to four riparian soil characteristics; soil hydric rating, drainage class, hydrological soil group, and flood frequency (Palik et al., 2004). A soil map unit considered to be riparian soil when all the soil parameters are true.

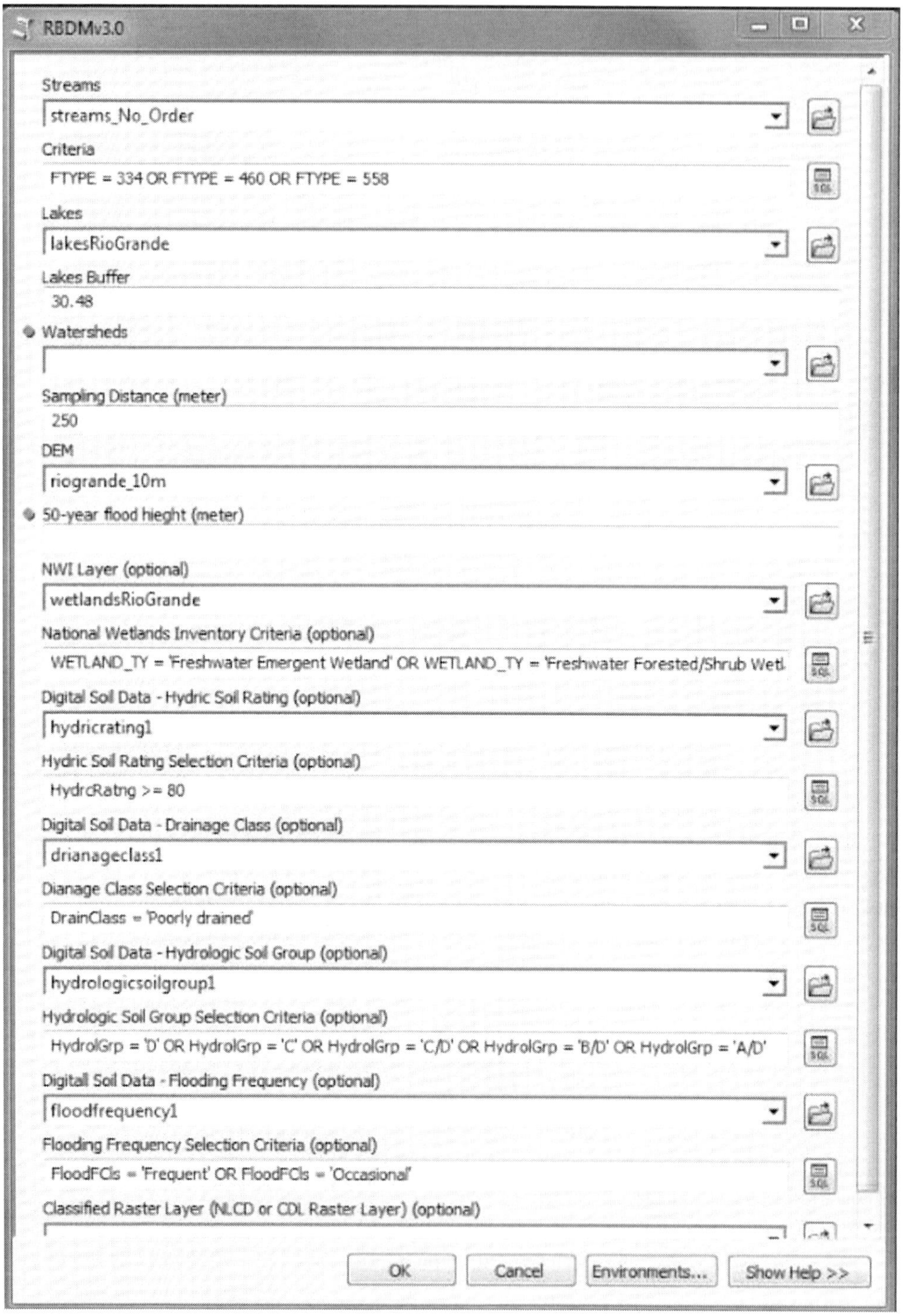

Figure 5. RBDMv3.0 model interface.

Table 2. Required RBDMv3.0 queries

Name	Query Input	Description
Streams Criteria	FTYPE = 334 OR FTYPE = 460 OR FTYPE = 558	FTYPE represents streams type 334 = connector 460 = streams (intermittent, perennial, and ephemeral) 558 = rivers or general flow waterbodies.
Lakes Buffer	Set at 30.48m (100 feet)	60%-80% of riparian functions occurs at 100 feet of waterbodies (Ilhardt et al., 2000)
Sampling Distance	Set to 500m.	This is a user defined input. The maximum sampling distance set at 3000m. This input controls the sampling distance away from the stream segment and the size of transects points cloud around stresms
50 year flood height	1.40m calculated according to Mason, 2007	This value is the hydrologic descriptor of riparian ecotones. (Ilhardt at al., 2000)
National Wetlands Inventory Criteria	WETLAND_TY = 'Freshwater Emergent Wetland' OR WETLAND_TY = 'Freshwater Forested/Shrub Wetland' OR WETLAND_TY = 'Riverine'	Wetlands divided into three major types: • Freshwater Emergent Wetland • Freshwater Forested/Shrub Wetland • Riverine Subclasses are according to Cowardin et al., 1979
Hydric Soil Rating Selection Criteria	HydrcRatng >= 90 (Palik et al., 2004)	Hydric soils are defined according to the National Technical Committee for Hydric Soils (NTCHS). Soils that formed under conditions of saturation, flooding or ponding long enough during the growing season to develop anaerobic conditions in the upper part (Soil Data Viewer 6.0 User guide, 2011)
Drainage Class Selection Criteria	DrainClass = 'Poorly drained' OR DrainClass = 'Somewhat Poorly Drained' OR DrainClass = 'Very Poorly Drained' (Palik et al., 2004)	Represents to the frequency and duration of wet periods under conditions similar to those under which the soil formed (Soil Data Viewer 6.0 User guide, 2011)
Hydrologic Soil Group Selection Criteria	HydrolGrp = 'D' OR HydrolGrp = 'C' OR HydrolGrp = 'C/D' OR HydrolGrp = 'B/D' OR HydrolGrp = 'A/D' (Palik et al., 2004)	Based on estimates of runoff potential and the rate of water infiltration when the soils are not protected by vegetation (Soil Data Viewer 6.0 User guide, 2011)
Flood Frequency Selection Criteria	FloodFCls = 'Very Frequent' OR FloodFCls = 'Frequent' OR FloodFCls = 'Occasional' (Palik et al., 2004)	Temporary inundation of an area caused by overwhelming streams, runoff from adjacent slopes, and tides (Soil Data Viewer 6.0 User guide, 2011)

RESULTS AND DISCUSSION

The (RBDM) v3.0 model offers a range of riparian areas delineation results. The results depend on the input data configuration. In this section all the required and optional inputs in Table 1 and 2 utilized in two main configurations to produce the final delineation and extent of riparian areas within our watershed; delineated riparian areas utilizing 50 year flood height and delineated riparian areas utilizing 50 year flood height, NWI data, and SSURGO data.

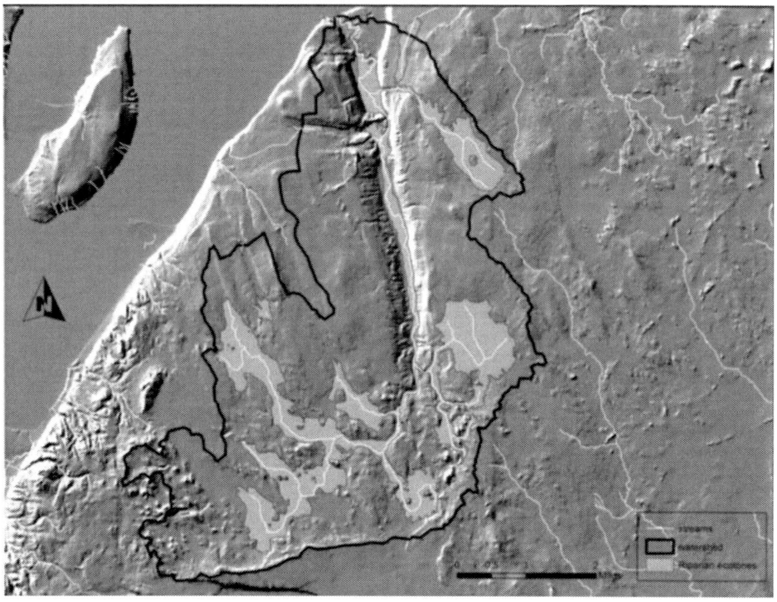

Figure 6. Delineated riparian ecotones utilizing 50 year flood height.

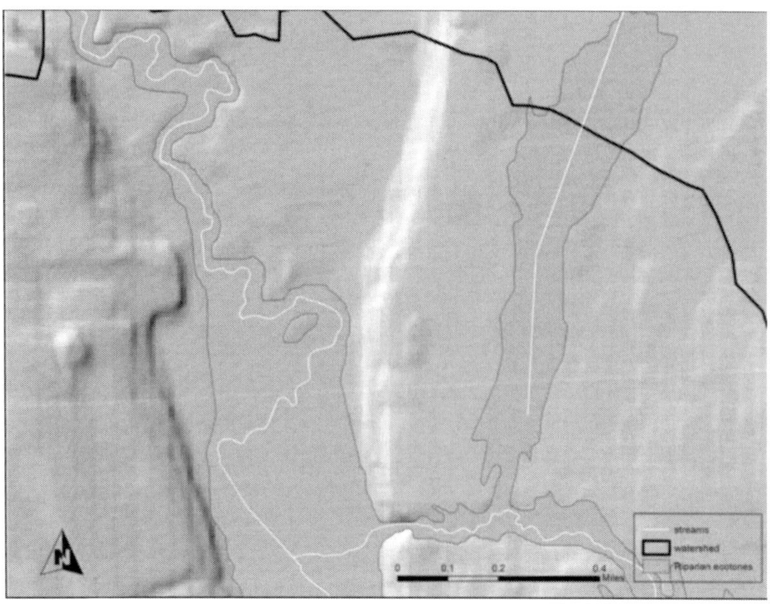

Figure 7. Closer look at riparian ecotones.

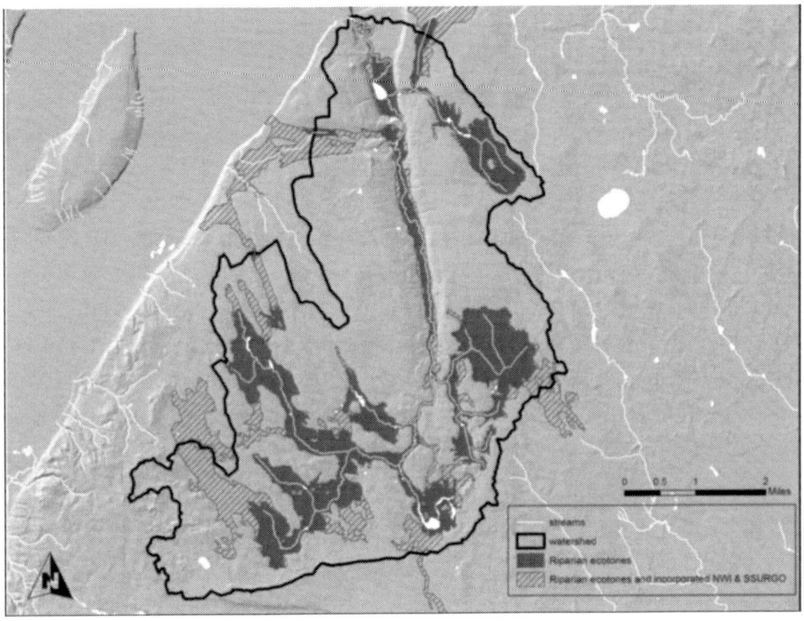

Figure 8. Difference in riparian areas extent between using the required inputs and incorporating optional data.

Figure 6 shows the final delineated riparian areas applying the calculated 50 year flood height. The model was successful in characterizing riparian areas for different landforms such as narrow and wide floodplains. Figure 7 illustrates how accurate is the final delineation utilizing the 50 year flood height as the hydrological descriptor of riparian ecotones. The delineated riparian areas were able to follow stream valley geomorphology retaining a continuous linear form as suggested by (Mitsch and Gosselink, 1993) even when the stream valley changes. Mason, 2007 and Abood et al., 2012a and b used the linear mixed effect model (Kutner et al., 2005) which shows that there is no linear correlation between the size of watershed area and delineated riparian area and also RBDMv3.0 delineation process is independent of landform.

Incorporating optional data NWI and SUSURGO in the delineation process offers a comprehensive estimate of riparian areas within our watershed. Figure 8 illustrate the difference in riparian areas extent between using the required inputs and incorporating optional data such as NWI and SSURGO in the delineation process (Table 3).

NWI and SSURGO data offer the advantage to extend riparian areas delineation beyond the mapped 50 year flood height floodplain. NWI data incorporates wetlands within or extends outward the floodplain and SSURGO data incorporates identified riparian soil units according to Palik et al. 2004 criteria within or extends outward the delineated riparian boundary. This spatial adjacency verifies that riparian ecotones are not limited to streams and rivers floodplains but includes lands associated with other kinds of surface water such as lakes and wetlands and also lands impacted by surface water such as riparian soils.

Table 4 illustrates wetlands distribution and type within the delineated areas utilizing required and optional data. 40.3% of delineated riparian areas are associated with wetlands where 95.4% freshwater emergent wetlands and the rest 59.7% are associated with streams. Figure 9 explain hydric soil distribution within delineated riparian areas. 76% of that has a hydric soil rating equal or more than 90% based on delineated riparian areas utilizing required and optional data.

Incorporating land use/cover attributes in the delineation process would result classified riparian ecotones such information can be used in monitoring riparian areas land cover distribution over time also to assess the land cover change. Figure 10 shows the incorporation of classified raster CDL 2011. The result offer detailed land use/cover of riparian areas at 30m spatial resolution generated from landsat satellite imagery. Such detailed information shows wide range of land cover classes within delineated riparian ecotones. For further focused land use/cover assessment riparian land cover data reclassified to six broad land cover classes. Table 5 shows the reclassified land cover classes according to Palik et al., 2004 suggestion with more subclasses details for the Natural/Semi Natural land cover attribute.

Table 3. Total riparian areas delineated utilizing required and optional data

Inputs	Total Riparian area (Hectares)
Required data (applying 50 year flood height)	1482.3 (21.6% of watershed area is riparian)
Required data + Optional data (applying 50 year flood height +NWI+SSURGO)	2672.6 (38.9% of watershed area is riparian)

Table 4. Wetlands type and distribution within delineated riparian areas

Wetlands Type	Area (Hectares)	% of Total Wetlands
Freshwater Emergent Wetland	37	3.4
Freshwater Forested/Shrub Wetland	1028.1	95.4
Freshwater Pond	10.2	0.9
Lake	2.0	0.18

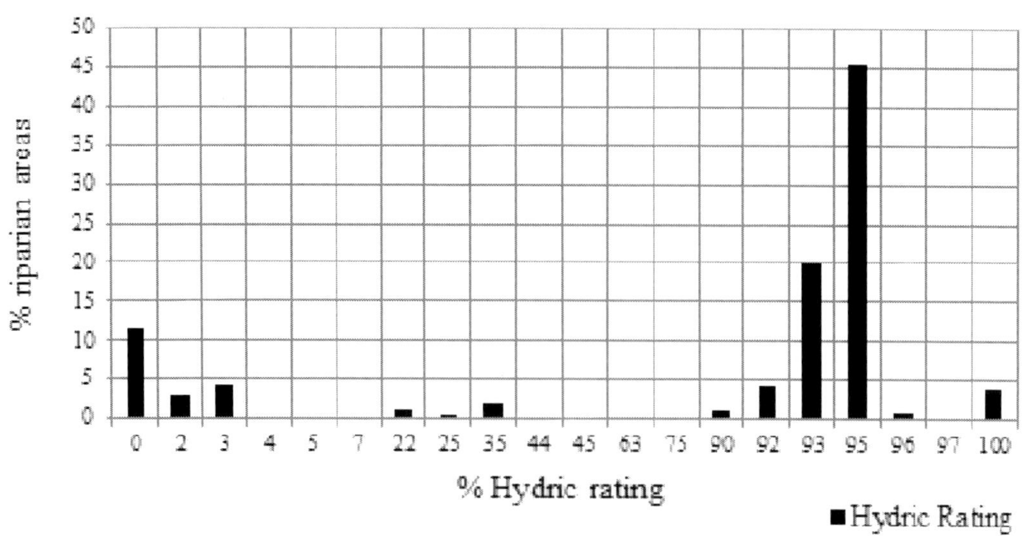

Figure 9. Hydric soils distribution within riparian ecotones.

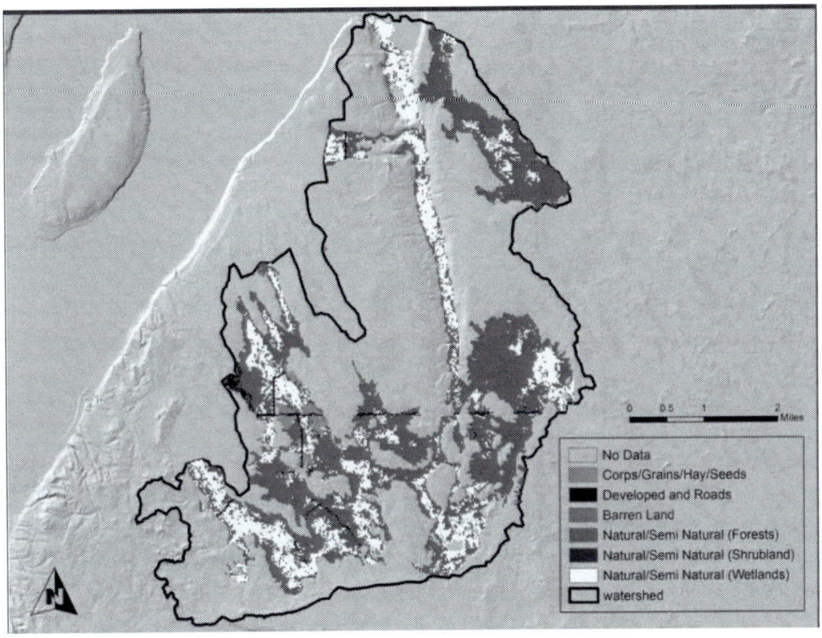

Figure 10. Riparian ecotones land cover distribution utilizing CDL 2010.

Figure 11 illustrates riparian ecotones land cover distribution from 2010 to 2014. The analysis shows no change in non-riparian land cover classes such as agriculture and developed area between 2010 and 2014. 98% of delineated riparian ecotones support natural/semi natural riparian land cover. 55% of delineated riparian ecotones associated with mixed and deciduous forest and 44% associated with herbaceous and woody wetlands. This shows riparian ecotones are not strictly associated with streams ecosystems but other surface water features such as wetlands (Palik et al., 2004) also shows that riparian ecotones maintained an averge of 98% natural/semi natural riparian land cover from 2010 to 2014.

Table 5. Recoded riparian ecotones land cover

New land cover classes	Original Land cover classes	Value
Crops/Grains/Hay/Seeds	Corn, sorghum, soybeans, sunflower, barley, spring wheat, winter wheat, rye, oats, millet, spletz, alfalfa, other hay/non alfalfa, sugar beets, dry beans, potatoes, clover/wildflowers, sod/grass seed, fallow/idle cropland, cherries, apples, grass/pasture, and celery	1
Developed and Roads	Developed/open space, developed/low intensity, developed/med intensity, and developed/high intensity	2
Barren Land	Barren	3
Natural/Semi Natural (Forests)	Deciduous Forest, Evergreen Forest, and Mixed Forest	4
Natural/Semi Natural (Shrubland)	Shrubland	5
Natural/Semi Natural (Wetlands)	Woody wetlands and herbaceous wetlands	6

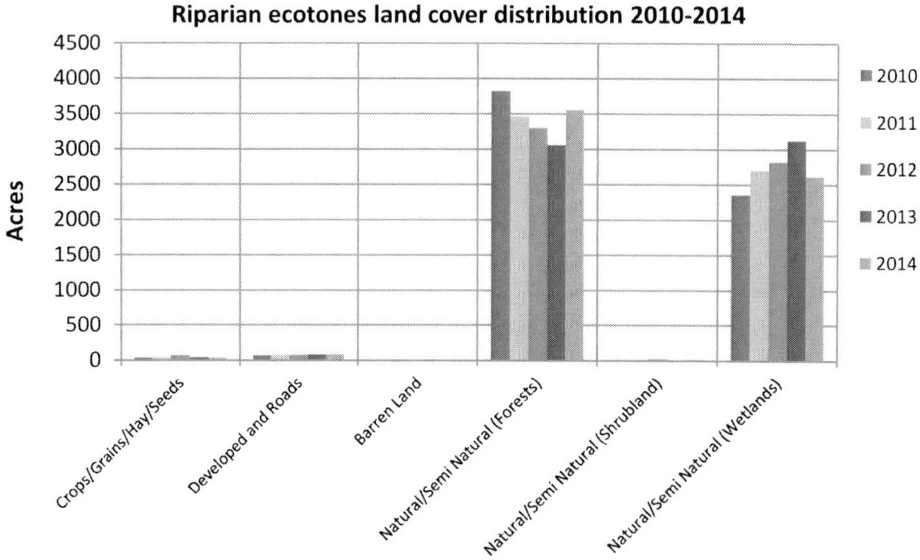

Figure 11. Riparian land cover distribution 2010-2014.

The generated annual riparian Land use/cover layers could be utilized for further change analysis between 2010 to 2014 to explain the interactions among riparian ecotones land use/cover. The majority of the delineated riparian land use/cover classes did not change during the period between 2010 to 2014. The mass intractions occurred between Natural/Semi Natural (Forests) and Natural/Semi Natural (Wetlands) classes (Figure 12). The change analysis showed that between 2010 to 2014 825 Acres of forest were changed to wetlands while 489 Acres of wetlands were changed to forests. Such information could help decision makers monitor riparian land use/cover within riparian ecotones setting over time also could guide conservation efforts for various uses.

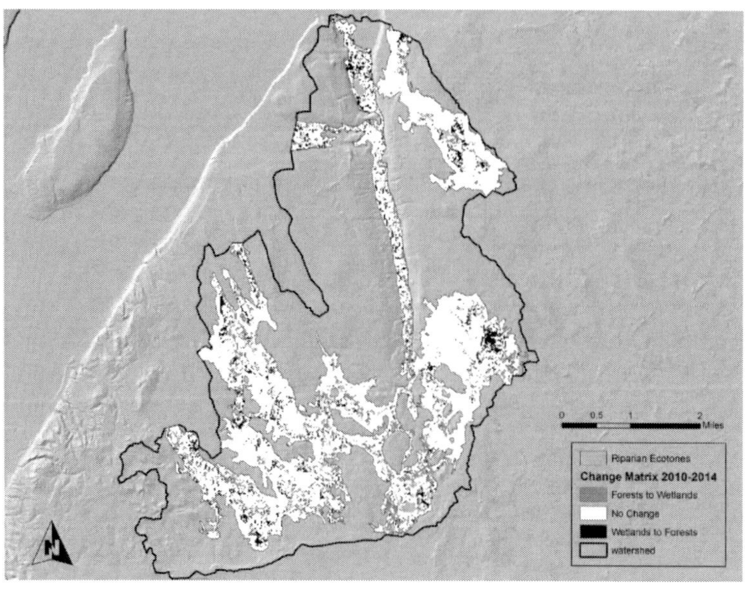

Figure 12. Riparian land cover change 2010 to 2014.

Model Sensitivity

There are three major parameters that impact the final delineation result, 50 year flood height, NHD streams positional inaccuracies, and DEM spatial resolution. Figure 11 shows the difference in riparian delineation using two different 50year floodplain values. Here, calculating an accurate hydrological descriptor for a given watershed is an essential step. A high 50 year flood height value would overestimate riparian ecotones and the probability to include uplands with the final mapped riparian ecotones is high on the other hand low 50 year flood height value would underestimate the full extent of riparian ecotones.

RBDMv3.0 uses NHD streams in the delineation process. Reviewing and correcting streams locations before starting the delineation process is an essential step. On the contrary this parameter would cause a shift in riparian areas location equal to the stream positional inaccuracy as reported by NHD.

DEM spatial resolution is an essential parameter and has a direct impact on mapping accuracy. A sensitivity analysis was conducted by Abood et al., 2011a and b involving four different DEM spatial resolutions 1m, 3m, 5m, and 10m. The analysis showed a linear relationship exists between the delineated riparian area and DEM spatial resolution (Figure 12). Analysis of variance showed that the change of delineated riparian area with different DEMs spatial resolution with P-value equal to 0.0617. Using a 10m DEM spatial resolution would introduce a 4% increase in riparian areas compared to 1m. Further Abood et al., 2012a and b conducted a statistical assessment comparing delineated riparian areas utilizing 10 and 30m DEMs using a linear mixed effect model developed by Kutner et al., 2005. The assessment showed that 30m DEM is inadequate to delineate geomorphology in a landscape heavily impacted by glaciation.

Conclusion

Riparian ecotones are dynamic transitional buffers between aquatic and terrestrial ecosystems with well-defined vegetation and soil characteristics. Riparian ecotones carry out many physical, ecological, and biological functions and pass energy and materials between the surrounding landscapes. Riparian ecotones are functionally and laterally connected to both aquatic and terrestrial ecosystems and maintain a linear form due to their proximity to streams and rivers. Mapping riparian ecotone at a fixed width buffer from streams and rivers is inadequate because such mapping approach does not consider the surrounding geomorphology and associated vegetation.

RBDMv3.0 is a robust automated GIS model developed as an ArcGIS for Desktop toolbox. The new version introduces many enhancements over previous versions such as streams criteria, an updated sampling technique, and updated soil criteria. Such enhancements introduce flexibility and provide more control for the user to better optimize the delineation process and decrease the computational time.

The task of delineating variable width riparian ecotones utilizing the 50 year flood height as the hydrological descriptor was successful. 50 year flood value is an essential model parameter and applying the wrong value could overestimate or underestimate the total area and extent of delineated riparian ecotones. The model incorporates optional data to achieve a complete and accurate riparian ecotones mapping. NWI and SSURGO data extends riparian mapping beyond the associated floodplain to close proximity wetlands and riparian soils according to a specific set of attributes. Furthermore the model incorporates classified land use/cover data that provides land cover attributes associated with delineated riparian ecotones. Such information provides quantitative attributed data that can help monitor riparian ecotones over time, show land cover distribution and change and guide decision maker's efforts to assess, and conserve riparian areas. This methodology recognizes that riparian ecotones occur at variable distances from streams and rivers by accounting for hydrologic, geomorphic and vegetation data.

Data preparation is an important step before starting the delineation process. The model uses many inputs and variables from different sources with inherited errors. Calculated 50 year flood height value, NHD stream positional inaccuracies, DEM spatial resolution have a direct impact on the final delineated riparian areas extent. Preparing and correcting input data is a critical step and necessary to achieve high mapping accuracy.

Incorporating land use/cover information permits assessment of land practices within delineated riparian ecotones. This can help decision makers monitor ecotones within a riparian setting over time, show land cover distribution and change, and guide conservation efforts for various uses. All can be achieved utilizing free nationally available data that provides data constancy, flexibility and known classification accuracies.

References

Abood, S. A. and Maclean A. L. (2015). Riparian Buffer Delineation Model v3.0. *2015 ESRI User Conference. Paper presentation UC85, San Diego Convention Center, San Diego, California*, July 20-24, 2015. Conference proceedings.

Abood, S. A. and Maclean A. L. (2012a). Modeling Riparian Zones Utilizing DEMs, Flood Height Data, Digital Soil Data and National Wetland Inventory via GIS. *Photogrammetric Engineering and Remote Sensing ASPRS 2011 Annual Conference, Milwaukee, Wisconsin.*, May 1-5, 2011. Conference proceedings.

Abood, S. A. (2011b). Modeling and Classifying Variable Width Riparian Zones Utilizing Digital Elevation Models, Flood Height Data, Digital Soil Data and National Wetlands Inventory: A New Approach for Riparian Zone Delineation, Ph.D. Dissertation, Michigan Technological University, Houghton, Michigan, 111 p.

Abood, S. A., Maclean A. L. and Mason L. A. (2012a) Modeling Riparian Zones Utilizing DEMs and Flood Height Data via GIS. *Photogrammetric Engineering and Remote Sensing.*, *78*(3), 259-269.

Abood, S. A. and Maclean A. L. (2012b). Modeling and Classifying Riparian Ecotones via GIS Utilizing Geophysical and Vegetative Inputs: A New Approach. *American Water Resources Association AWRA 2012 Summer Specialty Conference, Riparian Ecosystems IV: Advancing Science, Economics and Policy, Denver, Colorado*. June 27-29, 2012. Conference proceedings.

Albert, D. A. (1995). Regional landscape ecosystems of Michigan, Minnesota, and Wisconsin: a working map and classification. General Technical Report NC-178, U.S.

Department of Agriculture, Forest Service, North Central Forest Experiment Station, St. Paul, MN, 250p.

Aunan, T., Palik, B. J. and Verry, E. S. (2005). A GIS approach for delineating variable-width riparian buffers based on hydrological function, Research Report 0105, *Minnesota Forest Resources Council, Grand Rapids, MN*. 14 p.

Cowardin, L. M., Carter, V. and Golet, F. C. (1979). Classification of wetlands and deepwater habitats of the United States. U.S. Department of the Interior, Fish and Wildlife Service, Washington, D.C. 131p.

Drohan, P. J., Ciolkosz, E. J. and Petersen, G. W. (2003). Soil survey mapping unit accuracy in forested field plots in Northern Pennsylvania. *Soil Science Society of America Journal.*, *67*, 208-214.

ESRI, (2015). ArcDesktop 10. 1999-2015. Environmental Systems Research Institute. Redlands, CA.[CD-ROM]

Homer, C. C., Dewitz, J., Fry, J., Coan, M., Hossain, N., Larson, C., Herold, N., McKerrow, A., VanDriel, J. and Wickham, J. (2007). Completion of the 2001 National Land Cover Database for the Conterminous United States. *Photogrammetric Engineering and Remote Sensing*, *70*(4), 338-341.

Homer, C., Huang, C., Yang, L., Wylie, B. and Coan, M. (2004). Development of a 2001 National Landcover Database for the United States. *Photogrammetric Engineering and Remote Sensing*, *70*(7), 829-840.

Ilhardt, B. L., Verry, E. S. and Palik, B. J. (2000). Defining Riparian Areas, Riparian Management in Forests of the Continental Eastern United States. (Verry, E. S., J. W. Hornbeck and C. A. Dolloff, editors). Lewis Publishers, New York, NY, pp. 23-42.

Johnson, David M. and Hueller, R. (2010). The 2009 Cropland Data Layer. *Photogrammetric Engineering and Remote Sensing*, *76*(11), 1201-1205.

Kutner, M. H., Nachtsheim, C. J., Neter, J. and Li, W. (2005). Applied Linear Statistical Models, McGraw-Hill Irwin, Madison WI., 1396 p.

Mitsch, W. J. and Gosselink, J. G. (1993). Wetlands, John Wiley and Sons, Inc., New York, 454 p.

Mitsch, W. J. and Gosselink, J. G. (2000). Wetlands, John Wiley and Sons, Inc., New York, 920 p.

Mason, L. (2007). GIS Modeling of Riparian Zones Utilizing Digital Elevation Models and Flood Height Data, M. S. Thesis, Michigan Tech. University, Houghton, MI, 75 p.

National Land Cover Database. (1992). Multi-Resolution Land Characteristics Consortium (MRLC), URL: http://www.mrlc.gov/nlcd_product_desc.php, U.S. Department of Interior, U.S. Geological Survey (last date accessed July 15, 2015).

National Land Cover Database. (2001). Multi-Resolution Land Characteristics Consortium (MRLC), URL: http://www.mrlc.gov/nlcd.php, U.S. Department of Interior, U.S. Geological Survey (last date accessed June 15, 2011).

National Land Cover Database. (2006). Multi-Resolution Land Characteristics Consortium (MRLC), URL: http://www.mrlc.gov/nlcd_2006.php, U.S. Department of Interior, U.S. Geological Survey (last date accessed June 15, 2011).

National Agricultural Statistics Services. (2010). Cropland Data Layer, URL: http:// www.nass.usda.gov/research/Cropland/SARS1a.htm, U.S. Department of Agriculture (last date accessed June 20 2011).

Palik, B. S., Tang, M. and Chavez, Q. (2004). Estimating riparian area extent and land use in the Midwest, General Technical Report NC-248, U.S. Department of Agriculture, Forest Service, North Central Research Station, St. Paul, MN, pp. 28.

Palik, B. J., Zasada, J. and Hedman, C. (2000). Ecological considerations for riparian silviculture, Riparian Management in Forests of the Continental Eastern United States (Verry, E. S., J. W. Hornbeck and C. A. Dolloff, editors). Lewis Publishers, New York, NY, pp. 233-254.

Skally, C. and Sagor, E. (2001). Comparing riparian management zones to riparian areas in Minnesota: a pilot study, Research report RR-1001, Minnesota Forest Resources Council, 60 St. Paul, MN, 11 p.

Soil Survey Staff. (2008). Natural Resources Conservation Service, United States Department of Agriculture. Soil Survey Geographic (SSURGO) Database for [Survey Area, State]. Available online at http://soildatamart.nrcs.usda.gov. (last date accessed 15 January 2010).

United States Geological Survey. (1997). Standards for digital elevation models part I General, Digital Elevation Model Standards, URL: http://rmmcweb.cr.usgs.gov/nmpstds/demstds.html, National Mapping Program Technical Instructions, U.S. Department of Interior (last date accessed: 25 July 2015).

Verry, E. S., Dolloff, C. A. and Manning, M. E. (2004). Riparian ecotone: a functional definition and delineation for resource assessment, *Water, Air, and Soil Pollution: Focus, 4*, 67-94.

In: Riparian Zones
Editor: Oleg S. Pokrovsky

ISBN: 978-1-63484-613-4

Chapter 3

THE NEW BRAZILIAN FOREST LAW AND ITS ECOLOGICAL IMPACT ON RIPARIAN FORESTS: AN EXAMPLE NOT TO BE FOLLOWED

Vinícius Londe*
State University of Campinas, Campinas, Brazil

ABSTRACT

Brazil; one of the most biodiverse countries in the world and owner of two phytophysiognomies considered priority hotspots for conservation, yet it delayed to create, but mainly to apply legislation targeted to protect such biodiversity. The first law to protect flora was passed in 1934, and was amended in 1965, and it was called the Forest Code. This law could be considered robust and conservationist, and if it had been properly applied, it would have protected a great part of the Brazilian vegetation, including the Permanent Preservation Areas where the riparian forests which are so vital to biodiversity and ecologic, hydric, and geologic functions are a part. But in addition to the riparian forests having not been preserved, in 2012, the Forest Law was broadly altered. We should expect changes that take better care of our riparian forests, especially considering that Brazil has the largest amount of freshwater on Earth and we are experiencing a global environmental crisis. Nevertheless, such alterations tend to worsen the current situation of riparian zones and other ecosystems across the country. In this chapter, the reader will briefly learn about the evolution of the Brazilian Forest Law, the lack of application of this law and its consequences, and understand some modifications that occurred in the Law in 2012, verifying its ecological impact on undamaged and degraded riparian forests. I hope this discussion can influence critical thinking, help to preserve riparian forests and adequately restore them, and that our (bad) example can help to improve conservation laws in another places.

Keywords: biodiversity, Brazilian Forest Law, ecological impacts, environmental legislation, riparian forests

---

*Biologist, MSc in Ecology, PhD student in Ecology, State University of Campinas, Campinas, Brazil. Email: vlonde.ecologia@gmail.com.

CONTEXTUALIZATION

Brazil, as we know, is one of the most megadiverse countries on the planet and competes with Indonesia for the title of the most biologically rich Nation in the world (Mittermeier et al., 2005). To exemplify, in the moment that this chapter was written, 35.638 plant species are recognized - excluding algae and fungi - with angiosperms contributing to 92% of the total (List of Species of the Brazilian Flora, 2015). Considering an estimate of 320 thousand plant species on the Earth (Gaston, 2010), Brazil has about 11% of the total species richness. Also note that I did not consider here the diversity of fauna, fungi, or microorganisms, which greatly increase this importance. In this way, we expected large scale investments in deforestation control programs, environmental preservation, as well as robust legislation to protect such a great diversity of species, and the processes and functions that are linked to it on which all of us depend. Unfortunately, the conservation efforts were (and are) disproportionate, and the country has lost an enormous quantity of native vegetation, such that we have two of the 25 conservation priority hotspots, classified according to the loss of habitat in regions with high concentration of endemic species (Myers et al., 2000).

The phytophysiognomies (the predominant form of vegetation) found in Brazil are diverse, - varying from rupestrian complexes (small size vegetation found over rock formations) to exuberant forests, such as Amazonia and Atlantic Forests, which contain tall trees and a vast richness of other life forms (e.g., epiphytes) (Rizzini, 1997). Among the most relevant ecosystems, the riparian forests (or riparian zones) can be defined as areas of semi terrestrial transition with plant communities regularly influenced by freshwater (Naiman et al., 2009), or, in a simpler way, they are forest formations surrounding bodies of water.

The riparian forests stand out due to their species richness, genetic diversity, and their role in protecting hydric and edaphic resources, and terrestrial and aquatic fauna (Rezende, 1998). They are considered dynamic ecosystems, and perform diverse and important ecological, hydrological, and geomorphological functions, including rainwater storage during the rainy season and its release during dry ones, nutrient cycling, improvement of the water quality, and protection against erosion and sedimentation processes (Lima and Zakia, 2004; Naiman et al., 2009).

So, the relevance of riparian zones goes far beyond the ecological interface because they play a vital process in the provisioning of water, water regulation, and control of runoff, which is directly associated to the control of flooding, thus providing the ecosystem with services which are very important for human beings (Groot et al., 2002).

Nevertheless, even though so essential in several aspects, riparian forests have not escaped the overstated destruction of natural ecosystems by humans over time worldwide. What can we expect from the destruction of riparian forests? The first direct impact is the loss of functionalities of these ecosystems, together the loss of plant, animal, fungi, and microorganism diversity, loss of species interactions, altering of hydrological processes (such as water filtration in the ground, and the control of runoff), and an increase in erosion and sedimentation of water bodies, etc. Altogether, it produces a disruption of the eco, hydro, and geomorphological functions cited above.

Perhaps the closer and more visible example for most people, mainly in Brazil, is the occurrence of floods during rainstorms that flood streets, invade and destroy residences, drag out cars and kill people (Figure 1). All this happens because most cities were constructed next to water courses and the riparian forests were partial or totally removed (Martins, 2011), leading to loss of riparian functions. In watershed with good forest cover, that is, in preserved riparian zones, rainwater is partially filtrated in the soil and contributes to subsurface flow (Lima and Zakia, 2004). When vegetation is, for some reason, removed, soil becomes less aggregated and impermeable, increasing erosion and runoff during rainstorms (Montagnini and Jordan, 2005), which may cause destructive flooding to residences near rivers.

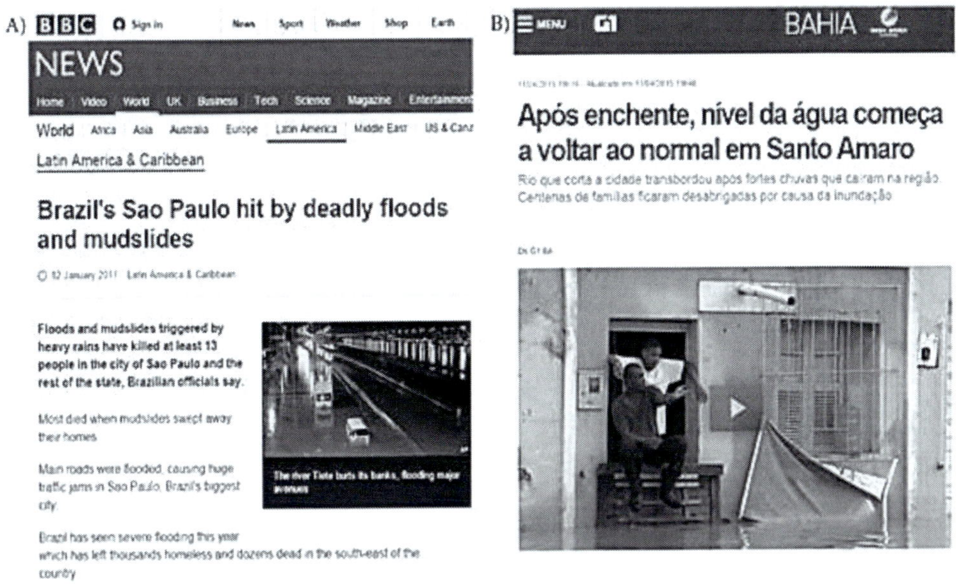
A) BBC NEWS

World | Latin America & Caribbean

Brazil's Sao Paulo hit by deadly floods and mudslides

12 January 2011 Latin America & Caribbean

Floods and mudslides triggered by heavy rains have killed at least 13 people in the city of Sao Paulo and the rest of the state, Brazilian officials say.

Most died when mudslides swept away their homes.

Main roads were flooded, causing huge traffic jams in Sao Paulo, Brazil's biggest city.

Brazil has seen severe flooding this year which has left thousands homeless and dozens dead in the south-east of the country.

The river Tiete bursts its banks, flooding major avenues

B) G1 BAHIA

Após enchente, nível da água começa a voltar ao normal em Santo Amaro

Rio que corta a cidade transbordou após fortes chuvas que caíram na região. Centenas de famílias ficaram desabrigadas por causa da inundação

Figure 1. A recurrent problem in Brazil are floods after/during rainstorms that impact many cities built near streams and rivers across the country. Here two newspaper headlines are shown about this: (A) a flood punishing São Paulo city in 2011 – from 12 January 2011, BBC News, and (B) residents of Santo Amaro (Bahia) hit by a flood in 2015 – from 11 April 2015, G1 News (the headline reads: “After floods, the water level starts back to normal in Santo Amaro” […] “hundreds of people were displaced by floods”). News source: (A) <http://www.bbc.com/news/world-latin-america-12164441> (B) <http://g1.globo.com/bahia/noticia/2015/04/apos-enchente-nivel-da-agua-comeca-voltar-ao-normal-em-santo-amaro.html>.

Therefore, natural ecosystems are dynamic, complex, composed of several biological interactions and physicochemical components (Andreasen et al., 2001), and once disrupted, their functions and processes are broken.

In this chapter, the main goal is to analyze the alterations made in the Brazilian Forest Law in 2012, and to discuss how these changes affect the riparian forests, and consequently all organisms that depend on its functioning, including, of course, human beings. I still have the perspective that through our (bad) example, other countries, and even probably Brazil itself, can better appreciate their riparian zones, and the ecosystem in general, and define guidelines that really protect its integrity and biodiversity on all scales, without solely focusing on the expansion of agriculture and livestock. Now, let’s know a little about the old Brazilian Forest Code.

LEGISLATION UPON BRAZILIAN VEGETATION: THE BEGINNING

The first Law to protect Brazilian ecosystems was published in 1934 and called the Forest Code (Brasil, 1934). According to Borges et al., (2011), this Code already contained some preservationists' features and classified some Brazilian ecosystems; however, the legislation was not properly applied due to carelessness by the authorities at that time. After 31 years, on 15 September 1965, the Law number 4.771 was passed, that instituted the second Forest Code, with many improvements and adjustments. This legislation could be considered robust, since it defined several parameters and measurements for protection of vegetation. As an example, one can cite the definition of areas for Legal Reserve, which are sites with native vegetation located on farms, and should be maintained in a sustainable manner without deforestation in order to maintain ecological processes and harbor biodiversity (Brasil, 1965).

The size of the Legal Reserve essentially depended on which biome the farm would be situated. For instance, a farm located in the Amazonia biome should preserve at least 80% of native vegetation, and ownership in the *Cerrado* (savanna) or Atlantic Forest biomes (the latter is the most deforested phytophysiognomy from Brazil, where less than 16% of vegetation remained (Ribeiro et al., 2009)), should preserve at least 20% of its original vegetation (Brasil, 1965).

Another important definition inherent to the Code from 1965 was the Permanent Preservation Areas (PPA), with one of them being the focus of this book: the riparian forests. Other PPA are the tops of hills and slopes with more than 45° of declivity, *restingas* (salt marshes), and all vegetation above an altitude of 1800m. One interesting point in this Law was determining minimal widths for riparian forests that should be kept in ownership according to the width of streams and rivers or other water bodies, and could not be suppressed without prior authorization by the Federal Government. Listed in the Table 1 are the types of water bodies and riparian widths defined by the old Code.

Table 1. The Brazilian Forest Code from 1965 determined distinct widths of riparian forests that should be preserved according to the type and width of water bodies

Streams and rivers	
Width of the water course (m)	Minimum width of the riparian forest (m)*
<10	30
10–50	50
50–200	100
200–600	200
>600	500
Water sources	
Minimum radius of 50 m width	
Lakes, ponds, and artificial reservoirs	
Undetermined	

* Measured from the highest level of water.
Source: Brazil, 1965.

I'd like to point out that according to the Forest Code from 1965, areas for Legal Reserve should not be the same as PPA (with some exceptions – Art. 16, §6, Brasil, 1965), thus, the owner should keep one area of Legal Reserve, which could be inserted in less productive sites, for instance, but should preserve the riparian forests too. This was a positive point in the Law because it helped to preserve more vegetated/forested areas, and enabled the maintenance of the connection and fluxes between them if adequate distances were kept.

The Forest Code from 1965 explained several other aspects regarding the protection and exploration of the Brazilian vegetation, however, we'll focus only on the riparian zones. It is worth highlighting that any type of damage or destruction of PPA, such as the use of fire or logging, was considered a penal misdemeanor and the owner was liable to penalties, such as the payment of fines and even imprisonment. Furthermore, it was the Federal Government – the highest level in the Brazilian policy hierarchy – or in agreement with States and Municipalities, who were responsible to supervise the real implementation of the Code.

Over the years, the Forest Code experienced some changes through Interim Measures, nº 2.166-67 (Brasil, 2001), e.g., and other Laws (Brasil, 1989, e.g.,), which modified isolated points in the Code. However, in the recent years, the desire for wide changes in the Code have enhanced and, after many discussions, alterations were consolidated in 2012, as we'll see below. But first let's see how the Code from 1965 was (not) enforced.

Disrespect of the Forest Code from 1965

The title of this section reflects that unfortunately, like so many Brazilian Laws, the Forest Code, despite being conservationist, wasn't definitively obeyed, and nowadays we have an already present environmental crisis, given the proportion of ecosystems' destruction. For the Atlantic Forest, for instance, the estimates of remnants of old-growth vegetation is only 12.5% from the original 1.3 million km^2, considering fragments over 3 hectares (SOS Mata Atlântica, 2013). Nevertheless, more than 80% of remnant fragments that have less than 50 ha are isolated, and the protected areas in fact protect only 9% of remnants (second-growth forests), and the scariest, only 1% of the original forest! (Ribeiro et al., 2009).

For the Amazonian Rainforest, perhaps the most famous Brazilian phytophysiognomy due its extension and exuberance, some studies have shown that its degradation and loss of habitat can be greater than described until recently, and depending on the improvement in forest monitoring methods. Analyses by remote sensing in high resolution showed that selective logging and other damage in its canopy were masked in previous satellite images and the forest degradation is, actually, much larger than previously thought (Asner et al., 2005; Foley et al., 2007).

The *Cerrado* (savanna) and *Caatinga* (dry forest) phytophysiognomies, which are seasonal, with well-defined dry and rainy seasons (Rizzini, 1997), also didn't escape from degradation. Recent estimates indicated that only from 1990 to 2010, the *Cerrado* lost 265.595 km^2 of native vegetation, while the *Caatinga*, lost 89.656 km^2 in this period (Beuchle et al., 2015). Another interesting study aimed to know if protected areas in the *Cerrado* were effectively contributing to maintain the 17% of old-growth vegetation in this phytophysiognomy, as defined by the Convention on Biological Diversity (CBD), and the result was surprisingly negative. Protected areas hold only 3% of the Brazilian *Cerrado*, and this of course is very far from the goal of 17% defined by the CBD (Françoso et al., in press). The authors highlighted that urgent measures must be taken to create further protected areas in the *Cerrado,* and to guarantee the representativeness and persistence of its high biodiversity (Françoso et al., in press).

As we have seen, we have lost and continue to lose giant areas of old-growth vegetation, and the Permanent Preservation Areas (PPA) are inserted among them; so important to preserve water resources, landscapes, geological stability, biodiversity, integrity of the climate system, and the maintenance of human well-being, as recognized by the Legislation itself (Brasil, 1965, 2012). Many other examples of habitat loss, fragmentation, and degradation from distinct Brazilian phytophysiognomies could be cited, but this is not the goal here. These examples show that despite having robust environmental legislation over the past 50 years, there has been a lack of implementation and inspection.

Figure 2. A fact that has been drawing media attention and has worried people in recent years is the water scarcity in many cities in Brazil when this problem was not common, especially in the Southeastern regions where there are large cities, such as São Paulo. Shown here are some international News headlines regarding this trouble. A) RT News, from 16 February 2014 (available on <http://rt.com/news/water-ration-drought-brazil-253/>); B) The Guardian, from 05 September 2014 (<http://www.theguardian.com/weather/2014/sep/05/brazil-drought-crisis-rationing>); C) BBC News, from 11 October 2014 (<http://www.bbc.com/news/world-latin-america-29581069>); D) The Guardian, from 11 February 2015 (<http://www.theguardian.com/global-development-professionals-network/2015/feb/11/brazil-drought-ngo-alliance-50-ngos-saving-water-collapse>).

Most likely if the proposals of the Code had been fulfilled, Brazil nowadays would be at a much better level regarding environmental conservation. However, to the contrary, there was disobedience of the environmental legislation and the consequences have already begun to be seen and felt. A fact that drew media attention in recent months was (and still is) the water crisis that is a concern across the Country, but mainly the Southeastern Region (Figure

2), even being the richest country in freshwater in the world (±12% of the total amount of freshwater worldwide) (Bedê, 2012). Unfortunately, it seems that people are only now paying attention to the importance of riparian zone vegetation and water preservation, and they have begun to demand action from the government.

In addition to the water crisis, which is intrinsically linked to riparian forests degradation, we can highlight other less familiar damage by humans in general, but which also affect them, such as biodiversity loss (due to habitat fragmentation), construction of dams for hydropower production (they interrupt fish migration and seed flow), and river flow change (caused by regional deforestation) (Costa et al., 2003; Silva et al., 2015). Even so, proposals of alterations in the Forest Code have permeated Brazilian National Congress over the years, with many debates between the farming sector, which defended lax in the Code, and the environmental sector, which did not accept these changes and wanted stricter implementation and control of the Law. However, the farming pressure was stronger, and in order to expand this sector, the Forest Code was altered.

Changes in the Forest Code

Due to the already present environmental crisis, and under such divulgated and dreaded climate change, major commitment from Brazil is expected in the sense of effectively protecting its natural resources and maintaining its high biodiversity. In other words, we are expected to follow a narrowing and real applicability of the Forest Code. If there were changes, as many wanted, they were to improve the Code and achieve a model to be followed. Nevertheless, in 2012 the changes to the Code became public and official and, unfortunately, as discussed below, the wishes of environmentalists were not met. The farming pressures of a developing country were viewed as being more important.

Now, under Law nº 12.651 from 12 May 2012, the Forest Code (or the New Forest Law, as it is known), with many alterations, aims at sustainable development and provides general rules to protect native vegetation, including the Permanent Preservation Areas (PPA) and Legal Reserves. Here, one should ask: Will the changes in the Code contribute to long-term perpetuation of biodiversity (on all scales), processes and functions? In other words, are these changes really sustainable? Does The New Forest Law effectively protect riparian forests? In the next section, we'll do an analysis of some of the alterations and their probably consequences.

Ecological Perspectives of the Changes to the New Forest Law Concerning Riparian Forests

In order to respond to the questions raised above and widen the discussion about the changes to the Forest Code, in this section I'll perform an analysis on some sections of the New Forest Law that affect directly the riparian zones. Comparisons between the Code from 1965 and the current one are necessary to verify if the changes in effect will improve or worsen the situation of riparian forests, from an ecological standpoint, with direct consequences on the social and economic scopes.

A positive point in the New Forest Law was the continuity of the width of riparian zones surrounding "native" water courses, that is, the width of remnant riparian zones, in rural and urban regions, was not altered (review Table 1). However, there are two important counterpoints to be considered here. First, these width measures become irrelevant if we consider that the great majority of riparian forests were already partially or totally removed, both in rural but mainly in urban regions, once cities were built surrounding water bodies. A close look through the South, Southeast, Midwest, and Northeast regions of Brazil allows for verification of this harsh reality. Moreover, the landscape analyses through satellite images clearly demonstrate this (Figure 3).

This section of the legislation is obviously more relevant for the Amazonian region, where large extensions of old-growth vegetation still exist. However, even they are threatened by the urban expansion, unsustainable economic development, deforestation, agribusiness (Laurance, 2015), and the construction projects of hydropower. There are currently construction projects of at least 58 new dams in Amazonia by the Federal Government, even with the negative experience of this type of energy generation in the region (Kahn et al., 2014), which is flat and does not provide enough water pressure for an efficient production of energy. I wonder why instead of building dozens of new dams with vast socioenvironmental impacts, the Government does not invest in the production of solar energy in this tropical region.

The second observation is concerning the point (distance to water body) to which the New Forest Law determines where the riparian forests begins. The Code from 1965 considered riparian forests as "*marginal strips along the river or any other water course since its highest level in marginal strip,*" but the New Forest Law considers riparian forests as "*marginal strips of any water course [...], since the gutter edge of the regular channel.*" What are the differences and implications of the underlined terms?

The Code from 65 considered the intrinsic variations in the output of water courses which occur between rainy and dry seasons, being the regular channel where water regularly goes during the year (the output is lesser) (Figure 4), and the highest channel when the output is high and the water course occupies more space in the channel, mainly in the rainy season (Figure 4). Even within the definition of riparian zone the differences of output are considered, because it is a natural process of these ecosystems. Naiman et al., (1993), for instance, clearly state: "*the riparian corridor encompasses the stream channel and that portion of the terrestrial landscape from the high water mark towards the uplands where vegetation may be influenced by elevated water tables.*" The New Forest Law, in order to restrict riparian zones from the regular channel, implicates in a considerable loss of riparian vegetation and its functionality, perhaps not so visible on a local scale, but consider the extension of Brazilian territory covered by bodies of water! This problem had already had been forewarned before the alterations to the Code by renowned Brazilian researchers (e.g., see Metzger et al., 2010 and Piedade et al., 2012); however, the warning was not heard by legislators.

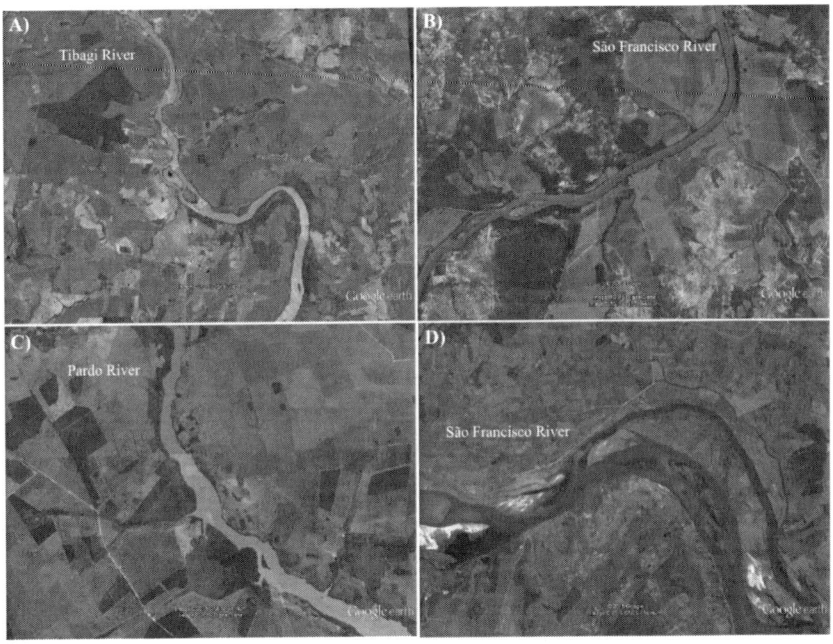

Figure 3. Satellite images examplifying damage in riparian zones and their vicinities in four Brazilian regions, caused by human activities as a result of the disobedience to the Forest Code from 1965. A) South region, Tibagi River, Paraná State, north of Maravilha city, image date: 1/29/2012, altitude of the viewpoint: 10.78 km; B) Southeast Region, São Francisco River, Minas Gerais State, north of Três Marias city, image date: 2/21/2010, viewpoint: 10.73 km; C) Midwest Region, Pardo River, Mato Grosso do Sul State, south of Ribas do Rio Pardo city, image date: 7/11/2013, viewpoint: 10.32 km; D) Northeast Region, São Francisco River, Sergipe State, north of Saúde city, image date: 8/14/2014, viewpoint: 9.2 km. Note the great lack of riparian forests and agricultural intensities around both rivers, including eroded lands in Figure B. All images are from Google Earth, 2015.

Figure 4. The differences between the regular and highest level of the river channel have strong consequences over the structure of riparian forests. Even though in the greater part of the year the water table remains low (regular level), during and after rainstorms the amount of water will greatly increase and occupy much area in the channel (highest level), so, only species adapted to this stressful condition will survive. Thus, this natural dynamic must be considered in the definition of riparian forests, which, in fact, encompasses the highest level. Photo: V. Londe - Pirapitinga River, Matutina, MG.

Furthermore, the New Forest Law allows that people and animals may enter riparian forests to obtain water, but it is not clear about the restriction of the vegetation's suppression. Likewise, it permits the realization of other activities of less environmental impact, such as the openness of ways to access water, build bridges, as well as the construction and maintenance of fences, etc. On the other hand, the previous Code was more specific and did not permit the suppression of vegetation, neither the impairment of the natural regeneration or maintenance of the riparian forests, which guaranteed the sustainability of the area over time.

Some studies have shown that cattle can damage riparian forests to compress the soil, which reduce the rainwater infiltration and increase the runoff, and cause damage to the vegetation structure by consuming plants from the underbrush, including native herbs and seedlings (Kauffman and Krueger, 1984). According to Mitchell and Kirby (1990), as important as the impact of large herbivores over seedlings is the damage that they may cause to bryophytes and in populations of invertebrates and vertebrates that depend on the vegetation structure. Lees and Peres (2008), for instance, highlighted that movement of cattle in riparian forests has inclusively had negative effects over terrestrial birds that inhabit these ecosystems, because cattle interrupt the natural regeneration of the forest, which is essential to structure and functional connectivity.

Another negative point that persisted in the New Forest Law is the admission, in some cases, of part of the PPA in the computation of Legal Reserve, because this reduces the portion of native vegetation in the properties, which in landscape terms can represent considerable losses. Furthermore, in properties with adequate amounts of riparian zones and legal reserve, the distance between fragments could be less, contributing to processes of propagule dispersal of and animals between them. Studies have evidenced that forest fragments exert influence over the adjacent ecosystem with respect to seedling recruitment of more than 230m distance (Bertoncini Rodrigues, 2008), and there is also evidence of greater propagule dispersal, with consequent evenness in species composition between fragments less than 200m apart (Rossi et al., in analysis).

Regarding exploitation of old-growth forests, the New Forest Law requires the owner to obtain approved licensing from the Environmental Department, upon the admission of the "Sustainable Forest Management Plan" (PMFS in Portuguese), in which the techniques of exploitation, management, and repositioning of the vegetation post-exploitation are previously defined. However, this Law becomes contradictory to require a paragraph on the obligation of vegetation repositioning in the PMFS (Art. 31, Chapter VIII, Brazil, 2012), but a posterior paragraph gives exemption of repositioning to owners whose forest raw material is coming from PMSF (Art. 33, Chapter VIII, Brazil, 2012). It is really surprising to require vegetation repositioning in the PMFS, and then exempt them from it! This means that we can exploit the remaining of our old-growth forests without the need for recovery!

Moreover, the Law exempts owners that exploit non-timber products, such as fruits, honey, hunting, medicinal plants, etc. of forest recovery. Many of these products are collected without costs, but have great value in productive use and are sold in local markets (Primack and Rodrigues, 2001). Therefore, it is interesting to ask: Are these products extracted in a sustainable way? If not, should the traders initiate forest recovery? Should there be a payment for these ecosystem services? These are open gaps as people understand that they can exploit forests without being concerned about inspection, let alone worry about natural resources.

Up to this point, questions of the legislation regarding unsuppressed areas were addressed. However, the majority of the old-growth vegetation, from the distinct Brazilian phytophysiognomies, has already been deforested (Figure 5), including, of course, riparian zones. Recognizing this, the New Forest Law incorporated an exclusive section about the degraded PPA.

The first standard was the continuity of human activities in the areas of riparian forest consolidated by July 22, 2008; that is, all owners that had deforested and settled in permanent preservation areas until this date (certainly the huge majority, with the exception of the Amazon (Figure 5)), have been legally consolidated/recognized. However, as already evidenced, today little remains for us to inspect, in comparison with the portion already deforested in most Brazilian phytophysiognomies (Figure 5).

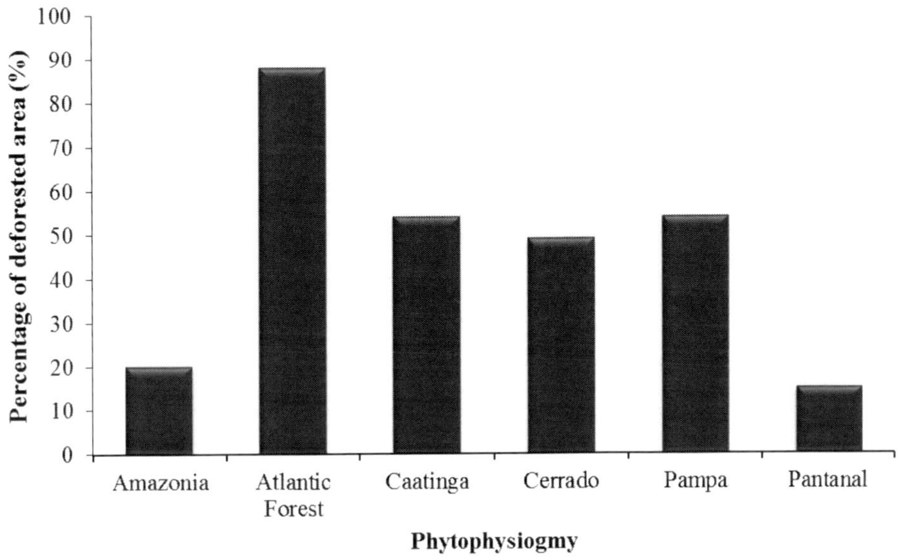

Source: IBGE, 2012.

Figure 5. Percentage of deforested area in the main Brazilian phytophysiognomies up to 2009 for Caatinga (dry forests), Pampa (grasslands), and Pantanal (wetlands), until 2010 for Atlantic Forest, and Cerrado (savanna), and 2011 for Amazonia.

What the Law requires of the owners is the recovery of marginal strips where there should be riparian forests - which are not present there because of the breach of the Code from 1965 - however, these strips are of minimum size (5, 8, or 15m), beginning from the regular baseflow level of water, and so are probably little or not functional. Moreover, the width to be recomposed is set according to the size of the rural property (Table 2), and not according to the width of the watercourse (Table 1, Forest Code from 1965), which represent a great regression in terms of conservation because the fiscal module (the measurement of the size of the area) greatly vary from a municipaly from another (Landau et al., 2012).

The question that arises however, is whether marginal strips of just five, eight, or fifteen meters wide are appropriate? Are they ecologically functional? Can they contain a minimum diversity of native species or they will be just habitat for exotic and aggressive species? Can they be equated to "native" riparian forests? Which methods can be applied to restore such narrow strips? Will we be able to actually recover these strips?

Let us take as an example a method of ecological restoration widely applied in Brazil, which is the total planting of seedlings. A spacing of 2 x 3m or 2 x 2m between lines of “filling” is usually employed between fast-growing species and lines of “diversity,” as slower growing species require specific habitat conditions to develop (Nave and Rodrigues, 2007). Thus, in the case of the owner having to recompose five meters of marginal strip, two parallel lines to the watercourse would be deployed. Two rows of seedlings, and one day trees (if the seedlings are not taken by exotic species, such as African grasses, for example), may not have any ecological function. It is therefore not possible to employ restoration methods in such short strips (maybe some technique of natural engineer), and most likely this “recovery” will be ineffective. We can expect, for example, that narrow strips may add sediment and silt to the watercourses. Let us assume that saplings are planted in five, eight, or even 15 meters in both margins of a river from the regular baseflow of the water level. One can imagine that these saplings and a good fraction of soil (plus invested money and work) will be leached away when the water level rises.

Table 2. According to the New Brazilian Forest Law, owners must be in charge of recovering the degraded riparian forests depending on the size of their rural properties

Water courses	
Farm size (fiscal module*)	Marginal strips (m)†
up to 1	05
1–2	08
2–4	15
>4	20–100
Water sources	
Radius minimum of 15m	
Lakes and ponds	
up to 1	05
1–2	08
2–4	15
>4	30

*Minimum area needed rural properties to be considered economically viable, varying from 5 to 110 hectares, according to the municipality (Landau et al., 2012).
†Measured from the regular level of water and independent of the width of the water bodies.
Source: Brazil, 2012.

We have examples of large restored riparian forests (300ha) of 5, 9, and 10 years post-plantation, which from the ecological perspective (recruitment of seedlings, dominance of grasses, inadequate dispersion of seeds) are not self-sustaining and can be compromised in the long term (Souza and Batista, 2004). Still, riparian forests restored around the Reservoir of the Hydroelectric power station of Volta Grande 20 years ago, in the States of Minas Gerais and São Paulo, of 30m and 100m wide, have low richness and density of regenerating wood species (Londe and Souza, unpublished), which represent the persistence of forests at the time because they will replace the adult trees.

These narrow strips should be considered as possessing edge effects, since these effects (increased incidence of solar radiation, temperature, wind, and reduction of humidity) are more pronounced at 200m of the forest borders, but may reach a distance of 500m (Laurance, 1991). Lees and Peres (2008), in studying several riparian forests in the State of Mato Grosso,

Brazil, found that those less than 200m wide were much more vulnerable to edge effects and did not contain cores of forest habitat. As the conditions of the border are different from the requirements necessary for the species adapted to the interior, the species composition is altered (Primack and Rodrigues, 2001), usually composed of common, invasive, and exotic species, especially grasses, which are highly combustible and can increase the likelihood of fires (Lamb and Gilmour, 2003). Therefore, the passage of fire in these narrow strips may kill trees and seedlings (if they exist), which in turn leads to a greater increase of invasive/exotic herbaceous biomass due to increased solar radiation on the ground, leaving the plant community in an alternate state dominated by herbs and grasses, mainly if the area is in an agricultural or urban matrix. It turns out that the maintenance of forested riparian zones in agricultural landscapes, for example, increases the abundance of wildlife species such as small mammals and reptiles, compared to areas dominated by herbaceous species (Maisonneuve and Rioux, 2001).

Furthermore, very narrow marginal strips do not contain a range of animal species, especially those that require larger areas and/or specific microhabitat conditions to survive. Semlitsch and Bodie (2003) did a compilation of data from several studies and found that amphibians require ciliary habitats from 159 to 290m wide to survive, although reptiles require from 127 to 289m. For salamanders, riparian forests of 30m or more may be efficient and ensure the survival of some species, while 9m ranges are as inefficient as cleared riparian forests (Peterman and Semlitsch, 2009). Narrow (<200m) and/or very disturbed riparian zones contain few species of birds and mammals, and yet they are common species to deforested habitats (Lees and Peres, 2008). In a study of riparian forests of different widths in the State of Paraná, Ramos and Anjos (2014) noted that species' richness and abundance of birds were greater in little disturbed riparian forests of 50m or more. Furthermore, species' richness and diversity of dung beetles were greater in wider riparian forests (>40m) than in narrow strips (<5m) in Southern Brazil (Viergas et al., 2014).

There was an interesting work which evaluated the effectiveness of riparian buffer strips with different widths in a region of New Zealand, and I believe this is a good example to present. Riparian forests of three sizes were evaluated (5-6m, 10m, and 15-20+m) to know if they were self-sustaining. The study warned that the narrower strips (5-6m and 10m) were not self-sustainable because the trees did not completely shade the ground, causing an invasion of weeds in the underbrush, leading to required periodic maintenance of them, and the natural regeneration of native species is also unlikely to occur (Parkyn et al., 2000). The larger strips (15-20+m) would decrease the likelihood of weeds to occupy all the underbrush, and this would increase the chances of native species recruitment, requiring little or no maintenance (Parkyn et al., 2000). The authors warned that the weed infestation is usually in 1-2m in the margins, and therefore, strips of only five meters are very susceptible.

Another example of a study that evaluated minimum width requirements of riparian forests for the protection of watercourses and biodiversity conservation occurred in Australia, in the State of Victoria. The authors did a good review of the subject and cited several examples, but in short, they suggested that the width of the riparian vegetation to be retained or restored should be greater for the intensity in land use and management goal (Hansen et al., 2010). They recommended establishing at least 30m of riparian vegetation on either side of the watercourse (Hansen et al., 2010). So, considering all the examples cited above, which summarize that much narrow riparian strips are unsustainable and do not contribute to biodiversity conservation, why should we continue with this approach in Brazil?

The New Forest Law stipulates that when recovering suppressed areas, landowners can use four approaches; conduction of natural regeneration of native species, planting of native species, conduction of natural regeneration plus the planting of native species, or planting native species interleaved with up to 50% exotic ones.

At first sight, the conduction of natural regeneration (which is to isolate the area, remove degrading factors and let the secondary succession occur naturally (Rodrigues and Gandolfi, 2007)), can be attractive, because it is a practical and low-cost method. However, it is likely that most areas to be recovered in Brazil do not fit in this method, especially if the native vegetation was suppressed a long time ago and the system has lost its resilience (their ability to resprout, stock of seeds in the soil, etc.), and/or if it has dominated by exotic and invasive species. In addition, natural regeneration is favored and accelerated when there are quality fragments nearby, such as a source of allochthonous propagules (Rodrigues and Gandolfi, 2007), however many riparian areas to be recovered in Brazil may be isolated from forest fragments and are usually surrounded by pasture or monocultures, such as sugar cane, hindering the regeneration process.

Yet, the indication of recovery by only planting native species is interesting, since the pool of regional species has to be respected, i.e., species adapted to phytophysiognomy and the region of interest. Sometimes this can go unnoticed, but it is important to encourage the use of local species in order to maintain the regional floristic characteristics, with an effect on animal diversity, and to avoid genetic miscegenation with species coming from remote regions. We must avoid, for example, the importation of species from very different places because each has a life-history and adapts to specific environmental conditions. In this way, biologically viable riparian forests will be recovered with guaranteed long-term persistence, and with typical species to reference ecosystems (Brancalion et al., 2010). Still, it can be a way of encouraging the production of regional saplings for planting in restoration projects, increasing the income and number of nurseries, together with the diversity of species produced. In the State of São Paulo, for example, the number of plant nurseries increased from 55 to 114, and the number of species produced increased from 30 to more than 80 after strong incentive of restoration practices of the Atlantic Forest (Brancalion et al., 2010). This is a good example to be followed and should be encouraged in other Brazilian States.

Regarding the planting of exotic species, it is important to note that many researchers point out problems caused by certain species in areas in restoration process, mainly because they can compete with and prevent the establishment of native ones (Lamb and Gilmour, 2003; Martins, 2011; Londe, 2013). Initially, the exotic and invasive species may not be problematic, because few individuals are present at the site, however, as more and more propagules are produced and dispersed by the adult plants, these species tend to become abundant, increasing their distribution, and cause local and even regional problems (Doren et al., 2009). Thus, the use of exotic species is expendable, because it reduces the chances of success of the restoration in the long term, and increases the likelihood of problems (Wang et al., 2013). Even for the restoration of small areas for the purpose of future exploitation, the planting of native species with economic potential is recommended, not exotic ones (Ivanaukas et al., 2007). The reduction or elimination of alien species in restored areas is inclusively suggested by the Society for Ecological Restoration International (SER, 2004), the reference organization for ecological restoration in the world.

In this sense, the Code from 1965 was more suitable because it prioritized recovery projects that used native species, and was not as malleable for smaller rural homeowners. The

New Forest Law, for example, favored the owners so that by July 2008, there were up to 10 fiscal modules of land and developed activities in areas where riparian vegetation was removed, by limiting the recovery of APP up to 20% of the total area of the property. This fact is very important, because in Brazil the number of small farms (less than 10 ha of area) correspond to more than 47% of the total number of rural properties and cover a total area of 7.798,607 ha (IBGE, 2006). Therefore, when the Law exempts or limits the recovery of riparian forests (and other physiognomies) in smaller rural properties, it favors nearly half of Brazilian landowners, while reducing the proportion of riparian forests of the national territory. Imagine if each of these owners had to keep 5 meters of riparian zones. Now imagine if they were 30 meters. Certainly, there would be more connectivity between areas and the landscape would be better.

And what does the New Forest Law determine about the settlements entered in the places that should be occupied by riparian forests? These settlements were regularized with caveats to PPA and identified as risk areas. Nevertheless, as the reader can foresee, almost all riparian forests can be considered as risk areas because they are naturally subjected to flooding. This happens because a portion of the rainwater which does not infiltrate into the soil is not transpired, thus contributing to the runoff, which supplies the aquatic ecosystems and provides water to people (Chapin III et al., 2011). So the amount of water in streams and rivers increases during the rainfall and they flood a portion of the riparian forests.

If riparian forests were replaced by settlements, it is expected that these places would often be flooded. To regularize these settlements, the New Forest Law allows people to continue in these flood areas. It should be remembered that the risks are not only restricted to flooding of homes and businesses, but also because they are subject to landslides caused by the charging of watercourses (Figure 6). If, on the one hand, it is easier to allow people to continue in these areas due to costs and problems with expropriation and allocation of these people to other sites, on the other hand, they are subjected to unworthy risk of life.

Figure 6. When riparian forests are deforested and replaced by settlements, a recurrent problem may be the erosion of the riverbank, which, in turn, can threat nearby residents. Here is shown an example of a stretch at the Das Velhas River, in Belo Horizonte, Minas Gerais, where water flux is destroying the riverbanks (A), including a strip where contention barriers were placed (B). Photos: V. Londe, 2011.

Furthermore, regarding riparian zones in urban areas, the Code sets a minimum width of 15 meters to be maintained without buildings, but does not determine what kind of vegetation must be recovered these areas. In fact, different types of vegetation may play different functions in riparian areas. For example, forested riparian zones tend to stabilize the channel better because their roots penetrate deep into the sediment below and adjacent to the watercourse for water and nutrient uptake (Likens, 1992). On the other hand, riparian forests occupied by herbaceous and grasses, which have a dense but shallow rhizosphere, may limit water infiltration in the topsoil and even divert its flow (Tabacchi et al., 2000). That is, a dense herbaceous cover can even increase the runoff, although the surface flow is less here than in exposed soil, and there is less loss of sediments (and consequently nutrients) as well (Hofmann and Ries, 1991). Regarding the definition of just 15 meters of riparian zones in urban areas, we have returned back to what was discussed earlier in this section. These very narrow strips are probably non-functional, and cannot shelter native species diversity, since as we have seen, even larger areas do not fulfill these functions.

CONCLUSION

Throughout this chapter, we have seen how the Brazilian Forest Laws changed over time and affected all ecosystems across the country. I focused on riparian forests, but we know that ecosystems exchange energy and matter with each other in many ways (Chapin III et al., 2011), so, conservation efforts should consider the system as a whole; that is, regarding the landscape perspective. When any substantial native vegetation is lost, a set of species also becomes extinct in the landscape (Fischer et al., 2009), and, as we saw, an enormous amount of native vegetation was (and still is) lost in Brazil. Imagine how many species went extinct and how many functions were lost with them!

In 1965, the second version of the Forest Code was deployed that, although without much scientific basis, was more robust and conservationist, but it was not properly enforced. In 2012, the New Forest Law was validated, and is a weak conservation policy, despite the available scientific information on the importance of natural ecosystems, an already present environmental crisis, and ongoing global climate change. Unfortunately, the majority of the resolutions in the New Forest Law tend to worsen the current situation of riparian zones (as well as other Brazilian ecosystems), and so I presented here some relevant points about the changes that directly affect them, such as the measurement of this system from the regular baseflow water level, the recovery of tiny marginal strips, and their ecological impacts, but also with social and economic effects.

The New Forest Law is only three years old, but studies have already shown that it contributes little to biodiversity conservation and provision of ecosystem services in the Atlantic Forest (Alarcon et al., 2015), and puts terrestrial and aquatic biodiversity, as well as ecological functions of riparian forests at risk (Viegas et al., 2014), as was warned by scientists before its deployment. Over time and with the development of new research, the negative impacts of the New Forest Law will be revealed, like many of them discussed here. New changes and improvements in the Law are unlikely (at least in the short term), but I truly hope that minds will be changed in Brazil and worldwide, and that other Countries do not follow this example, and maybe we will live in a place where nature is not seen only as an area to deforest, build or plant. Because in times of water scarcity, loss of biodiversity and global climate change, riparian forests are always good options.

REFERENCES

Alarcon, G. G., Ayanu, Y., Fantini, A., Farley, J., Filho, A. S., Koellner, T. 2015. Weakening the Brazilian legislation for forest conservation has severe impacts for ecosystem services in the Atlantic Southern Forest. *Land Use Policy*, 47, 1-11.

Andreasen, J. K., O'Neil, R. V., Noss, R., Slosser, N. C., 2001. Considerations for the development of a terrestrial index of ecological integrity. *Ecol. Ind.*, 1, 21-35.

Asner, G. P., Knapp, D. E., Broadbent, E. N., Oliveira, P. J. C., Keller, M., Silva, J. N., 2005. Selective logging in the Brazilian Amazon. *Science,* 310, 480.

Bedê, L., 2012. In Brazil, working to safeguard 1/8 of the world's fresh water. Human Nature Conservation International Blog. Available in: <http://blog.conservation.org/2012/06/in-brazil-working-to-safeguard-one-eighth-of-worlds-fresh-water/>. Acessed in: 26 May 2015.

Bertoncini, A. P., Rodrigues, R. R., 2008. Forest restoration in an indigenous land considering a forest remnant influence (Avaí, São Paulo State, Brazil). *For. Ecol. Manag.*, 255, 513-521.

Beuchle, R., Grecchi, R. C., Shimabukuro, Y. E., Seliger, R., Eva, H. D., Sano, E., Achard, F., 2015. Land cover changes in the Brazilian Cerrado and Caatinga biomes from 1990 to 2010 on a systematic remote sensing sampling approach. *Apllied Geography*, 58, 116-127.

Borges, L. A. C., Rezende, J. L. P., Pereira, J. A. A., Júnior, L. M. C., Barros, D. A., 2011. Areas de preservação permanente na legislação brasileira. *Ciência Rural*, 41, 1202-1210.

Brancalion, P. H. S., Rodrigues, R. R., Gandolfi, S., Kageyama, P. Y., Nave, A. G., Gandara, F. B., Barbosa, L. M., 2010. Legal instruments can enhance high-diversity tropical forest restoration. *Rev. Árvore,* 34, 455-470.

Brasil, 1934. *Decreto Federal 23.793*, de 23 de janeiro de 1934. Aprova o Código Florestal. Brasília, DF. Disponivel em: <http://www.planalto.gov.br/ccivil_03/decreto/1930-1949/d23793.htm> Acesso em: 16 Maio 2015.

Brasil, 1965. *Lei 4.771*, de 15 de setembro de 1965. Institui o Código Florestal. Brasília, DF. Disponível em: <http://www.planalto.gov.br/ccivil_03/Leis/L4771.htm> Acesso em: 13 Abril 2015.

Brasil, 1989. *Lei 7.803*, de 18 de julho de 1989. Altera a redação do Código Florestal de 1965. Brasília, DF. Disponível em: <http://www.planalto.gov.br/ccivil_03/leis/L7803.htm#art2> Acesso em: 25 Maio 2015.

Brasil, 2001. *Media Provisória 2.166-67*, de 24 de agosto de 2001. Altera alguns artigos do Código Florestal de 1965. Brasília, DF. Dispoível em: <http://www.planalto.gov.br/ccivil_03/MPV/2166-67.htm#art1> Acesso em: 25 Maio 2015.

Brasil, 2012. *Lei 12.651*, de 25 de maio de 2012. Dispõe sobre a proteção da vegetação nativa. Brasília, DF. Disponível em: <http://www.planalto.gov.br/ccivil_03/_Ato2011-2014/2012/Lei/L12651compilado.htm> Acesso em: 13 Abril 2015.

Chapin III, F. S., Matson, P. A., Vitousek, P. M., 2011. *Principles of terrestrial ecosystem ecology*. 2nd ed. Springer: New York. 529 pp.

Costa, M. H., Botta, A., Cardille, F. A., 2003. Effects of large-scale changes in land cover on the discharge of the Tocantins River, Southeastern Amazonia. *J. Hydro.*, 283, 206-217.

Doren, R. F., Richards, J. H., Volin, J. C., 2009. A conceptual ecological model to facilitate understanding the role of invasive species in large-scale ecosystem restoration. *Ecol. Ind.*, 9s, 150-160.

Fischer, J., Lindenmayer D. B., Hobbs, R, 2009. Landscape pattern and biodiversity. In: Levin, S. A., (ed.), *Princeton Guide to Ecology*. Princeton University Press, Princeton. p. 431-437.

Foley, J. A., Asner, G. P., Costa, M. H., Coe, M. T., DeFries, R., Gibbs, H. K., Howard, E. A., Olson, S., Patz J., Ramankuttt, N., Snyder, P., 2007. Amazonia revealed: forest degradation and loss of ecosystem goods and services in the Amazon Basin. *Front. Ecol. Environ.,* 5, 25-32.

Françoso, R., Brandão, R., Nogueira, C. C., Salmona, Y., Machado, R. B., colli, G. R., 2015. Habitat loss and the effectiveness of protected areas in the errado biodiversity hotspot. *Nat. Conservação*, in press.

Gaston, K. J., 2010. Biodiversity. In: Sodhi, N. S., Ehrlich, P. R. (ed.), *Conservation Biology for All*. Oxford University Press: Oxford. p. 27-44.

Groot, R. S., Wilson, M. A., Boumans, R. M. J., 2002. A typology for the classification, description and valuation of ecosystem functions, goods and services. *Ecol. Econ.*, 41, 393-408.

Hansen, B., Reich, P., Lake, P. S., Cavanaro, T., 2010. Minimum width requirements for riparian zones to protect flowing waters and to conserve biodiversity: a review and recommendations. Report to the Office of Water, Department of Sustainability and Environment, Monash University. 151 pp.

Hofmann, L., Ries, R. E., 1991. Relationship of soil and plant characteristics to erosion and runoff on pasture and rage. *J. Soil Water Conserv.*, 46, 143-147.

IBGE – Instituto Brasileiro de Geografia e Estatística, 2006. *Censo agropecuário 2006: Brasil, grandes regiões e Unidades da Federação.* IBGE: Rio de Janeiro. 777 pp.

IBGE – Instituto Brasileiro de Geografia e Estatística, 2012. *Indicadores de desenvolvimento sustentável.* IBGE: Rio de Janeiro. 350 pp.

Ivanauskas, N. M., Rodrigues, R. R., Souza, V. C., 2007. The importance of the regional floristic diversity for the restoration successfulness. In: Rodrigues, R. R., Martins, S. V., Gandolfi, S. *High diversity forest restoration in degraded areas: methods and projects in Brazil.* Nova Science Publisher: New York. p. 61-76.

Kahn, J. R., Freitas, C. E., Petrere, M., 2014. False shades of green: the case of Brazilian Amazonian hydropower. *Energies,* 7, 6063-6082.

Kauffman, J. B., Krueger,W. C., 1984. Livestock impacts on ripariam ecosystems and streamside management implications: a review. *J. Range Manag.*, 5, 430-438.

Lamb, D., Gilmour, D., 2003. *Rehabilitation and restoration of degraded forests.* IUCN, Gland, Switzerland and Cambridge, UK and WWF, Gland, Switzerland. 110 pp.

Landau, E. C., Cruz, R. K., Hirsch, A., Pimenta, F. M., Guimarães, D. P. 2012. *Variação geográfica do tamanho dos módulos fiscais no Brasil.* Documentos 146. Embrapa Milho e Sorgo, Sete Lagoas. 199 pp.

Laurance, W. F., 1991. Edge effects in tropical forest fragments: application of a model for the design of nature reserves. *Biol. Conserv.*, 2, 205-219.

Laurance, W. F., 2015. Emerging threats to tropical forests. *Ann. Missouri Bot. Gard.,* 100, 159-169.

Lees, A. C., Peres, C. A., 2008. Conservation value of remnant riparian forest corridors of varying quality for Amazonian birds and mammals. *Conserv. Biol.*, 22, 439-449.

Likens, G. E., 1992. *The ecosystem approach: its use and abuse*. Ecology Institute: Germany. 166 pp.

Lima, W. P., Zakia, M. J. B., 2004. Hidrologia de matas ciliares. In: Rodrigues, R. R. and Leitão-Filho, H. F. (Eds.). Matas ciliares: conservação e recuperação. 2 ed. 1ª reimpr. São Paulo: Editora da Universidade de São Paulo, FAPESP, 2004.

Lista de Espécies da Flora do Brasil. 2015. Jardim Botânico do Rio de Janeiro. Disponível em: <http://floradobrasil.jbrj.gov.br/>. Acesso em: 13 abr. 15.

Maisonneuve, C., Rioux, S., 2001. Importance of riparian habitats for small mammals and herpetofaunal communities in agricultural landscapes of southern Québec. *Agric. Ecosyst. Environ.*, 83, 165-175.

Martins, S. V., 2011. *Recuperção de matas ciliares*. 2ª ed. Aprenda Fácil: Viçosa. 255 pp.

Metzger, J. P., Lewinsohn, T. M., Joly, C. A., Casatti, L., Rodrigues, R. R., Martinelli, L. A., 2010. Impactos potenciais das alterações propostas para o Código Florestal Brasileiro na biodiversidade e nos serviços ecossistêmicos. *BIOTA-FAPESP e ABECO*, 1, 1-13.

Mitchell, F. J. G., Kirby, K. J., 1990. The impacts of large herbivores on the conservation of semi-natural woods in the British uplands. *Forestry*, 63, 333-353.

Mittermeier, R. A., Fonseca, G. A. B., Rylands, A. B, Brandon, K., 2005. Uma breve história da conservação da biodiversidade no Brasil. *Megadiversidade*, 1, 14-21.

Montagnini, F., Jordan, C. F., 2005. *Tropical Forest Ecology: the basis for Conservation and Management*. Springer: Berlin. 295 p.

Myers, N., Mittermeier, R. A., Mittermeier, C. G., Fonseca, G. A. B., Kent, J., 2000. Biodiversity hotspots for conservation priorities. *Nature*, 403, 853-858.

Naiman, R. J., Décamps, H., McClain, M. E., 2009. *Riparia: ecology, conservation and management of streamside communities*. Elsevier.

Naiman, R. J., Décamps, H., Pollock, M., 1993. The role of riparian corridors in maintaining regional biodiversity. *Ecol. Appl.*, 3, 209-212.

Nave, A. G., Rodrigues, R. R., 2007. Combination of species into *filling* and *diversity* groups as forest restoration methodology. In: Rodrigues, R. R., Martins, S. V., Gandolfi, S. *High diversity forest restoration in degraded areas: methods and projects in Brazil*. Nova Science Publisher: New York. p. 103-126.

Parkyn, S., Shaw, W., Eades, P., 2000. *Review of information on riparian buffer widths necessary to support sustainable vegetation and meet aquatic functions*. Prepared by NIWA for Auckland Regional Council. Auckland Regional Council Technical Publication Munber 350. 38 p.

Peterman, W., Semlitsch, R. D., 2009. Efficacy of riparian buffers in mitigating local population declines and the effects of even-aged timber harvest on larval salamanders. *For. Ecol. Manag.*, 257, 8-14.

Piedade, M. T. F., Junk, W., Sousa Jr., P. T., Cunha, C. N., Schongart, J., Wittman, F., 2012. As áreas úmidas no âmbito do Código Florestal brasileiro. In: Souza, G., Jucá, K., Wathely, M., (org.). *Código Florestal e a Ciência: o que nossos legisladores ainda precisam saber*. Comitê Brasil: Brasília. 115 p.

Primack, R. B., Rodrigues, E., 2001. *Biologia da conservação*. Editora Planta: Londrina. 328 pp.

Ramos, C. C. O., Anjos, L., 2014. The width and biotic integrity of riparian forests affect richness, abundance, and composition of bird communities. *Natureza and Conservação*, 12, 59-64.

Rezende, A. V., 1998. Importância das matas de galeria: manutenção e recuperação. In: Ribeiro, J. F. (ed)., *Cerrado: matas de galeria*. Embrapa: Planaltina. 164 pp.

Ribeiro, M. C., Metzger, J. P., Martensen, A. C., Ponzoni, F. J., Hirota, M. M., 2009. The Brazilian Atlantic Forest: how much is left, and how is the remaining forest distributed? Implications for conservation. *Biological Conserv.*, 142, 1141-1153.

Rizzini, C. T., 1997. *Tratado de fitogeografia do Brasil: aspectos ecológicos, sociológicos e florísticos*. 2 ed., Âmbito Cultural Edições Ltda, Rio de Janeiro.

Rodrigues, R. R., Gandolfi, S., 2007. Restoration actions. In: Rodrigues, R. R., Martins, S. V., Gandolfi, S. *High diversity forest restoration in degraded areas: methods and projects in Brazil*. Nova Science Publisher: New York. p. 77-102.

Semlitsch, R. D., Bodie, R., 2003. Biological criteria for buffer zones around wetlands and riparian habitats for amphibians and reptiles. *Conserv. Biol.*, 5, 1219-1228.

SER – Society for Ecological Restoration International, Science and Police Working Group, 2004. *The SER International Primer on Ecological Restoration*. www.ser.org and Tucson: Society for Ecological Restoration International.

Silva, C. J., Sousa, K. N. S., Ikeda-Castrillon, S. K., Lopes CRAS, Nunes, J. R. S., et al., 2015. Biodiversity and its drivers and pressures of change in the wetlands of the Upper Paraguay-Guaporé Ecotone, Mato Grosso (Brazil). *Land Use Policy,* 47, 163-178.

SOS Mata Atlantica. 2013. Atlas da Mata Atlantica. Disponivel em: <http://sosma.org.br>. Acesso em: 18 abril 2015.

Souza, F. M., Batista, J. L. F., 2004. Restoration of seasonal semidecidous forests in Brasil: influence of age and restoration design on forest structure. *For. Ecol. Manag.,* 1-3, 185-200.

Tabacchi, E., Lambs, L., Guilloy, H., Planty-Tabacchi, A. M., Muller, E., Décamps, H., 2000. Impacts of riparian vegetation on hydrological processes. *Hydrological Processes*, 14, 2959-2976.

Viegas, G., Stenert, C., Schulz, U. H., Maltchik, L., 2014. Dung beetle communities as biological indicators of riparian forest widths in southern Brazil. *Ecol. Indic.*, 36, 703-710.

Wang, X., Wang, Y., Wang Y., 2013. Use of exotic species during ecological restoration can produce effects that resemble vegetation invasions and other unintended consequences. *Ecol. Engineering*, 52, 247-251.

In: Riparian Zones
Editor: Oleg S. Pokrovsky

ISBN: 978-1-63484-613-4

Chapter 4

BIOGEOCHEMICAL FUNCTIONING OF AMAZONIAN FLOODPLAINS: THE CASE OF LAGO GRANDE DE CURUAI

M-P. Bonnet[1], J. Garnier[2], G. Barroux[1], G. R. Boaventura[2] and P. Seyler[3]
[1]IRD, GET UMR 5563 CNRS, Toulouse, France
[2]UnB, Universidade de Brasilia, Instituto de Geociencias, Brasilia-DF, Brasil
[3]Laboratoire HydroSciences Montpellier, France

ABSTRACT

The large floodplains associated with the Amazon River and its main tributaries affect water transport, influence sediment and chemical budgets and support highly diverse ecosystems and productive fisheries. Approximately 25% of the Amazon River discharge is routed through these systems along a 2,000 km reach between Sao Paulo de Olivença and Óbidos (Richey et al., 1989), with an extension of over 100,000 km^2 (Melack and Hess, 2010). A major challenge in understanding the Amazon fluvial system is evaluating how and to what extent biogeochemical floodplain processes influence the hydrological, chemical and biological dynamics of the Amazon River. We focus our study on one of the largest floodplain systems of the Amazon, "Lago Grande de Curuai," which is representative of the central and lower reaches of the Amazon basin. A ten-year study of the "Várzea do Lago Grande de Curuai" floodplain allowed us to obtain a comprehensive understanding of how hydrological and geochemical processes occurring in floodplains alter the dissolved and particulate flux of the Amazon River. In terms of internal hydrological functioning, we demonstrated that floodplain water balances are influenced by direct rainfall, local runoff and seepage as well as flooding from the river and that the relative importance of different inputs varies seasonally. Regarding the sediment and particulate carbon budget of the Curuai floodplain, the annual volume of sediment trapped in the floodplain was of the same order of magnitude as the mean annual sediment fluxes outflowing from the floodplain into the Amazon River, and the floodplains act as an important source of particulate and dissolved organic carbon. The water passing through the floodplains undergoes important biogeochemical transformations under the influence of biotic processes, sorption and redox reactions. Floodplains also play an important role in the organic carbon balance of the Amazon basin. Coupling the hydrological model with the database of elemental concentrations, we discuss the conservativity or non-conservativity of certain selected elements and present the mass balance of these elements between the floodplain and the main stream.

INTRODUCTION

Wetlands cover ca. $2.43 \cdot 10^6$ km² globally (Adams and Faure, 1998) and constitute important sites for the production, storage and export of dissolved and solid matter. Floodplains, or "varzeas" (Amazonian floodplains), are dynamic, complex wetland systems that periodically oscillate between terrestrial and aquatic phases, with constant sediment exchange between river channels and floodplains. The floodplains of large rivers are generated through the formation of bars and the accumulation of sediment carried in diffuse overbank flows and channelized flows (Dunne et al., 1998). Floodplains affect erosion, transport and sedimentation flux budgets in the watershed system and are of special importance to the carbon cycle due to their high productivity (Junk, 1997).

The Amazon River is of particular importance because its floodplain is the largest in the world, covering an estimated area of approximately 800,000 km² (Melack and Hess, 2010) and harboring large stocks of water and sediment along the main rivers of the basin (Figure 1). Many processes occur in flooded areas, including the recycling of various elements that are naturally present in the waters and produced elsewhere. Flooded areas are usually very productive under the influence of floods, which recurrently provide large amounts of nutrients. Richey et al., (1989) indirectly estimated that 25% of the flow of the Amazon River is routed through the floodplain along a 2010 km long reach, inundating an estimated area of approximately 100×10^3 km^2 (Sippel et al., 1998). The Amazonian floodplain is composed of thousands of lakes, which are often grouped into sub-systems of inter-connected lakes, and forms a complex mosaic of freshwater systems with contrasting morphologies (Sippel et al., 1998), resulting in different inundation patterns (Mertes, 1997; Hamilton et al., 2002; Alsdorf et al., 2007), water chemistry and ecological characteristics (Hamilton et al., 2007, Junk et al., 2010). The floodplain lakes act as a temporary storage system for dissolved and particulate elements and as system that exports these elements into the main stream during large floods (Junk et al., 1989; Seyler and Boaventura, 2003; Bourgoin et al., 2007). During the residence of water in floodplain lakes, various processes, such as adsorption, desorption, redox chemical reactions, and uptake by biota influence the chemistry of the surface water of the Amazon River (Junk et al., 1989; Richey et al., 1989; Barroux et al., 2006). These systems are highly productive and are considered to be hotspots of biodiversity, which are characteristics that are primarily controlled by the river flood pulse (Junk et al., 1989), favoring the renewal of nutrients and the creation of various micro-habitats that differ in space and time throughout the course of the water year. Primary productivity of the floodplain is estimated to be 110 TC d.w. (dry weight) ha^{-1}, 73% of which originates from phytoplankton and terrestrial and aquatic macrophytes, whereas 27% originates from the inundated forest (Junk and Furch, 1993). The inflow of the turbid, nutrient-rich waters of the Amazon River helps sustain the floodplain's high fertility (Melack and Forsberg, 2001), which in turn sustains human activities such as agriculture, animal husbandry and forestry (Dufour, 1990).

This chapter summarizes a ten-year study conducted in one of the main floodplains of the medium and lower courses of the Amazon River, the "Varzea do Lago Grande de Curuai" ("várzea" is the local name of Amazonian floodplain lakes), where hydrological, chemical and sedimentological analyses were conducted. Based on observations, we built a water and mass balance model to understand and quantify the impact of floodplains on the dissolved and solid fluxes of the Amazon main stream.

STUDY SITE

The "Várzea do Lago Grande de Curuai" floodplain (Figure 2) is located on the right bank of the Amazon River, across from the city of Óbidos, 900 km from the mouth of the river (between 56.10°W and 55.00°W from upstream to downstream and 2.3°S and 1.9°S). It is a complex system of more than 30 interconnected lakes, with a maximum inundated area of 2,300 km^2, representing 1% of the total flooded area of the Central Amazon basin. The catchment of the Várzea Lago Grande de Curuai covers an area of 3,660 km^2, including open water areas. It comprises several lakes characterized by high suspended sediment loads as well as lakes characterized by high concentrations of dissolved humic acids and low concentrations of suspended sediment, which are interconnected with each other and permanently connected to the Amazon main stream by various small channels. During the rising stage, the open-water area varies between 575 and 2,090 km^2, which represents approximately 13% of the total flooded area of the Amazon River, between Manaus and Óbidos (Hess et al., 2003; Martinez and Le Toan, 2007). This floodplain is typical of the eastern Amazon floodplains; it is extensive and flat and is vegetated during the low-water season with savannah, low vegetation and alluvial forest (RADAMBRASIL, 1976). The soils are mainly composed of recent alluvial deposits and ferralitic soils, which mostly developed on the Alter do Chão formation (Lucas, 1989) since the Cenozoic period.

METHODS

Sample Collection and Analytical Procedures

Twelve field cruises were organized between 2002 and 2008 in the lakes of the Curuai floodplain (Figure 2). The sampling campaigns were distributed across three different hydrological periods: March–May, which corresponds to the rising water stage; June–July, which corresponds to the flood peak of the Amazon; and October–December, during the low-water stage. The sampling, filtration and the chemical analysis techniques were very similar to the methods used during our previous studies (Seyler et al., 2003; Barroux et al., 2005; Viers et al., 2005; Moreira-Turcq et al., 2013). Water samples were collected from multiple stations in the floodplain and in the Amazon River. For major and trace element analysis, an ultraclean sampling procedure was used throughout the manipulations in the field. The concentrations of dissolved organic carbon (DOC), Cl, SO_4, cations and trace elements as well as alkalinity were measured using routine methods. DOC was measured using a

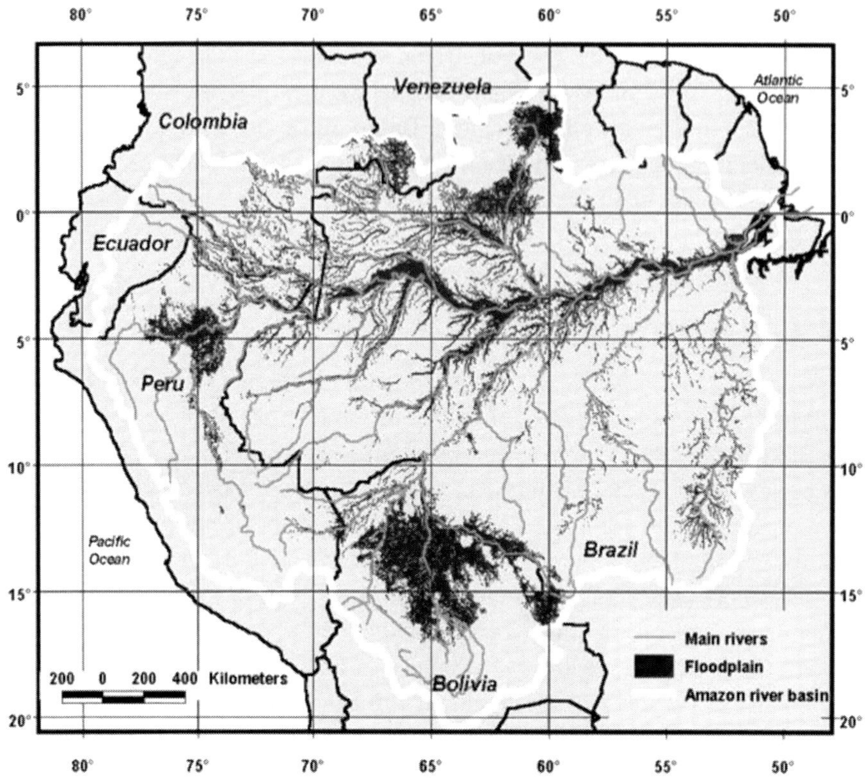

Figure 1. Extension of floodplains in the Amazon basin. Wetlands extend over 800,000 km^2 approximately 14% of the Amazon basin (Melack and Hess, 2010).

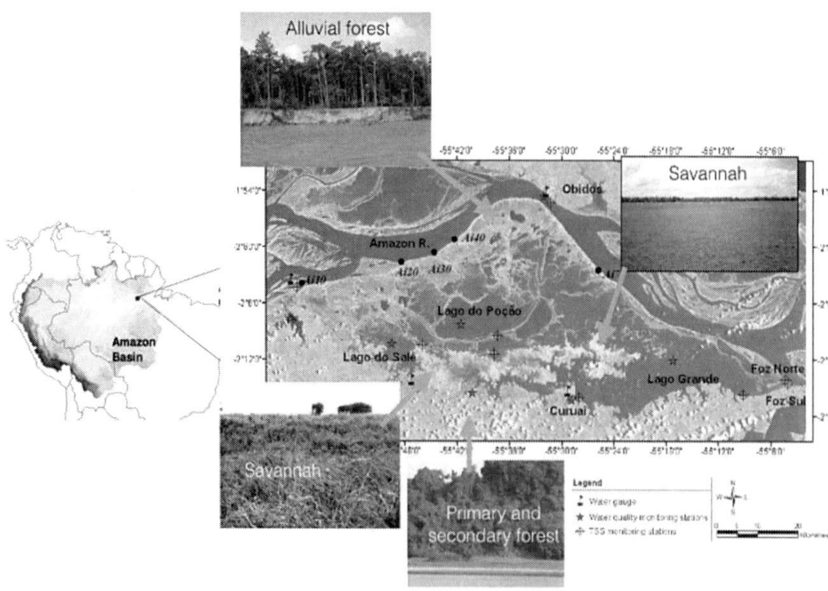

Figure 2. Study area. The area encompasses several lakes in the south and is bordered to the north by the Amazon R. In the south, the local drainage basin land cover consists essentially of primary and secondary forests. Water quality was studied in the principal lakes, marked with stars.

Shimadzu TOC 6000 with an uncertainty of 5%, and anions using a Dionex ICS-90 Ion Chromatography System. Trace elements were measured via inductively coupled plasma-mass spectrometry (ICP-MS, Perkin Elmer Elan 6000 and Agilent 7500), with indium and rhenium as the internal standards and a precision better than ±5%. The international geostandard SLRS-4 (Riverine Water Reference Material for Trace Metals certified by the National Research Council of Canada) was used to check the accuracy of the ICP-MS analyses, based on the average of four measurements; the relative difference was 8% for all elements.

Hydrological Modeling

The water level in the Curuai floodplain was surveyed for several consecutive water years. Water level gauges were installed at several locations in the floodplains at Curuai and in Sale Lake (Figure 2). Water levels were manually read twice a day, and discharge measurements in the principal connecting channels were regularly measured using an Acoustic Doppler Current Profiler (ADCP) at 1200 Hz. The basis of the model used to compute the water exchange between the floodplain system and the Amazon River is detailed in Bonnet et al. (2008, 2011). This model computes the variation of the stored volume as a function of the different sources and losses of water in the floodplain, i.e., direct rainfall, evaporation, seepage, runoff from the upland local watershed and exchanges with the main stream via diffusive overbank flows and through the connecting channels. Despite its relative simplicity, the model was recently shown to provide rather realistic estimations of the various contributions, at least at an annual scale (Rudorff et al., 2014) and a weekly scale (Bonnet et al., in prep). It was extended to represent the floodplain's internal lakes according to a simplified geometry (Figure 3) to help interpret chemical heterogeneity in the floodplain.

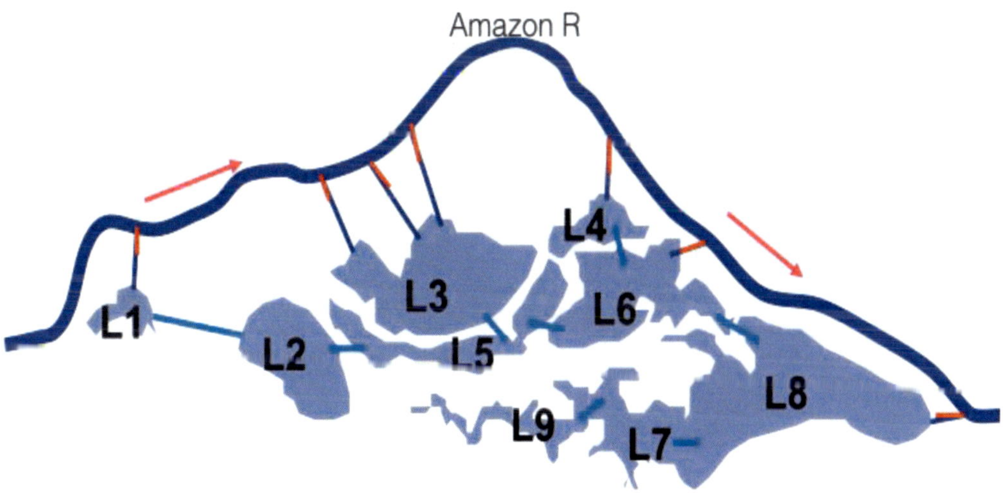

Figure 3. Proposed geometry for modeling the floodplain as a set of interconnected lakes. The model considers nine lakes connected by "channels" (light blue lines).

For each lake, the water balance can be written as follows:

$$V_j^{t+1} = V_j^t + \left(Q_{R_j} - Q_{E_j} + Q_{UpWs_j} + Q_{attz_j} \pm Q_{G_j} \pm \sum_i Q_{ij} \right) \quad (1)$$

where subscript j refers to each considered lake. Each flux term is computed at time t.

Q_{R_j} and Q_{E_j} denote the rainfall and evaporation fluxes obtained as the product of the rainfall and evaporation rates with the j lake open area, respectively. The term Q_{UpWs_j} is the runoff from the upland local watershed, and Q_{attz_j} is the runoff from the seasonally inundated lands drained by the considered lake. Both terms are computed from the non-linear reservoir equation $Q_{UpWs_j} = a\left(R_j^{t-d}\right)^b$, where a, b, and d are three calibration parameters. Parameter b was fixed at 0.5, which is a common value (Wittenberg and Sivapalan, 1999), and d accounts for the lag in time between rain events and runoff and was assumed to be 0 for the seasonally inundated lands.

Q_{G_j} is deduced from the total groundwater flux, Q_G, estimated using the lumped model to describe the floodplain as a unique lake as follows: $Q_{G_j} = Q_G \frac{V_j}{V}$, where V is the floodplain volume.

$\sum_i Q_{ij}$ represents the fluxes exchanged with the Amazon R. and/or between individual lakes through channels. The geometry of the river-to-lake channels was the same as that found using the lumped model (Bonnet et al., 2008), whereas the inter-lake channels required calibration.

- **Data used:**

The model requires volume-to-water level and surface-to-water level relationships for each lake. These relationships were built from the available DEM. The lakes' watersheds were obtained from the hydrotools extension implemented in ArcGIS®.

Rainfall rates were deduced from Thiessen polygons and, thus, differed among lakes, whereas the evaporation rate was uniformly distributed. Monthly chloride concentrations measured in each of the nine regions were used to constrain the model and calibrate the inter-lake channel geometry. The model was validated against data collected during the field trips organized during the 2002-2003 water year.

Modeling of Solid and Dissolved Fluxes

Once the water fluxes of the different channels are known, TSS and elemental fluxes are deduced, assuming that inflowing water exhibits the composition of the Amazon River at the locations of connections, whereas outflowing water exhibits the composition of the floodplain water. Daily solid fluxes were computed using daily TSS concentrations calculated from the correlation between the TSS data obtained each 10 days and the daily discharge data. Dissolved fluxes were computed using monthly measured concentrations. Data for the Amazon River are available at the HYBAM website.

RESULTS

Water Balance of the LAGO Grande de Curuai Floodplain

The results obtained from the hydrological model allow us to better understand the temporal dynamics of water exchange between the várzea and the Amazon. Each year, the filling phase extends from December-January to May-June, whereas the draining phase extends from June-July to November-December. The beginning and duration of the filling and draining phases are controlled by the hydrology of the Amazon and local precipitation. The results show that the Amazon River provides the main water input for the Curuai floodplain, which represents approximately 70 to 90% of the total water input to the lakes, (i.e., 4.5 times the maximum volume of the várzea during the hydrological cycle), whereas the outflow represents 4.7 times this volume. The local net precipitation and runoff from the local watershed constitute 0.36 and 0.24 times this volume, respectively, and the ground water inputs contribute less than 5%. The river water residence time in this floodplain is 5.0± 0.8 months, whereas the residence time of the other sources (ground water, runoff and rainfall) is 3.0 ± 0.2 months. The várzea represents a net source of 2 km^3 of fresh water for the Amazon.

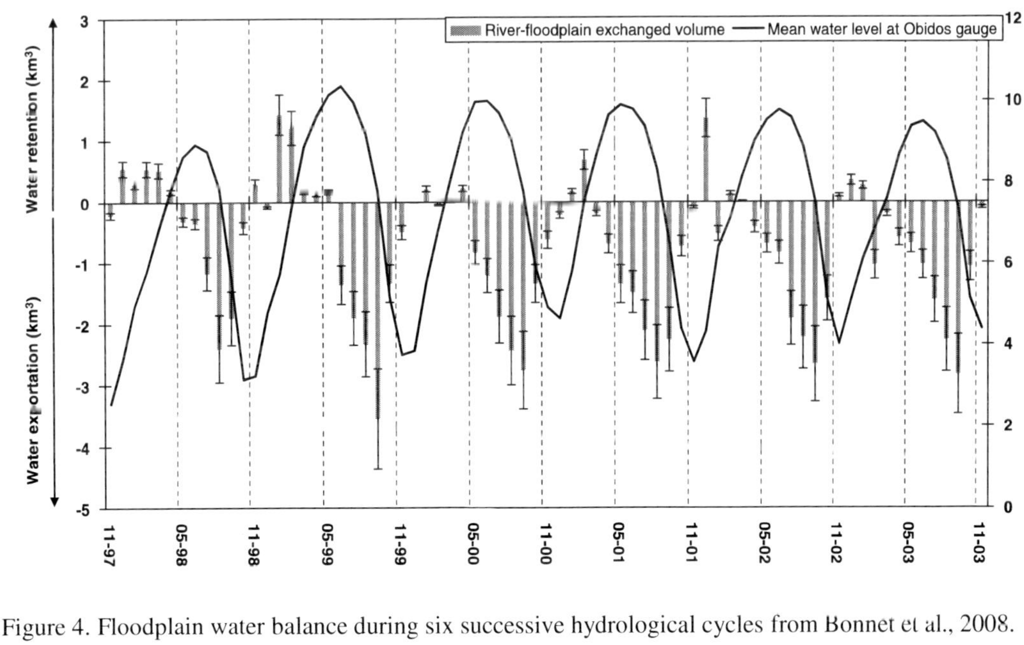

Figure 4. Floodplain water balance during six successive hydrological cycles from Bonnet et al., 2008.

The respective influence of local waters (precipitation and runoff from the local watershed and seepage) and regional water from the main stream varies during the course of the year. For example, at the beginning of the 2001-2002 water year until mid-December, the floodplain mixture was dominated by the water from the previous year. In early January, the Amazon River dominated the mixture (64%). From this date until the beginning of April, the contribution of river water decreased slightly, whereas the contributions from watersheds and direct rainfall increased. By the end of the year, river water represented 78% of the mixture. By mid-April, the contribution from rainfall water was as much as 17%, whereas the contributions from local upland watersheds and watersheds located in the ATTZ reached their maximum levels by the end of February, constituting 14% and 15% of the mixture, respectively. The contribution from the groundwater reservoir reached its maximum at the end of December, at 5% of the mixture. However, there were large inter-annual variations, mainly as a function of the celerity and amplitude of the Amazon flood wave.

Global Biogeochemical Functioning of the Curuai Floodplain

The lakes under the direct influence of the Amazon constitute the majority of lakes in the Curuai floodplain. The temperature measured during the campaign was approximately 30° ± 5°C throughout the year. The pH was generally close to neutral, except in October and June, when it was slightly acidic (6.3), though not during the phytoplankton blooms occurring in the falling water stage (up to 8.7 at the surface). The conductivity of these lakes varied between 30 to 100 µS/cm and globally followed the values measured in the Amazon River.

- **Major elements**

Carbonate alkalinity presented an average value of 22.82 $mg.l^{-1}$ and ranged from 10 to 40 $mg.l^{-1}$. The major dissolved anions were Cl^- and SO_4^{2-}, with average values of 1.91 and 2.91 $mg.l^-$, respectively. The major cations are classically Na^+, K^+, Mg^{2+} and Ca^{2+}, which exhibited average values of 2.12, 0.82, 1.03 and 5.48 $mg.l^{-1}$, respectively. For SO_4^{2-}, K^+, Mg^{2+} and Ca^{2+}, the minima occurred in June and the maxima in March, whereas for Cl^- and Na^+, the minima occurred in June and the maxima in October. Nitrates presented low concentrations throughout the year (0.180 ± 0.050 $mg.l^{-1}$), with the maxima being measured during low water stages. The average concentration of PO_4^{3-} in the lakes was 0.150 $mg.l^{-1}$, with a minimum of 0.05 and a maximum of 0.200 $mg.l^{-1}$. In Figure 5, we report the mean concentrations of each major anion and cation (in $meq.l^{-1}$) measured in the varzea lakes.

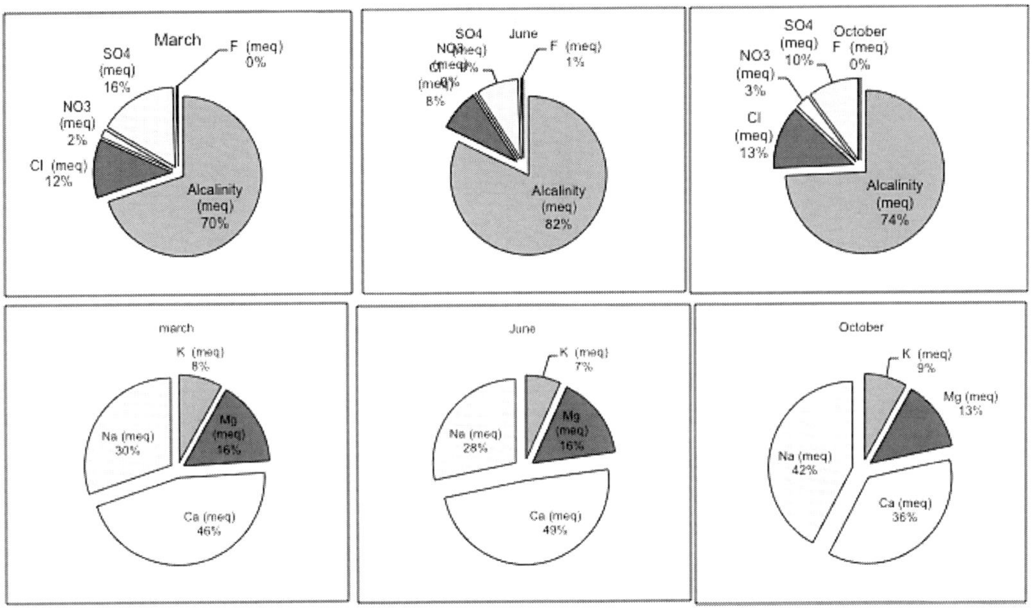

Figure 5. Mean concentrations of major elements in the Curuai varzea lakes. (probleme sur la figure du haut au milieu).

- **Trace elements**

Mn, Fe, Sr and Ba exhibited concentrations between 10 and 200 ppb and were the most important elements in the dissolved load. Al, Cu, Zn and Rb presented values between 1 and 10 $\mu g.l^{-1}$; V, Cr, As, Y, Mo, Sb and Ce showed concentrations in the range of 10 $\mu g.l^{-1}$; Co and REEs (except for Ce, Pb, Th and U) exhibited concentrations in the range of 10 $\mu g.l^{-1}$; and Cs and Lu displayed concentrations lower than 10 $\mu g.l^{-1}$. The trace elements measured in the dissolved phase (<0.22 µm) of the lake water are shown in Figure 6 as average concentrations. The lakes under the influence of the Amazon are in blue. The elements with the highest concentrations were Al (from 2 to 20 $\mu g.l^{-1}$), Fe (10 to 100 $\mu g.l^{-1}$), Sr and Ba (10 to 70 $\mu g.l^{-1}$). Lower concentrations were reported for Cd (from 0.001 to 0.01 $\mu g.l^{-1}$), rare earth elements (REEs) (0.001 to 0.1 $\mu g.l^{-1}$), and Pb and Th (0.001 to 0.1 $\mu g.l^{-1}$). The average abundance of the elements for all lakes presented the following order: Fe > Sr > Ba > Al > Mn > Cu > Rb > As > V > Sb > U> Pb > Y and TRs > Co> Cd > Th. The elements with the greatest variability were Mn> Pb> La> Al (> 100%). The variabilities of Th> Cd> Y and REEs were between 50 and 80%, whereas the variabilities of Fe > V > Sb > Cu> Rb > Co > Sr > U> Ba > As were between 15 and 40%.

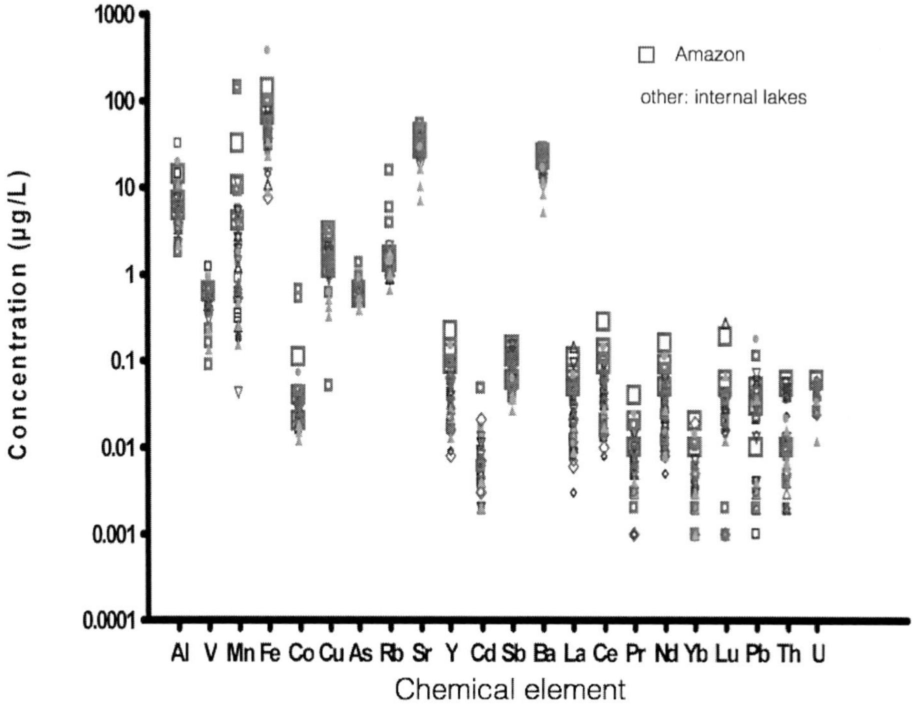

Figure 6. Trace element concentrations in the internal lakes of the Curuai floodplain – The concentration in the Amazon is also reported for comparison.

- **Sediment and POC fluxes**

The mean TSM was 135 mg.l^{-1}. The highest concentrations (up to 800 mg.l^{-1}) were observed during low-water periods (November-December). To estimate the TSM mass balance, we used the results of the hydrological model described above. The annual sediment storage entering through connected channels into the Curuai floodplain reached approximately 710 × 10^{3} t year^{-1} (± 19%), which represents between 41% and 53% of the annual flux of sediments entering from the main stream. The mean residence time of the river water was on the order of 3 months, whereas for water of any origin, it was approximately 2 months (Bourgoin et al., 2007). The mean POC content was approximately 8.2%, ranging from 1.69 to 4.45%. During periods of rising water, the mean POC content in floodplain 5 was 10.9%. The mean POC content was lowest during low-water periods (4.1%), whereas in floodplain 6, the contents were 8.48 and 8.61% during high- and falling-water periods, respectively. In the Amazon River and the Curuai floodplain, POC was inversely related to the TSM. The average concentration of COD in the varzea lakes directly connected to the Amazon ranged from 5.5 ± 2.4 mg.l^{-1}, with a maximum of 7.0 mg.l^{-1} during low waters and a minimum of 2.0 mg.l^{-1}. To estimate the carbon mass balance, we used the results of the hydrological model described above. According to the model results, DOC flowing into the várzea ranged from 0.63 to 29 Gg per month, and POC inflows varied from 0.15 to 6 Gg C per month. The greatest input fluxes occurred during high-water periods, when water from the Amazon River entered the lakes. DOC outflows ranged from 0.09 Gg C in January to 28

Gg C in June. Outflowing POC fluxes were at their minimum in January (0.06 Gg C) and at their maximum in August (16 Gg C). The efflux of total organic carbon represented approximately 0.3% of the total carbon flux at the Óbidos (which was estimated to be 32.7 ± 3.3 Tg/year by Moreira et al., 2003). The particulate organic carbon (POC) and dissolved organic carbon (DOC) fluxes exported by the Curuai floodplain represented 1.3% and 0.1%, respectively, of the annual POC and DOC fluxes at Óbidos.

Discussion

Internal Heterogeneity of Elemental Behavior

- **A distributed model for conservative elements**

Our observations revealed strong heterogeneity across the different lakes that make up the floodplain system. For example, Figure 8 shows the spatial distribution of the chloride concentration observed during the filling period (March 2003). There are clearly two distinct regions: one with chloride concentrations ranging between 2.1 and 2.3 $mg.l^{-1}$ close to the chloride concentration found in the Amazon R., and one with lower concentrations, between 0.7 to 1.6 $mg.l^{-1}$.

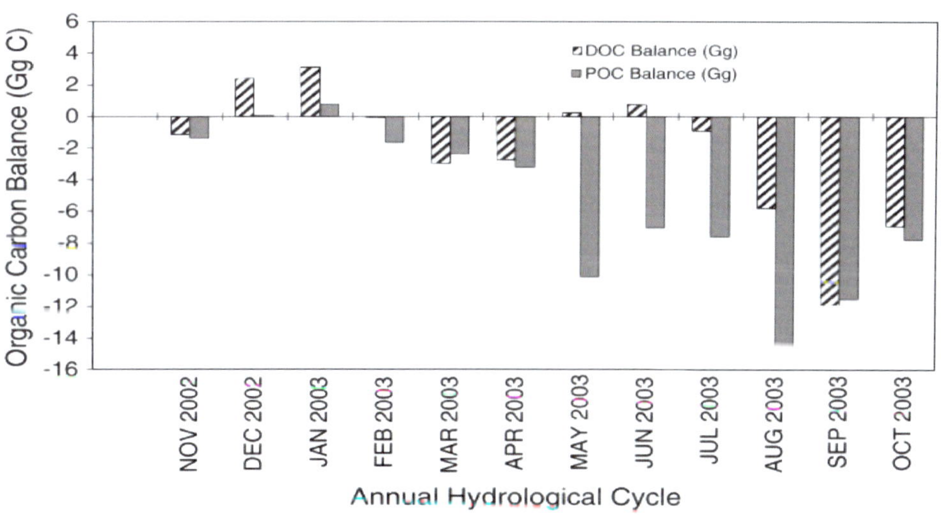

Figure 7. Monthly fluxes of particulate organic carbon (POC) and dissolved organic carbon (DOC) between the Amazon River and the floodplain system for the 2003 hydrological year. Fluxes are positive during the filling of the floodplain lakes (inflow) and negative during drainage (outflow). From (Moreira-Turcq et al., 2013).

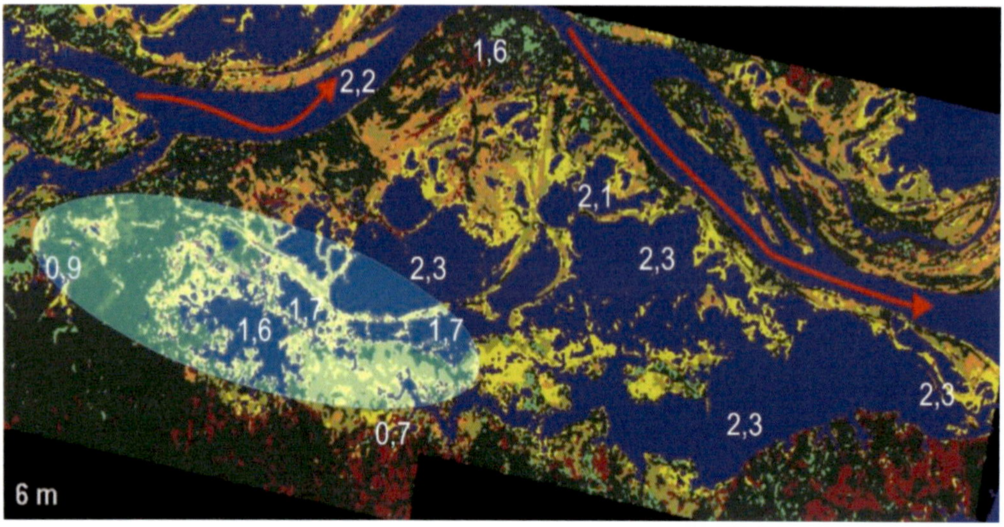

Figure 8. Distribution of chloride concentrations in the floodplain in March, 2003.

As chloride behaves as a non-reactive element (Hem, 1970) and presents different concentrations in the Amazon R. and in rain, groundwater and local runoff, this spatial distribution demonstrates a dissimilar proportion in the inflow paths to the different lakes of the floodplain. To assist in the interpretation of variations in water chemical properties among lakes, we proposed to model individual lake water and mass balances. The model was adapted from (Bonnet et al., 2008) to introduce inter-connected lakes within the floodplain and to allow computation of the mass balance as presented in the Methods section (distributed lakes model).

The distributed lake model was calibrated using water level data recorded at two locations in the floodplain and the chloride concentrations measured during the period from March 2002 to December 2003. This version of the model does not consider exchanges with the water table. The modeled and measured concentrations of dissolved chloride (<0.22 μm) in Lakes Curumucuri, Salé, Poçaõ and Grande are reported in Figure 9. Generally speaking, the seasonal evolution of dissolved chloride was well reproduced by the model, but the model slightly under-estimated the concentration during low-water periods and over-estimated it during high-water periods. The average absolute differences between the simulated and measured chloride concentrations were 0.14, 0.18 and 0.31 $mg.l^{-1}$ for Lakes Salé, Poçaõ and Grande, respectively. These differences correspond to relative absolute errors of 8.1, 9.6 and 13%, which are of the same order of magnitude as the measurement accuracy (approximately 10%). The absolute mean difference was slightly greater for Cumurucuri Lake (0.16 $mg.l^{-1}$, which represents a relative error of 16.4%). For this lake, even if the model was correctly reproducing the seasonal trend, the simulated chloride concentrations were systematically slightly over-estimated.

The model enabled the filling modalities for each represented lake to be analyzed. In particular, we confirmed that the Sale and Curumucuri Lakes are mainly filled by local waters, whereas Lake Grande is mainly filled from the Amazon and Lake L6 (in Figure 3). Poção Lake was essentially flooded from the Amazon R.

- **Identification of chemical elements with non-conservative behavior**

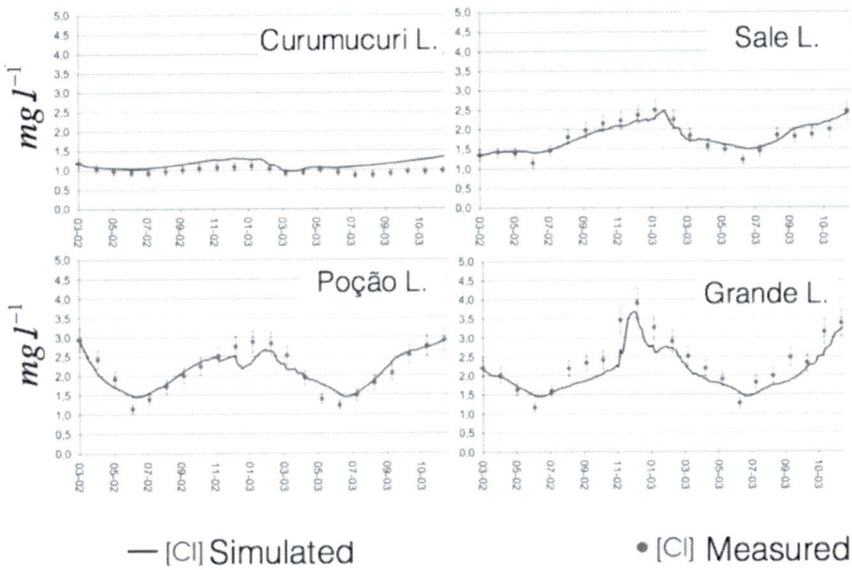

Figure 9. Observed and simulated Cl concentrations in the four main internal lakes of the floodplain.

Here, we aimed to identify the chemical elements that do not behave as non-reactive tracers by comparing the modeled and observed concentrations. As the model assumed conservative behavior for all the considered elements, large differences between its results and the observations indicated deviation from the conservative expectation of the considered element. We focused on Ca, Sr, K, Cu and U, which exhibited the highest-confidence measurements and show interesting behavior in the Amazon. The Ca, K, Sr and Cu concentrations varied inversely with the Amazon discharge (Stallard and Edmond, 2000; Seyler and Boaventura, 2003), whereas U presented a rather constant concentration year-round. Here, we focused only on Lakes Sale, Poção and Grande, for which the conservative model was successful in representing chloride concentrations.

To determine whether a considered element behaved conservatively, we analyzed the relative differences between the reported modeled and measured concentrations and those computed for the chloride concentration. If these relative differences are of the same order observed for chloride, the element behaves conservatively in the floodplain. To detect seasonal changes in the behavior of each element, we considered the following ratio:

$$R = \frac{E_m - E_c}{\Delta_{cl}} \qquad \text{Eq. 1}$$

where Δ_{cl} is the mean absolute difference computed for the chloride concentration over the whole study period, and E_m-E_c is the difference between the measured and computed concentration of the considered element E (in $mg.l^{-1}$ or $\mu g.l^{-1}$).

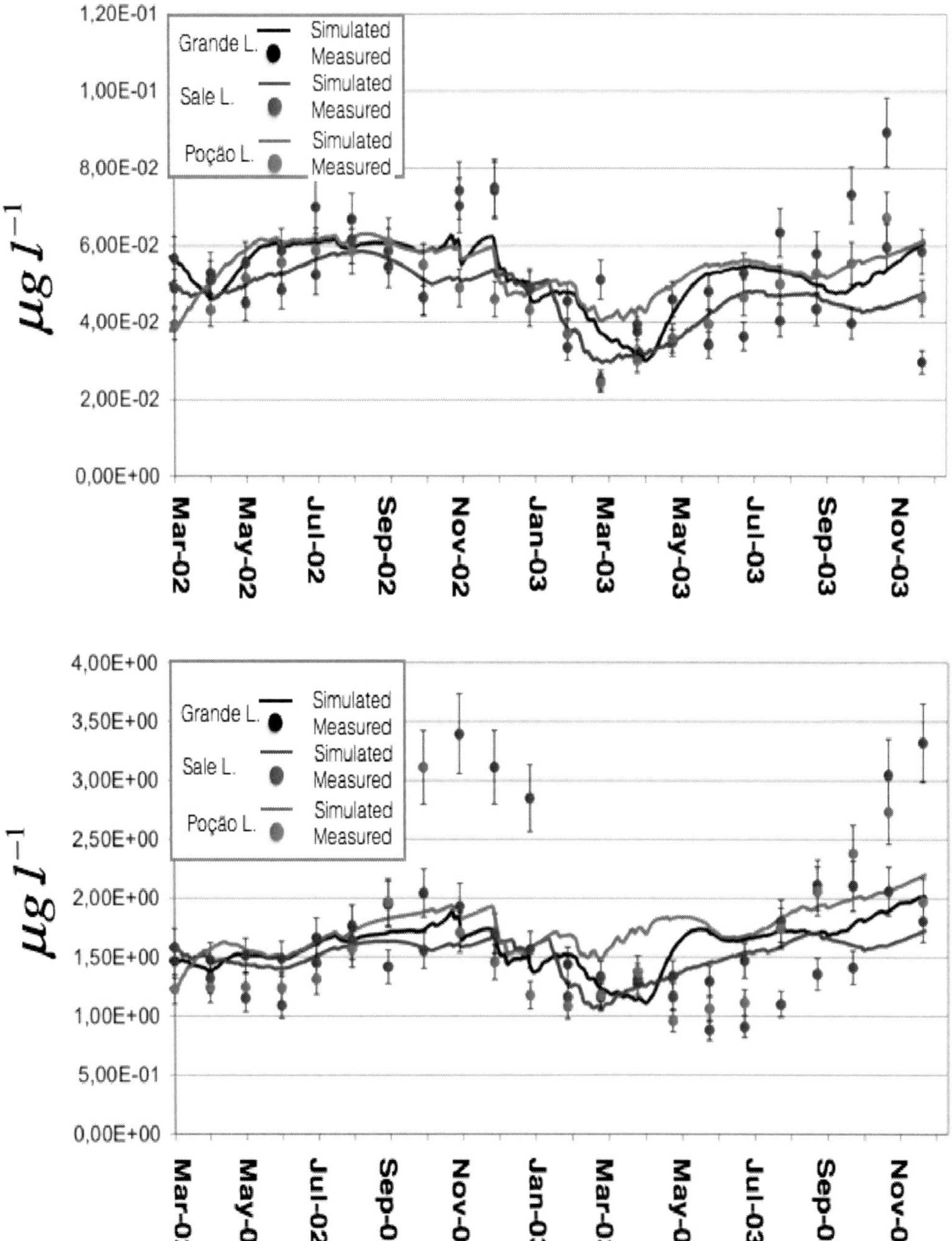

Figure 10. Observed and simulated dissolved U and Cu concentrations in the internal lakes of the floodplain.

This ratio highlights periods of time when an element's behavior departs from conservative expectations. When R is greater than 1, there is a source term for E that may be linked to biological activities or to an abiotic source that is not addressed by the model (such as sediment release or inputs from the water table). When R is lower than -1, there is a sink (such as a change into a solid phase or uptake) that is not addressed by the model.

Elements That Present a Relatively Constant Concentration throughout the Water Year

The dissolved Cu and U concentrations varied only slightly during the water year, except during the low-water period, when the concentrations sharply increased by a factor of two (Figure 10). The model was globally successful in reproducing the temporal trends in the three studied lakes but did not reproduce these concentration peaks well. The agreement of the model with the observations was better when considering the input from the water table. In the case of U, our computation showed that a cumulative flux of 8 tons over the 2002-2003 water year was required to better fit the observations. Considering the water flux computed previously, this U flux supposed a U concentration of 1.10^{-3} $mg.l^{-1}$, which is approximately 20-fold higher than the concentration measured in the surface water. This concentration value is consistent with data from Godoy and Godoy (2006), who observed U concentrations ranging between 1.10^{-5} and 5.10^{-3} $mg.l^{-1}$in the northern Brazilian water table. Thus, we concluded that U behaves conservatively in the floodplain, with the model-observation agreement being significantly improved by realistic water table inputs. The same conclusions were drawn for Cu. We found that a Cu flux of 79 tons from the water table was necessary to significantly improve the model results. This flux led to a Cu concentration in the water table of $2{\cdot}10^{-2}$ $mg.l^{-1}$, which is one order of magnitude greater than in the lake waters.

Elements That Vary during the Water Year

The dissolved Ca concentration followed similar seasonal trends among Lakes Sale, Pocão and Grande, marked by an increase during the rising period and diminution during the flushing period (Figure 11). The concentrations in the lakes clearly differed from those in the Amazon R., especially from November 2002 to April 2003. There were also some differences between the lakes. The increase in Ca in Lake Grande was sharper at the beginning of the rising period (from December) but lasted only until February, whereas the increase was regular until May in Lake Sale. In Lake Pocão, the Ca concentration increased from December to March. The K concentrations presented similar trends to the Ca concentrations.

The R ratio clearly indicated deviation from conservative expectations for these two elements. In particular, the model under-estimated the Ca and K concentrations during the rising period in the three studied lakes and over-estimated their concentrations during the flushing period.

Vegetation uptake could explain a portion of the differences between the model and observations during the flushing period. Some semi-aquatic plants (such as Echinocloa) growing in this period of the year and present a large biomass. These plants are able to uptake nutrients directly from the water (Furch et al., 1983; Furch and Klinge, 1989; Piedade et al., 1991; Junk and Piedade, 1993). During the rising period, we believe that the differences between the model and observations could be explained by the release of semi-aquatic plant degradation and by washing out of the emerged soils and sediments.

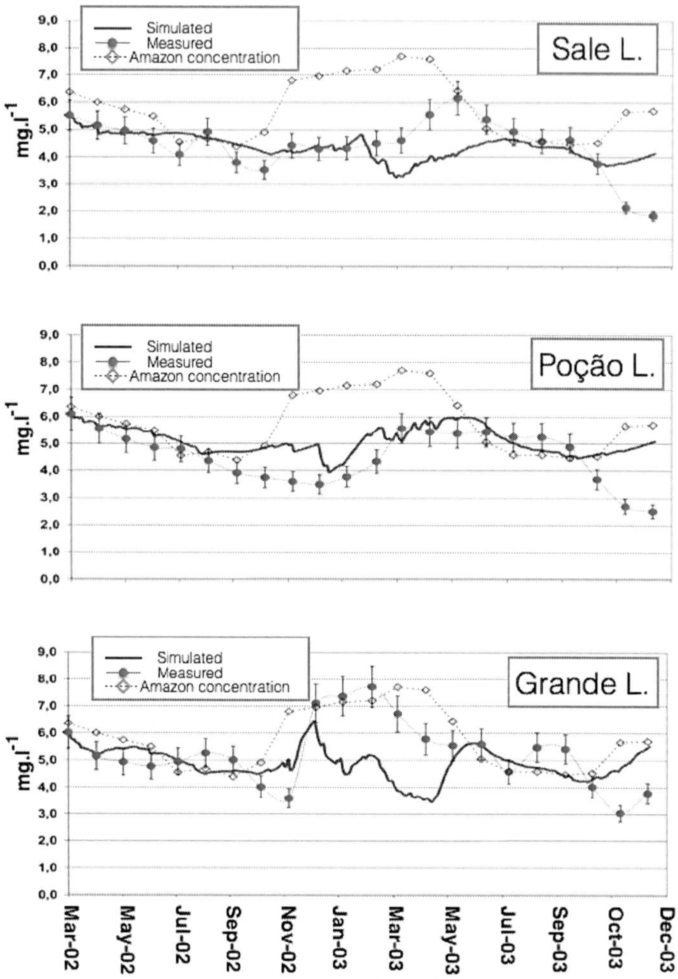

Figure 11. Observed and simulated Ca concentrations in the internal lakes of the floodplain. The observed concentration in the Amazon measured at Obidos is also reported for comparison.

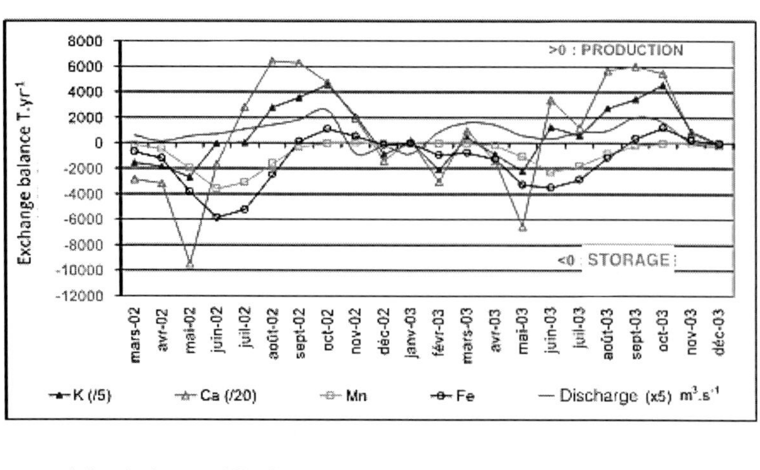

Figure 12. Elemental flux balance of K, Ca, Mn, and Fe between the floodplain and the Amazon. The water balance is also reported for comparison.

Influence of Internal Processes on the Biogeochemical Behaviors of Elements

Combining the observed concentrations and the water balance model made it possible to compute the elemental mass balance of K, Ca Mn and Fe during the 2002-2003 water year (Figure 13). The computed water balance is also reported in the same figure. By convention, a positive balance indicates exportation to the Amazon River. As mentioned above, the floodplain was exporting water from May to October in 2002 and from February to the end of October in 2003, whereas it was storing water the rest of the time. Generally speaking, Ca and K behaved relatively similarly: both elements were exported during the flushing period from June (August) to November 2002 for Ca (K) and from June to December 2003. Both elements showed a small amount of storage during the flushing period, likely under the influence of vegetation uptake. Fe and Mn also behaved similarly: significant storage was observed for both elements from March to September 2002 and for Fe (Mn) from February (April) to August 2003. Fe was slightly exported during the low-water period. In contrast to Ca and K, the storage period was longer, until late in the flushing period. Mn was only very slightly exported compared with the other studied elements. The storage of these elements was explained by their precipitation as oxides and hydroxides, explaining the Fe and Mn enrichment observed in the sediment of the floodplain, and by their uptake by vegetation (Viers et al., 2005).

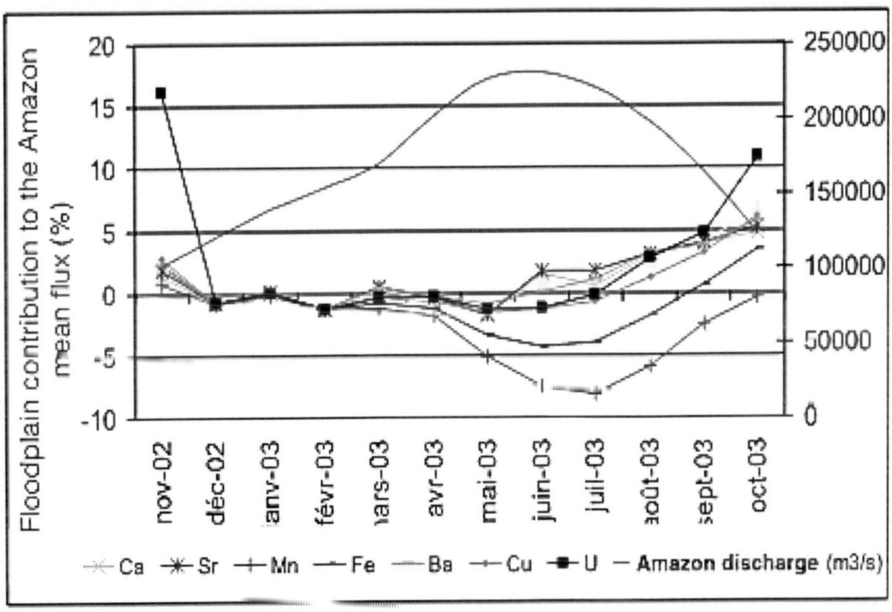

Figure 13. Influence of the floodplain expressed as a % of the Amazon flux for K, Ca, Sr, Mn, Fe, Ba, Cu and U. The Amazon discharge measured at the Obidos gauge station is also reported for comparison.

The cumulative fluxes observed during the 2002-2003 water year are reported in Table 1. On an annual scale, Ca, K, Ba and Sr were exported to the Amazon. The export represented between 10 to 30% of the input flux. Thus, Ca and K storage through vegetation uptake was not significant, reinforcing the hypothesis that this uptake is essentially ensured by annual semi-aquatic plants, rather than perennial vegetation. This result also confirms that the annual semi-aquatic plant production is released in dissolved form, as stated in (Abril et al., 2014). On the other hand, dissolved Mn and Fe were clearly trapped on an annual scale. This storage represented 47% and 97% of the input flux for Fe and Mn, respectively. These elements likely underwent chemical phase changes under precipitation as hydroxide or oxide, and a portion remained trapped in the sediment, as indicated by our data. According to Viers et al., 2005, Mn storage is mainly due to uptake by perennial vegetation, whereas Fe is stored mainly in annual semi-aquatic plants. This difference could explain the differences in the residence times of Fe and Mn, which were 2 and 10 times greater than their respective water residence times.

Table 1. Dissolved cumulative fluxes of selected elements deduced from the hydrological model during one hydrological cycle. A negative balance indicates storage in the floodplain. The water fluxes (Q) are also reported

	K T/yr	Ca T/yr	Sr T/yr	Mn T/yr	Fe T/yr	Ba T/yr	Cu T/yr	U T/yr	Q km^3/yr
Annual cumulative flux entering the floodplain	24245	161666	987	744	2798	679	52	2	947
Annual cumulative flux exported from the floodplain	27874	185479	1090	17	1467	778	58	2	1110
Annual Exchange balance	3629	23813	103	-727	-1331	99	6	0	163
Residence time in the floodplain in days	126	133	133	1471	234	122	141	122	122

Influence of Internal Processes on Transfer to the Ocean

The extrapolation of our results was based on the ratios between the Curuai floodplain area (2,500 km^2) and the surface of the floodplains located along the Solimões River (95,000 km², Sippel et al. (1998)). The results are summarized in Table 2.

Table 2. Influence of the floodplains on the elemental flux of the Amazon River deduced from the extrapolation of the Curuai mass balance to floodplains located along the Solimoes/Amazon Rivers

	K 10^3 T/yr	Ca 10^3 T/yr	Sr 10^3 T/yr	Mn 10^3 T/yr	Fe 10^3 T/yr	Ba 10^3 T/yr	Cu 10^3 T/yr	U 10^3 T/yr	Q km^3/yr
Annual mean flux of the Amazon at Óbidos	4846	33011	201	79	480	119	10	0,3	5361
Annual influence of the regional floodplain expressed as % of the Amazon flux	3%	3%	2%	-36%	-11%	3%	2%	8%	4%

At an annual scale, the floodplain located along the Solimões/Amazon River represented between 3 and 36% of the flux of the Amazon as a function of the element considered. The influence of the floodplain was more significant at the seasonal scale, regardless of the element. For Ca, K, and Sr, it contributed 5 to 7% of the Amazon flux, and Mn and Fe storage represented between 5 and 10% of the Amazon flux from April to September.

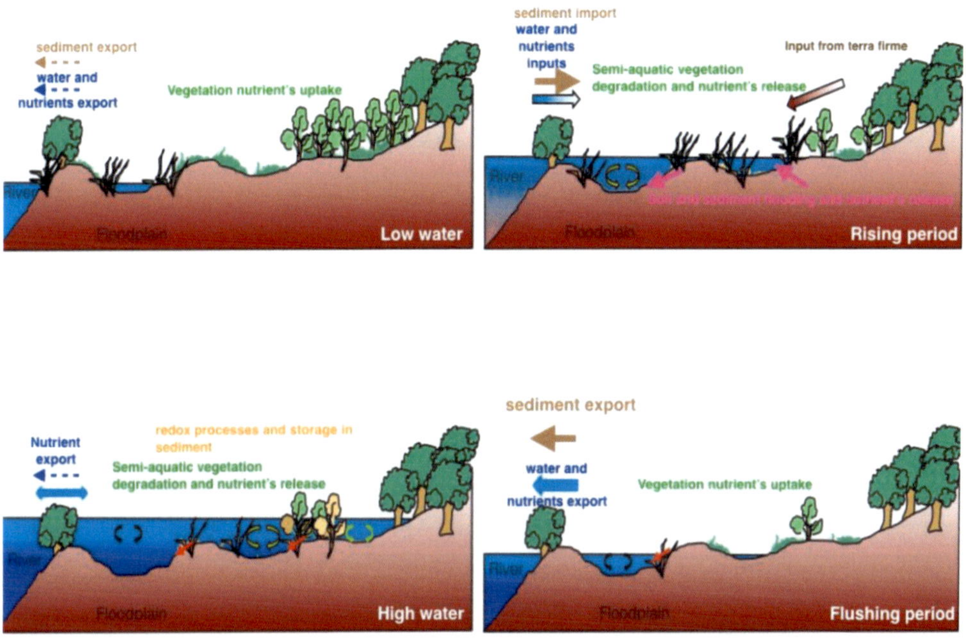

Figure 14. Conceptual scheme of the biogeochemical functioning of the floodplain.

Conclusion

Based on six years of research conducted in a representative floodplain from the medium reach of the Amazon, we were able to highlight the main hydrological and chemical characteristics of this emblematic ecosystem in the Amazon Basin. The hydrological regime was mainly driven by the associated main stream, which contributed more than 70% of the total inflow. Nevertheless, we observed contrasts in water quality between the Amazon River and the floodplain. Too interpret these contrasts, we built a water balance model based on the available hydrological measurements. The model enabled a better understanding of the temporal dynamics of water exchange between the floodplain and the main stream and calculation of the cumulative influxes and outfluxes on a monthly basis, in addition to the water residence time. Each year, the floodplain stored approximately 4.5 times its maximal volume and exported 4.7 times this volume. The residence time was estimated to be 5 ± 0.8 months.

Based on analysis coupled with the sediment database, we estimated that 710 x 10^3 tons/year of sediment was stored in the floodplain, representing between 40 and 50% of the annual flux entering from the Amazon. Organic matter is imported into the floodplain from the Amazon River mainly during the rising water period and is produced in the floodplain and exported to the river during high- and falling-water periods (Moreira-Turc et al., 2013). Based on the Cl mass balance, we computed the fluxes exchanged with the Amazon for several elements. Nutrients such as Ca and K and oligo-elements such as Mn and Fe were taken up by vegetation. In particular, Ca, K and Fe were more likely to be influenced by non-perennial vegetation (Echinocloa), whereas Mn was more likely to be stored in flooded forests. This result is in agreement with the findings of Viers et al., 2005, regarding the Marchantaria floodplain lake.

On an annual scale, the floodplain mass balance of elements contributed less than 1% of the flux of the Amazon to the ocean. However, once extrapolated to the whole inundated surface of the Amazonian floodplains located along the Solimoes/Amazon River, the fluxes became significant. In particular, floodplain storage contributed to reducing the total Amazonian flux of Mn and Fe to the ocean by 36 and 11%, respectively. In addition, on a seasonal scale, the floodplain introduces a time lag, as described by Seyler et al., 2003.

REFERENCES

Abril, G., Martinez, J., Artigas, L., Moreira-Turcq, P., Benedetti, F., Vidal, L., Meziane, T., Kim, J., Bernardes, M. C., Savoye, N., Deborde, J., Souza, E. L., Albéric, P., Landim de Souza, M. and Roland, F. (2014). Amazon River carbon dioxide outgassing fuelled by wetlands. *Nature*, *505*, 395–398.

Adams, J. M. and Faure, H. (1998). A new estimate of changing carbon storage on global land ecosystem reconstruction. *Global and Planetary Change*, 16–17, 3–24.

Alsdorf, D., Bates, P., Melack, J., Wilson, M. and Dunne, T. (2007). Spatial and temporal complexity of the Amazon flood measured from space. *Geophys. Res. Lett.*, *34*, L08402.

Alsdorf, D. E. (2003). Water Storage of the Central Amazon Floodplain Measured with GIS and Remote Sensing Imagery. *Ann. Assoc. Am. Geogr.*, *93*, 55–66.

Barroux, G., Sonke, J. E., Viers, J., Boaventura, G., Godderis, Y., Bonnet, M. P., Sondag, F., Gardoll, S., Lagane, C. and Seyler, P. (2006). Seasonal dissolved rare earth element dynamics of the Amazon River main stem, its tributaries, and the Curuaí floodplain. *Geochemistry*, *Geophys. Geosystems*, *7*.

Bonnet, M. P., Barroux, G., Martinez, J., Seyler, F., Moreira-Turcq, P., Cochonneau, G., Melack, J., Boaventura, G., Maurice-Bourgoin, L., León, J. G., Roux, E., Calmant, S., Kosuth, P., Guyot, J. L. and Seyler, P. (2008). Floodplain hydrology in an Amazon floodplain lake (Lago Grande de Curuaí). *J. Hydrol.*, *349*, 18–30.

Bonnet, M. P., Lamback, B., Boaventura, G. and Oliveira, E. (2011). Impact of the 2009 exceptional flood on the flood plain of the Solimões River, in: Conceptual and Modelling Studies of Integrated Groundwater, Surface Water and Ecological Systems. IAHS Publ, Melbourne, Autralia, 200–206.

Bourgoin, L. M., Bonnet, M. P., Martinez, J. M., Kosuth, P., Cochonneau, G., Moreira-Turcq, P., Guyot, J. L., Vauchel, P., Filizola, N. and Seyler, P. (2007). Temporal dynamics of water and sediment exchanges between the Curuaí floodplain and the Amazon River, Brazil. *J. Hydrol.*, *335*, 140–156.

Dufour, D. L. (1990). Tropical Rain Native forests by Amazonians as models of sustainable agroecosystems, in: Ecosystem Science for the Future. 652–659.

Dunne, T., Mertes, L. a. K., Meade, R. H., Richey, J. E. and Forsberg, B. R. (1998). Exchanges of sediment between the flood plain and channel of the Amazon River in Brazil.

Furch, K., Junk, W. J., Dieterich, J. and Kochert, W. (1983). Furch_etal_ 1983_amazonia.pdf. Amazonia VII, 75–89.

Furch, K. and Klinge, et. H. (1989). Chemical Relationships between Vegetation, Soil and Water in Contrasting Inundation Areas of Amazonia. Mineral Nutrients in Tropical Forest and Savanna Ecosystems. J. Proctor. Oxford, Blackwell Scientific Publications: 189-204.

Godoy, J. M. and Godoy, et M. L. (2006). "Natural Radioactivity in Brazilian Groundwater." *Journal of Environmental Radioactivity*, *85*(1), 71-83.

Goulding, M., Barthem, R. and Ferreira, E. (2003). The Smithsonian Atlas of the Amazon. Smithsonian Institution Press, Washington DC. 256 p.

Hamilton, S. K., Kellndorfer, J., Lehner, B. and Tobler, M. (2007). Remote sensing of floodplain geomorphology as a surrogate for biodiversity in a tropical river system (Madre de Dios, Peru). *Geomorphology*, *89*, 23–38.

Hamilton, S. K., Sippel, S. J. and Melack, J. M. (2002). Comparison of inundation patterns among major South American floodplains. *J. Geophys. Res. Atmos.*, *107*, 1–14.

Hem, J. D. (1970). "Study and Interpretation of Chemicals Characteristics of Natural Waters." U.S. Geol. Surv. Water-Supply-Paper 1473.

Hess, L., Melack, J., Novo, E. M. M. C. B. C. F. and Gastil, M. (2003). Dual-season mapping of wetland inundation and vegetation for the central Amazon basin. *Remote Sens. Environ.*, *87*, 404–428.

Junk, W., Bayley, P. and Sparks, R. (1989). The flood pulse concept in river-floodplain systems. Can. Spec. Publ. *Fish Aquat. Sci.*, *106*, 110–127.

Junk, W. J. and Piedade, et M. T. F. (1993). "Biomass and Primary-Production of Herbaceous Plant Communities in the Amazon Floodplain." *Hydrobiologia*, *263*, 155-162.

Junk, W., Piedade, M. and Wittmann, F. (2010). Amazonian Floodplain Forest. Ecophysiology, biodiversity and sustainable mamagement, Springer. Springer.

Junk, W. J. (1997). General aspects of floodplain ecology with special reference to Amazonian floodplains, in: Junk, W. J. (Ed.), The Central-Amazonian Floodplain: Ecology of a Pulsing System. Springer Verlag, Berlin Heidelberg New York, 3–22.

Junk, W. J. and Furch, K. (1993). A general review of tropical South American floodplains. *Wetl. Ecol. Manag.*, *2*, 231–238.

Lucas, Y. (1989). Systèmes pédologiques en Amazonie brésilienne. Equilibres, déséquilibres et transformations. PhD dissertation, Univ. of Poitiers (France).

Martinez, J. and Le Toan, T. (2007). Mapping of flood dynamics and spatial distribution of vegetation in the Amazon floodplain using multitemporal SAR data. *Remote Sens. Environ.*, *108*, 209–223.

Melack, J. M. and Forsberg, B. R. (2001). Biogeochemistry of Amazon floodplain lakes and associated wetlands, in: M., M. E., Victoria, R. L., Richey, J. E. (Eds.), Biogeochemistry of the Amazon Basin. Oxford university press, New York, p. 365.

Melack, J. M. and Hess, L. (2010). Remote sensing of the distribution and extent of wetlands in the Amazon basin, in: Junk, W.J., Piedade, M. (Eds.), Amazonian Floodplain Forests: Ecophysiology, Biodiversity and Sustainable Management. Ecological Studies Springer.

Mertes, L. A. K. (1997). Documentation and significance of the perirheic zone on inundated floodplains. *Water Resour. Res.*, *33*, 1749.

Moreira-Turcq, P., Bonnet, M. P., Amorim, M., Bernardes, M., Lagane, C., Maurice, L., Perez, M. and Seyler, P. (2013). Seasonal variability in concentration, composition, age, and fluxes of particulate organic carbon exchanged between the floodplain and Amazon River. *Global Biogeochem. Cycles*, *27*, 119–130.

Piedade, M. T. F., Junk et, W. J. and Long, S. P. (1991). "The Productivity of the C_4 Grass *Echinochloa Polystachya* on the Amazon Floodplain." *Ecology*, *72*, 1456-1463.

RADAMBRASIL. Projeto RADAMBRASIL, 1976. Levantamento de Recursos Naturais, 10 (Folha SA 21, Santarém). *Departamento Nacional da Produção Mineral*, *Rio de Janeiro*, pp. 522.

Richey, J., Mertes, L. A. K., Dunne, T., Victoria, R. L., Forsberg, B., Tancredi, A. C. N. S. and De Oliveira, E. (1989). Sources and routing of the Amazon River flood wave. Global Biogeochem. *Cycles*, *3*, 191–204.

Rudorff, C., Melack, J. and Bates, P. (2014). Flooding dynamics on the lower Amazon floodplain: 2. Seasonal and interannual hydrological variability. Water Resour. Res. Accepted a.

Seyler, P. T. and Boaventura, G. R. (2003). Distribution and partition of trace metals in the Amazon basin. *Hydrol. Process.*, *17*, 1345–1361.

Sippel, S. J., Hamilton, S. K., Melack, J. and Novo, E. M. M. (1998). Passive microwave observations of inundation area and the area/stage relation in the Amazon River flooodplain. *Int. J. Remote Sens.*, *19*, 3055–3074.

Stallard, R. F. and Edmond, J. M. (2000). Oltman, Recent studies Meade et al. sediment. *J. Geophys. Res.*, *88*, 9671–9688.

Viers, J., Barroux, G., Pinelli, M., Seyler, P., Oliva, P., Dupré, B. and Boaventura, G. R. (2005). The influence of the Amazonian floodplain ecosystems on the trace element dynamics of the Amazon River mainstem (Brazil). *Sci. Total Environ.*, *339*, 219–232.

Wittenberg, H. and Sivapalan, M. (1999). Watershed groundwater balance estimation using streamflow recession analysis and baseflow separation. *J. Hydrol.*, *219*, 20–33.

In: Riparian Zones
Editor: Oleg S. Pokrovsky
ISBN: 978-1-63484-613-4

Chapter 5

An Approach to the Integrated Management of Exotic Invasive Weeds in Riparian Zones

***J. Jiménez-Ruiz*[1] *and M. I. Santín-Montanyá*[2,*]**
[1]Technical Directorate for Evaluation of Plant Varieties and Plant Protection Products
[2]Plant Protection Department, National Institute of Agricultural and Food Research (INIA), Madrid, Spain

Abstract

The riparian zones have been for millennia a fundamental element for the settlement of the population due to the availability of water, the existence of meadows and fertile valleys as roads. Their ecological importance and livelihood, however, have not prevented their degradation. For centuries this degradation has been gradual and slow, exponentially increasing from the twentieth century by human transformation capabilities of the medium. In mountain areas, the less human presence, coupled with an abrupt relief, a more hostile climate and lower agricultural potential, have allowed a better conservation of rivers, in morphological and biological aspects. However, in middle and lower zones, the degradation has been intense and it is increasingly difficult to find a riparian zone with its original and unaltered riparian vegetation. The main causes of riparian zones degradation are the agriculture on the floodplain, large hydraulic pipes, diversions and dams, channels that have destroyed and altered their morphology, biological and hydrological function. The rapid urban expansion in the riparian zones, in recent years, have caused the channelling of rivers to prevent flooding of building areas. The degradation of riparian zones join to undesirable presence of the exotic species, which will replace the native vegetation in these zones eventually. The riverbanks act as the last refuge of natural vegetation, but it has been replaced by agricultural crops or forest plantations. This situation presents a unique duality: the widespread degradation of the riverbanks and, at the same time, their high ecological value as stronghold for natural vegetation for wildlife and their role as biological corridors. In Mediterranean riparian habitats, uncontrolled presence of invasive and exotic species can reduce native plant species richness, which would be devastating for our environments. Based on the

* Email: isantin@inia.es.

historical management, there are very limited control options available. The presence of weeds or exotic species in riparian zones creates controversial because these species invade very sensitive habitats. In this context, we suggest a methodology to address the management and control of one of the most invasive riparian weeds in the world, the grass *Arundo donax* L. (Giant reed), from an integrated perspective, improving the ecological status of the riparian zones and producing the diminishing impact on the biodiversity. We combined mechanical, physical, chemical and biological techniques in a management programme of riparian zones, taking account the ecological factors of plant communities and social context. The findings of this research project should encourage further studies on the integrated management of invasive weeds in riparian zones, and environmental conditions that may influence field efficiency.

Keywords: exotic, control, riverbanks, aquatic ecosystems, integrated control methods

Conceptual Framework of the Ecology of Invasive Species

Invasive species are defined as any organism, introduced or established in an ecosystem or natural habitat. In addition, it is an agent of change and threat for the native biological diversity either for their invasive behavior or by the risk of genetic contamination (Vilá et al., 2008). The problems of introduction of species, whether animal or vegetable, with invasive behavior is considered the second leading cause of biodiversity loss, by the order of magnitude, affecting global biosphere (Genovesi and Shine, 2004), only behind the loss and habitat destruction (Lowe et at., 2000). For this reason, the issue is gaining in recent years further consideration by researchers and managers of administrations and government agencies responsible for conservation of nature. This has led to a series of agreements, and action programs, both on local level and on the global scale, in order to improve the knowledge of these invasive species. Thus, it is necessary to assess properly the risks involved and the evaluation of pathways to obtain their control, taking account the possible adverse impacts for biodiversity, the economy, health and welfare of the population. Most biological invasions have been caused by accidental introductions; in other cases they have been intentional. The species transfer between regions is increasing with human movements, growing trade, disruption of ecosystems and further development have accelerated the process (Meyerson and Mooney, 2007; Hulme, 2009). Considering, for example, that maritime commercial fleet is doubling every 10 years, and the transport calculation is close to 1,000 tons of water and near to 7,000 to 10,000 species daily.

Historically there have been various reasons and motivations that produced the introductions of invasive alien species (IAS): economic purposes (agriculture, horticulture, ornamental plants, forestry, fishing, hunting activity, biological pest control, etc.), scientific or educational (zoos, botanical gardens, etc.) and aesthetics (landscaping, pets, gardening, etc.). The negative effects that could have such introductions on the environment and in the economy are yet not taken into account. The quality of life enjoyed by many countries is largely consequence of plants and animals, in particular of crops and livestock (Baker, 1986). This human dimension is an essential element in determining which legal, financial and criminal brakes should be imposed to discourage trade and transport activities involving high risk (Jenkins, 2001).

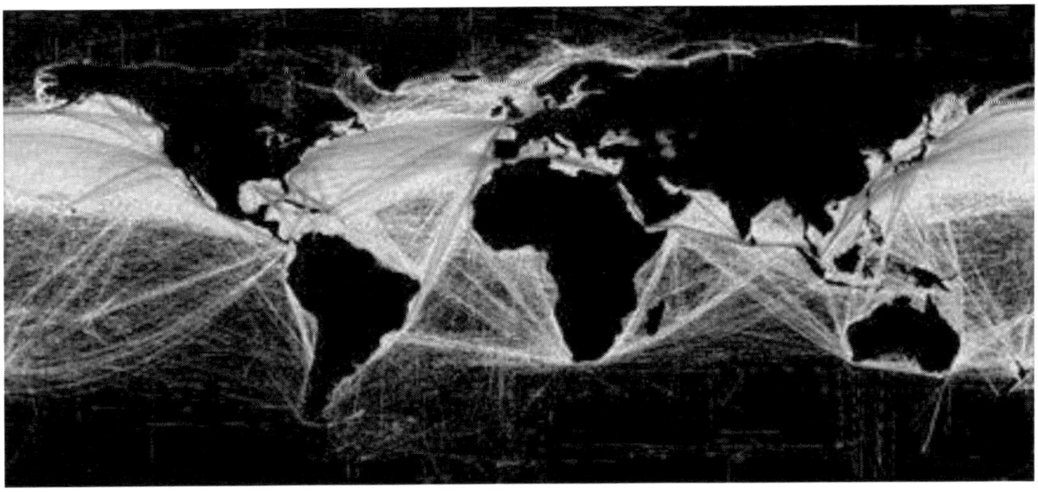

Figure 1. Shipping routes, showing relative density of commercial shipping around the world. More details available at: http://spatial-analyst.net/worldmaps/shipping.rdc.

From this perspective, it is recognized that cities are focal areas of the global economy and entry points of many invasive species, where there is rarely an awareness of the problems that invasive species can cause in natural ecosystems. Settlement patterns of human populations determine the distribution of many invasive species introduced by means of transport or brokers.

Types introductions: i) Intentional: made by the human, consciously and with specific purposes, ii) Unintentional: accidentally produced by the transport, iii) Negligent, always occurring through human actions. The latter may be of two types: a) Escapes species kept in captivity for lack of the necessary security measures and b) Introductions occur through input channels known as high risk for which they have not taken preventive measures. It is estimated that in marine ecosystems - which are quite permeable to the entry of new species the human intervention has accelerated the introductory rates on the order of magnitude of 10^6. Actually, a species that needed 5000 years to reach a new biogeographic region, today only needs one day. In terrestrial ecosystems, invasion rates have been no less spectacular.

Main Characteristics of the Invasive Species

Exotic invasive plants that are able to survive, grow and invade new ecosystems, represent the major ecological experiment, happening on a global scale (Stuart et al., 2000). Understanding how the success of invasive plants that a priori are not adapted to the new ecosystem can be a challenge for scientists and managers for to know how to cope and try to reduce the serious impact they produce. Inherent characteristics of invasive species are the following:

- High rate of growth and reproduction. These two biological characteristics lead to effective monopolization of resources and displacement of native species by competitive exclusion (Cadotte et al., 2006). For example, some of the invasive trees of banks of rivers and ditches, as ailanthus (*Ailanthus altissima*), black locust

(*Robinia pseudoacacia*), Siberian elm (*Ulmus pumila*) or negundo maple (*Acer negundo*), grow rapidly, produce abundant seed crops each year and/or generate abundant vegetative regrowth (Richardson et al., 2006). The small aquatic snail *Potamopyrgus antipodarum,* from New Zealand, has spread with great success by Australia, Europe and America. One of the characteristics that explain their success is the parthenogenetic reproduction, i.e., the female can have children without being fertilized by a male. Thus, the introduction of a single female to a river or lake can result in a large population, as their reproductive rate is very high.

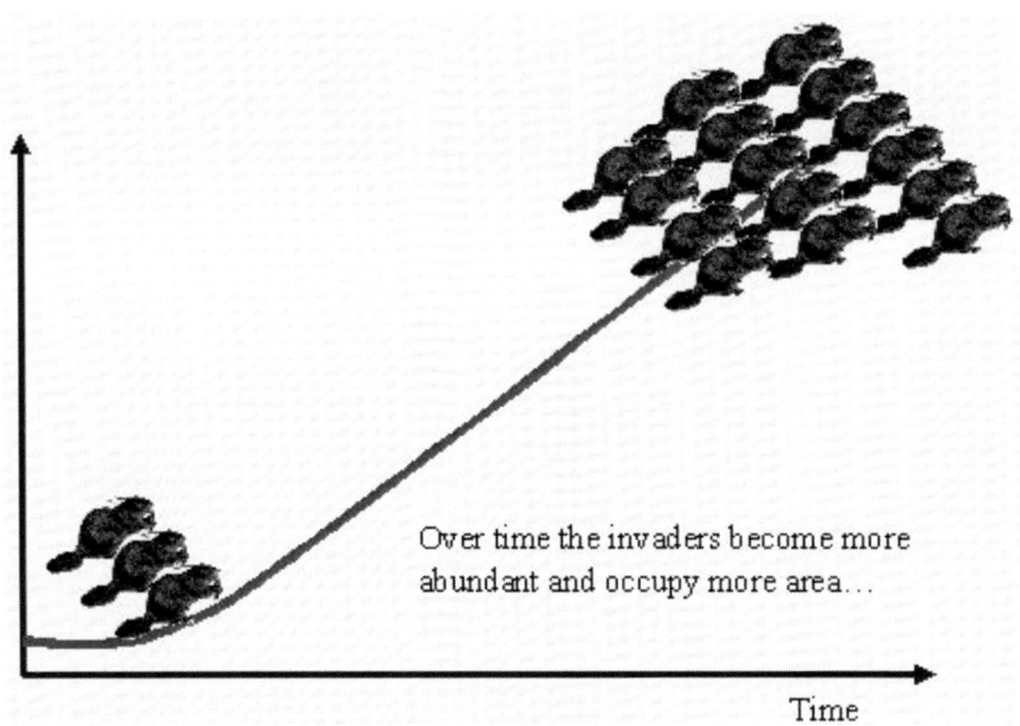

Figure 2. The figure shows how the increase in the number of individuals directly affects the capacity of the ecosystem.

- Flexibility and phenotypic plasticity. Several authors have suggested that invasive species are able to acclimate better than native species to new or changing environmental conditions (Pysek et al., 2007; Rejmanek et al., 1996 and Thompson et al., 2001). This acclimation capacity may be due to a high phenotypic plasticity when a certain genotype results in very different phenotypes in response to the environment, or high functional flexibility, ie, the phenotype can vary over time in response to environmental fluctuations. Moreover, phenotypic plasticity can vary not only between invasive and noninvasive, but between populations of the same species show different invasiveness (Niinemets et al., 2003). Such is the case of relict indigenous population of hornbeam (*Rhododendron ponticum*) in southern Spain, which shows a low phenotypic plasticity versus invasive populations of this species in localities of central Europe, whose plasticity is very high.

Figure 3. *Caprobrotus edulis* is an important invasive species presented in the Mediterranean zone and contain local adaptations to the ecological factors in the ecosystems invaded.

- Easy hybridization. Some species have very easily to hybridize with each other, allowing them to increase their genetic variability. This capability can give them a great invasive potential, since it favors the establishment of stable populations in new areas from a few specimens introduced (Ellstrand and Schierenbeck, 2000). As an example, include the case of ruddy duck (*Oxyura jamaicensis*), a species introduced in England in the forties, which has hybridized with our native species, the White-headed Duck (*Oxyura leucocephala*). Hybrids are fertile and can cross one another and to the parents, which contributes to the expansion of this invasive species at the expense of a loss of the original individuals of the indigenous species (Muñoz-Fuentes et al., 2007).

Conditions That Favor the Biological Invasions

The planet's ecosystems have a strong and growing anthropogenic component which is increasing due to the globalization of the economy. Therefore, the problem of invasive alien species (IAS) has a considerable human dimension. According Perrings et al., (2000) there are two hypotheses about the relationship between economic activities and biological invasions. The first is that *invasions increases depending on the extroversion of the economy*; the second is *that invasions happen depending on the degree of disturbance of habitats by economic activities.*

Many scientists have asked this question and have searched for the answer from invasive species presence in different territories. Published data shows that there is no single explanation. It is necessary to take in account that always it depends on both ecosystem and species considered. However, some general patterns have also noticed that it seem to recur in various places on the planet (Cadotte et al., 2006). Among the reasons behind the success of invasions, we can differentiate between those characteristics proper of the receiving ecosystem and the characteristics of invasive species.

Although there is no ecosystem on earth that does not have invasive species, it is also true that there are environments with a higher proportion of exotic than other species. For example, the islands are considered as "fragile" ecosystems where the introduction of exotic species usually have more drastic consequences that in the continents (Planty-Tabacchi et al., 1996). Also, disturbed environments are more likely to hold exotic species that those stable and well-preserved.

Two hypotheses help to explain this pattern: a) The empty niche hypothesis postulates that certain ecosystem functions cannot be performed by any kind due to phylogenetic or bio-geographic constraints. For example, there ecosystems that lack of trees and temperate grasslands, such as New Zealand, where there are hardly any native annual grasses, the weeds introduced grown and spread with great success (Mack, 2003). And b) the hypothesis of the absence of enemies posits that success of some invasive species is due to the absence of predators, parasites or diseases, able to slow its expansion in ecosystems (Colautti et al., 2004). The low diversity of species, such as some islands, or those caused by some disturbance, reduce the likelihood of an invasive species found resistance from competitors, predators or diseases that regulate their invasion in the ecosystem.

Effects of the Invasive Species in the Riparian Ecosystems

Exotic plants are common in rivers and riparian environments for several reasons. On the one hand, the bottom of riparian basin sediments accumulate and they are naturally rich in water and nutrients, whose concentration is further increased by human activities. Furthermore, because it is altered by human activity, introducing deliberately or accidentally these species in an environment that favors dissemination of the invaders, and finally, because the disturbances that occur in these ecosystems provide liberated spaces (Planty-Tabacchi et al., 1996) ready to be colonized. When all these elements come together it is easy to understand why these environments are, in the words of Hood and Naiman (2000) "ecosystems disproportionately susceptible to invasion." In these cases, invasions are carried out by plants able to exploit resources opportunistically (Jiménez et al., 2011). In fact, this type of behavior coupled with a tendency to establish monotypes, vegetation where exotic species dominate over native species and exclude or subordinate the features shared by many of the plants that invade wetlands (Zedler, 2004).

Effects on the Biological Diversity (Species Richness)

Invasive species can reduce biodiversity in ecosystems. They affect not only specific native species, but may also reduce native biodiversity at the level of the community or ecosystem (Mack et al., 2000). When the invading organism is extremely competitive, it can even be monospecific in the area. In the marine environment the most obvious case is that of *Caulerpa taxifolia* which produces homogenization of habitats and particularly the habitat for *Posidonia oceanica*. Vegetal formations of water hyacinth (*Eichhornia crassipes*) floating in the river, also affect directly the whole community so because its root system linked can drag turtles, snakes, snails and may even uproot emerging native species (Capdevila et al., 2006). In most cases, the impact on the community is indirect and occurs when invasive species alter the networks of interaction between native species (Muñoz-Fuentes et al., 2007; Vilá et al., 2006). In waterways, invading organisms can alter the functioning of the ecosystem and result in loss of local biodiversity. For example, crawfish alters the turbidity of the water due to over exploitation of macrophytes, which in turn means reducing the light entering and prevents the recolonization of macrophytes. Primary carnivores and herbivores (fish larvae, molluscs) are affected indirectly, but also directly as part of the diet of crawfish. The high density of mosquito fish can cause a series of knock-on effects. The disappearance of cladocerans and predation of macroinvertebrates may generate increased protozoa, rotifers and phytoplankton. Decomposition of phytoplankton by the large amount of excrement produced by the mosquito fish leads to a clouding of the water, which in turn promotes eutrophication and algae growth. This can reduce the available oxygen and cause the disappearance of sensitive organisms. Examples of similar effects can be found among vegetables. The dense canopy of water hyacinth (*Eichhornia crassipes*) can limit the arrival of light underwater species and decrease the concentration of oxygen in the water, which can also adversely affect fish stocks. The same effect is caused by the water fern (*Azolla filiculoides*) currently expanding in the Doñana National Park, and film forming on the water surface preventing gas exchange between the air and water and promoting anoxic processes (Vilá et al., 2006).

Figure 4. The photo of the left shows the invasion of the riverbank by *Arundo donax* such as a monospecific vegetal formation. The photo of the right shows the aquatic invasive species *Azolla filiculoides*.

Effects on the Functionality of the Ecosystems

Introduction of IAS alters the structure of the native communities, as it involves changes in wealth, diversity and species dominance (Hopper et al., 1997). Such structural changes generally affect the cycles of matter and energy flow in the ecosystem, and often represent a significant alteration of its functioning (Craine et al., 2002).

Currently there is no consensus in establishing relationships between structure and function of ecosystems. Some authors believe that each species has a unique role, so that the entry of any invasion would impact the ecosystem functioning. Others studies argue that the species of a community can be grouped into a few "functional groups" within which each species would have similar effects on ecosystem processes (Lavorel and Garnier, 2002).

Figure 5. Photo shows an aquatic ecosystem totally invaded by different invasive species. *Arundo donax* in the riverbank, and in the surface of water *Eichhornia crassipes* and *Ludwigia grandiflora*.

According to Hopper et al., (1997), ecosystem functioning depends more on the diversity and identity of functional groups of species diversity, and functional redundancy among species helps to ensure the maintenance of these functions, against the loss of species and their fluctuations of abundance. Other authors, however, attach special significance to certain species, called "keystone species," capable of performing unique functions and therefore controlling the principal ecosystem processes. Some cases of invasions often support this prediction, as they show very different effects on ecosystem processes invaded. For example, considering the cycling of nitrogen species can be grouped between fixing and not fixing atmospheric nitrogen; or between plants C3 and C4 metabolism depending on the efficiency of water use; or between herbaceous and woody, depending on the capacity to store carbon; in the case of animals, they can be classified between herbivores and predators depending on the type of the diet (Yelenik et al., 2007; Grigulis et al., 2005).

MANAGEMENT OF THE INVASIVE SPECIES IN RIPARIAN ECOSYSTEMS

Initial Considerations

Population management of invasive species is influenced by multiple variables (Genovesi and Shine, 2004). Decisions on population management of invasive species can be conditioned by several factors. Some have to do with their biology and ecology. Others are related to the scope of work, the type of action or the ecological context in which the processes occur. Therefore, to define the objectives and work plan for the control of invasive species it is necessary to obtain information on these factors (Andreu and Vilá, 2007). This preliminary work is important and can influence on the outcome of the proceedings for the management of the invasive species applied. For example, the information obtained is essential to choose the methodology to eliminate invasive weeds, because different methods differ in effectiveness, implementation period, costs, impacts and adaptation to the various situations in which these species grow (Deltoro et al., 2012). The success of the initiative does not only mean eliminating the invasive species, but producing the minimal impact on the environment without compromising the recovery of the ecosystem.

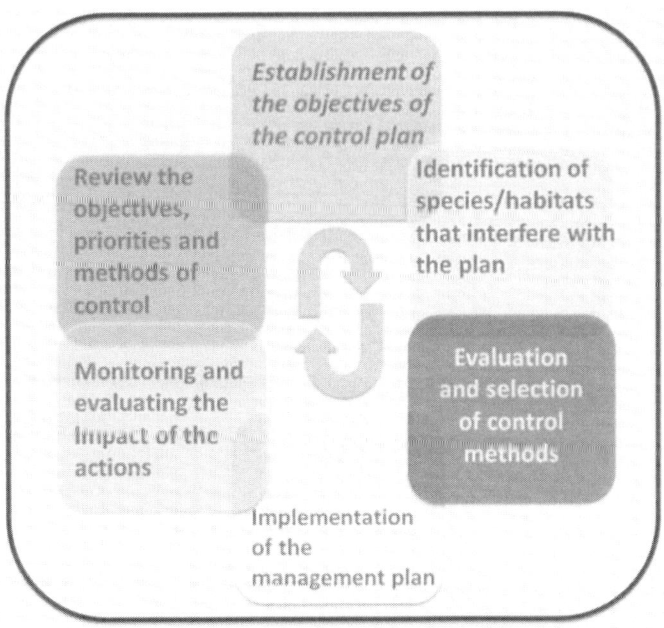

Figure 6. According to adaptive management, continuous assessment of the effectiveness and impact of the methods allows review of management objectives established at an earlier stage.

Despite this, even in well-planned actions, the obtained results may be below expectations or mishaps occur. For example, the method for eliminating or controlling invasive species may be less effective than expected and may cause unanticipated impacts. The monitoring to be carried out on actions allows reviewing objectives and strategy, reassessing the methods used and suggesting alternatives as part of a process. Thus, it is necessary: a) Track the impact as part of a strategy of adaptive management, b) Choose the method according to the type of intervention that is framed within the tasks and c) Prevent the possible impacts.

Objectives and Management Strategy

The International Union for Conservation of Nature and Natural Resources (IUCN) recommends that, whenever feasible, eradication is the best management strategy to deal with invasive alien species. This is possible in the early stages of invasion, when populations are small or localized.

A proper way to proceed in the case of interventions removal of invasive species at medium scales and large is to establish homogeneous sectors on the type of plant formation and the physical environment that supports them from mapping conducted as part of the previous work (Deltoro et al., 2012). Once concretized, you must assign a target to each based on: i) the likelihood of success of the action, ii) the benefit for biodiversity and iii) the cost.

To estimate the probability of success of a management program of invasive species, the eradication factors that facilitate and hinder the actions must be take into account. In our experience, we can cite the following:

- Type of plant formation.
- Access to the work area: the easy access facilitates the work and improves its performance, especially when transporting large amount of material.
- Environmental Features: work in banks of the gentle slopes and homogeneous substrate is easier and improves the performance of actions.
- Applicable environmental standards: applicable regulations can prevent the use of a method of controlling a priori species suitable for other more costly or less effective.
- Duration of the actions: the possibility of rehearsals in later years contributes to the success of any initiative to control invasive species.
- Budget: the budget constraints will mean the impossibility of some methods of higher cost relative to other, cheaper, but perhaps less suitable for normative social aspects such as rejection or prohibition.
- Social aspects: the commitment to the initiative of local leaders and social support to the initiative favors the achievement of objectives, to the point of being able to count on the help of volunteers in carrying out certain tasks.

To determine the benefit for biodiversity of a performance control the invasive plants should be assessed what changes the medium that sustains if there is no intervention would be. The cost of the performance should be measured accurately and is determined by the method employed, taking into account the factors that determine the likelihood of success, which can increase it significantly. In equal chance of success and benefit to biodiversity between two performances, the performance should opt for lower cost. The objective of the actions should aim at eradication in situations where the probability of success and benefit to biodiversity are high and the cost is reasonable within the budget of the initiative. On the contrary, when the probability of success is low and the cost is high, even if the benefit is higher, the biodiversity reasons to act and to do so should be carefully evaluated, because the chances of failure are high. For intermediate situations, the goal should always be oriented towards the likelihood of success. In this way, one will avoid investing resources in unattainable goals.

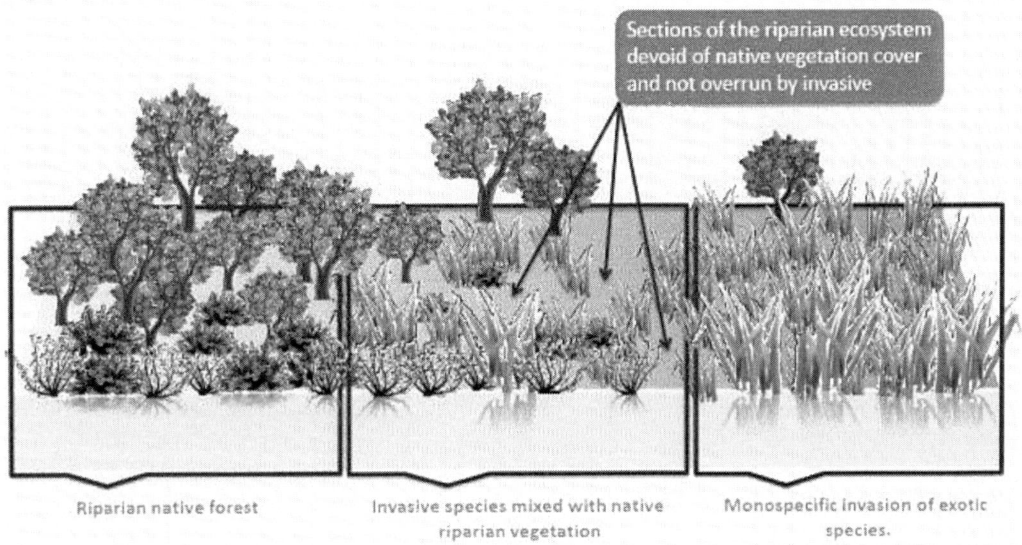

Figure 7. Schematic transition from a native riparian forest to an area invaded by exotic vegetation. Adapted from Deltoro et al., 2012.

The complexity of the work of removal and control of invasive species increases with the scale of the intervention. Therefore, it is necessary to develop a strategy in which the following aspects are taken into account: The intervention should begin in less invaded areas and gradually move toward the core of the invasion, but not before having achieved the restoration of the first. Bradley (1988) suggests in his approach to control invasive plants in a riparian environment, that the control work should begin in areas that conserve the best remnants of native vegetation formations, eliminating the emerging populations as part of the goal of eradication. It has been shown that small focused invasive species expand faster than larger ones (Moody and Mack, 1988). Therefore, because its removal is easier, this action represents a small investment that is profitable in the long term. Hence, we have to try prioritize the actions most likely to succeed, having lower cost and greater benefit for biodiversity, and subsequently initiate actions that would have been outstanding.

A CASE STUDY: MANAGEMENT AND CONTROL OF *ARUNDO DONAX* IN MEDITERRANEAN CONDITIONS

The degradation of riparian vegetation environments is one of the main causes affecting the surface water quality, and the presence of invasive species is considered an environmental problem that affects global biodiversity, and poses a major threat to the ecological integrity of river ecosystems. The main reasons for plant invasions are, on the one hand, the ability to avoid attack by natural enemies where native species struggle with this, and, on the other hand, the increase in resources for invasive plants.

The riparian ecosystems are especially prone to invasion by exotic plants, probably due to extreme changes in the water system, which naturally occur within the riverbanks (Hood and Naiman 2000; Quinn et al., 2007; Vila et al., 2007). Several studies have shown that alien species can reduce the native plant species richness in these communities (Hulme and Bremner 2006; Maskell et al., 2006; Truscott et al., 2008). Therefore environmental managers, government agencies, and conservation groups struggle to deal with the ongoing threat of invasive species.

Giant reed (*Arundo donax* L.) is considered native to eastern Asia (Polunin and Huxley 1987). It is a perennial and rhizomatous grass, the largest of six species in the genus *Arundo* and is one of the tallest grasses (up to 10 m). It has been cultivated across Asia, southern Europe, North Africa, and the Middle East for thousands of years (Perdue 1958). It was extensively planted along ditches for erosion control. Plants have been maintained in rural areas for fence material, roof thatching, construction of baskets and other artisanal products. Because of its rapid biomass production, *A. donax* is a proposed candidate for biofuel development in North America (Barney and DiTomaso 2008; Mack 2008) and other warm regions of the world (Seca et al., 2000).

It is a successful invasive, non-native weed of riparian zones in Spain, as well as other Mediterranean areas and semiarid riparian zones worldwide (Sanz Elorza et al., 2004). In the Mediterranean region, Giant reed has colonized riparian systems (Cushman and Gaffney 2010) forming extensive monospecific clumps that aggressively exclude competitors for light, nutrients and water (Davis et al., 2010; Lambert et al., 2010) and has altered wildlife habitats and displaced native species. *Arundo* spreads primarily through rhizome elongation and fragmentation (Decruyenaere and Holt, 2001; DiTomaso and Healey, 2007; Khudamrong-sawat et al., 2004) and not by seed, despite having a large inflorescence (Johnson et al., 2006).

A. donax invasion has many ecological and environmental impacts, (Bell 1997; Dudley 2000). It has been classified as a noxious weed in North America (USDA 2009). Its invasion in riparian areas alters the native vegetative structure (Herrera and Dudley 2003) and rapid growth after floods or wildfire leads to competitive displacement of native riparian vegetation (Coffman 2007). This dominance reduces arthropod diversity and abundance (Herrera and Dudley 2003) and also leads to decline in avian diversity and abundance (Kisner 2004). However, there is no comprehensive treatment of this species' ecology in its invasive range. The lack of data concerning this hinders management of *A. donax* and protection of essential riparian habitat for endangered species.

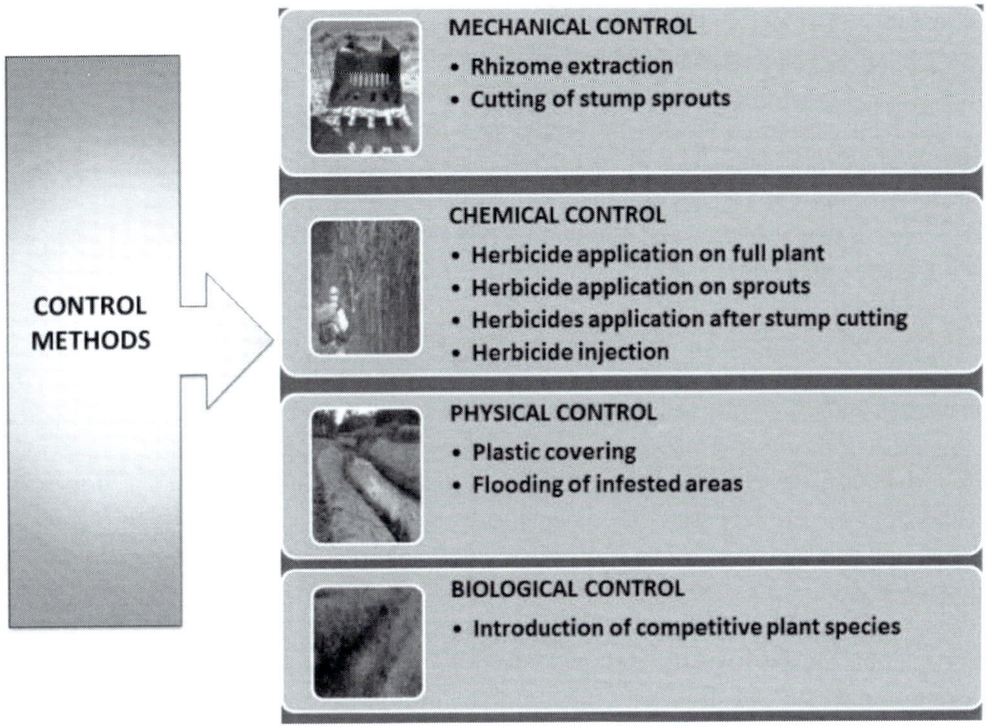

Figure 8. Methods to control *Arundo donax* L.

The propagation of Giant reed is difficult to control for several reasons. First, it rapidly produces large amounts of biomass, both above and below ground (Sharma et al., 1998). Second, it propagates very readily by regrowth from rhizome and stem pieces (Decruyenaere and Holt, 2001), which are easily dispersed by river currents. Finally, because a large proportion of its biomass is sequestered underground in the rhizome, the removal of the whole plants is difficult. In order to more successfully manage infestations of Giant reed in a given site, information is required regarding its potential invasiveness and its responses to different control methods. Integrated weed management programs (IWMP) provide a multidisciplinary approach to address the problems posed by invasive species. The most effective tools to control its spread include mechanical, chemical and biological methods (Figure 8).

Mechanical Methods

Heavy rains in the wet winter season can lead to flash flooding, which can cause rhizome fragments to break off and be carried downstream. As a result, reed populations tend to increase downstream (Else 1996). Once established, it is highly drought tolerant, and rhizomes deposited at the upper reaches of floodplain often thrive even in the absence of flowing water. Plants also tolerate high salinity levels, inhabiting estuarine and coastal strand environments, even colonizing marine islands after flooding that transports rhizomes from rivers across ocean waters. The reed has a very high photosynthetic rate (Rossa et al., 1998); it is known that *A. donax* is able to grow in absence of light for up to 100 days (Decruyenaere and Holt 2005) or from rhizomes covered by up to 1 meter of soil (Boose and Holt 1999; Else 1996). This allows new sprouts to become independent of the rhizome reserves when they emerge.

Giant reed can be controlled by mechanical treatments, such as cutting the stems or excavating the entire plant (Lowrey and Watson 2004). The main method commonly used to eliminate *A. donax* is mechanical removal by means of the complete removal of aerial biomass once a year (hereafter one stump treatment). This is problematic, however, due to the plant's tendency to re-sprout which is a highly efficient strategy in response to the loss of above-ground biomass following the disturbance. Stump treatment only once a year is usually an unsuccessful control method. Continued long-term stump re-cutting, especially in the period after this species has spent some of its reserves during its annual growth pulse, could conceivably compromise the chances of regeneration.

Despite the fact that *A. donax* has not evolved in the Mediterranean climate, with its persistent summer droughts, and unpredictable soil water content, it has managed to invade the ecosystem surprisingly well. Theoretically, *A. donax* should not be able to outcompete native woody species, which have evolved structural and physiological mechanisms to cope with these environmental constraints. Furthermore, the plant attempts to compensate control treatments applied, and consequent loss of biomass by re-sprouting. This often leads to soil water depletion and a derived increase in competition with native vegetation.

Control methods which access and utilize the physiological positive and negative feedback systems could reduce the competitive ability of this species against native vegetation. The direct negative feedback pathways are for example biomass loss via mechanical methods and indirect negative feedback routes include reduced water availability and reduced competitively compared to native plants.

Because of these possible complex and at times counter-intuitive effects, it is important to improve our knowledge about the traits that allow *A. donax* to invade Mediterranean environments – in particular its water use strategy – and how control treatments affect these adaptations. This is critical if we are to understand its potential effects on Mediterranean ecosystems, how it competes with native woody plants, and how we can work towards preventing it from becoming a dominant woody invader.

Despite limited evidence, it seems as though mechanical treatments applied twice per year (and thereafter double cut stump treatment) during the growing period could improve the success of control methods, particularly in the long-term. Some experiences demonstrate that between five and nine cut-stump treatments successively every 20 days lead to the reduction of plant density by 80% and limit height of new sprouts as well (Mota 2009; Godé 2008). Also, it is important that the periods between cutting stumps are short; so as to remove the sprouts which have been generated from the rhizome reserves. In this way, it will lead to the depletion of its reserves quickly.

Furthermore, recent research in temperate environments has identified the fact that the joint application of mechanical and chemical (herbicide) treatments is more effective at eradicating this species than mechanical treatment alone. These joint methods have not, however, been tested under Mediterranean conditions. The mid- and long-term effects of repeated application of these treatments on this re-sprouting species are virtually unknown, and therefore studies to assess the long-term prognosis for control are needed in order to test the usefulness of these techniques.

Chemical Methods

The employment of herbicides for weed control is crucial in weed management programs and herbicide selection is key to obtain effective control. Presently, there is a lack of information on proper timing of herbicide application for maximum control of giant reed with respect to its growth stage. Giant reed has shown tolerance to several herbicides (Odero and Gilbert 2012). In addition, control with herbicides may be restricted because giant reed invades very herbicide-sensitive habitats such as riparian areas. Glyphosate has been the most widely used non selective systemic herbicide for control of invasive non-native plants in Mediterranean rivers over the last four decades (Puértolas et al., 2010; Spencer et al., 2009). It is also an integral component of programs which aim to restore native vegetation (Finn and Minnesang 1990; Bell 1997 and Jackson 1994).

Although data on dosage and efficacy of glyphosate on giant reed is limited, some authors have carried out the evaluation of glyphosate on giant reed management (Lowrey and Watson 2004). Bell (1997) reported that 2 to 5% glyphosate solution was effective in controlling giant reed when applied to cut stems, and Jackson (1994) recommended a glyphosate solution of 1.5% for the control of giant reed. More recently, a number of authors have suggested different formulations of glyphosate and other active ingredients for giant reed management (Odero and Gilbert 2012; Spencer et al., 2008) the literature contains little more information in this respect. Bottoms et al., (2011) evaluated the activity of glyphosate, cyhalofop-butyl and penoxsulam on the growth and vegetative reproduction of creeping river-grass (*Echinochloa polystachya* Kunth A. S. Hitchc), an invasive aquatic grass. It was found that herbicides cyhalofop, glyphosate and penoxsulam reduced the fresh weight 25-50% at 14 days after treatment (DAT) and 63-80% at 28 DAT compared with the non-treated plants.

Recent research has focused on combining glyphosate application with stem cutting in order to control and reduce the competitive ability of the invasive species, and it should be used together with native species restoration programs. Given an infested area (with invasive plants), the first step to re-establish a native population would be to treat with herbicide in order to reduce the invasive plant biomass. The next step is to restore the area with native species. Renz and DiTomaso (2006) found that one particular pre-treatment (mowing and ploughing) enhanced efficacy of glyphosate on perennial pepper-weed (Lepidium latifolium L.). Also a similar pre-treatment with tillage improved the control of Canada thistle (Cirsium arvense L. Scop.) with glyphosate (Hunter, 1996). In one long term study of Alianthus altissima L., under Mediterranean conditions (Constán-Nava et al., 2010), it was indicated that joint stem-cutting and herbicide application is the only effective treatment to reduce this species in the long-term. Although Miller and D´Auria (2011) concluded that in areas where tillage is not possible, glyphosate treatment could provide satisfactory control of the perennial weed wild chervil (Anthriscus sylvestris L.)

There are limited chemical control options available for giant reed. The high invasive potential of giant reed is based on its widespread distribution and inherent weedy characteristics, which greatly increases the likelihood of its post-treatment survival, and subsequent environmental damage (Barney and DiTomaso 2008).

Unfortunately, determining herbicide efficacy in the field or the greenhouse, can take several weeks and requires suitable infrastructure, meaning that simpler methods or techniques that are faster than conventional methods are needed for herbicide efficacy evaluation.

The chlorophyll fluorescence analysis (CFA) technique has been used for early detection of physiological herbicide effects on plants (Spencer et al., 2009; Puértolas et al., 2010). The CFA technique can provide a less laborious means of testing the effect of herbicides on giant reed and other invasive plants. CFA is a quick, easy, and reproducible method; the confirmation of its efficacy would provide a useful tool for land. In previous experiments carried out in greenhouse, we confirmed that CFA techniques could be used to easily monitor the effects of herbicide treatments on new giant reed sprouts following cutting (Santín-Montanyá et al., 2013).

We conducted one greenhouse study and one field study, independently, to evaluate the giant reed response to several herbicides using CFA. Therefore, the objectives of these studies were to evaluate the response of giant reed to four systemic herbicides after initial cutting using CFA. The active ingredients (a.i) were: profoxydim, azimsulfuron, cyhalofop-butil and penoxsulam, commonly used for control of weeds adjacent to surface water, and glyphosate which was chosen because it is the standard herbicide used by many land managers for giant reed management.

Greenhouse Study

The *Arundo donax* rhizomes used in this study were collected from the banks of the River Harnina in Almendralejo (Badajoz, Spain). These rhizome fragments were planted in pots and were placed in a greenhouse with natural light. The temperature was maintained between 17°C and 40ºC. All pots were watered frequently in accordance with the needs of this species.

Figure 9. Overview of the test conducted in greenhouse. It is observed the first sprouts of *A. donax*.

Five systemic herbicides commonly used against grasses and perennials weeds were applied at these different doses: 0 (control), 1/2X, 1X and 2X the on-label application rate (OLA). Thc active ingredients (a.i) were azimsulfuron 50% (OLA 50 g ai.ha-1), cyhalofop-butyl 20% (OLA 1.5 L ai.ha-1), glyphosate 36% (OLA 10 L ai.ha-1), penoxsulam 2.04% (OLA 2 L ai.ha-1) and profoxydim 20% (OLA 0.75 L ai.ha-1). The effect of the herbicides was also examined by CFA, measuring the quantum yield of fluorescence (QY). Values of QY were recorded 4, 6, 8, 13, 17, 21, 25, 30, 40, 50 and 60 DAT.

The results show that the effects of the different doses of herbicide on QY mirror their effects on sprout relative height and biomass. Thus, CFA provides a less laborious means of field testing the post-cutting effects of herbicides on *Arundo donax.* QY values were obtained (expressed as a percentage with respect to controls) at 4, 6, 8, 13, 17, 21, 25 30, 40, 50 and 60 DAT.

The plants treated with glyphosate (from the 1/2X dose upwards) and profoxydim (from the same dose upwards) showed 100% and 30% reductions in relative QY at 60 DAT (Figures 10a and 10b).

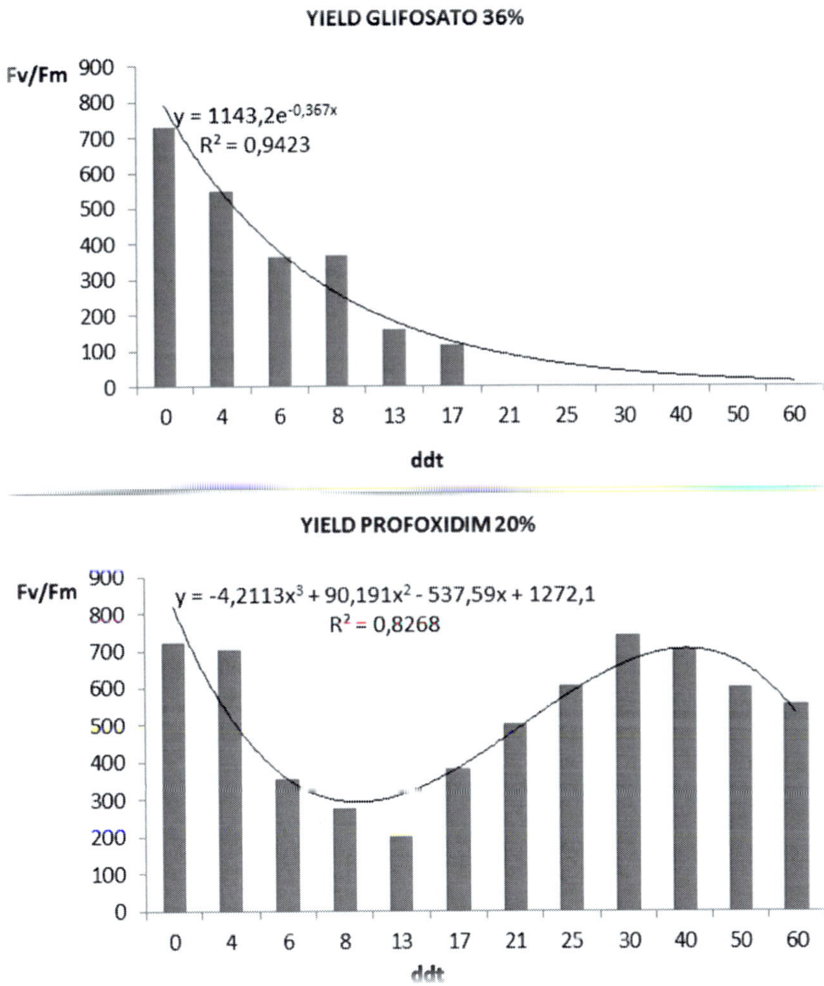

Figure 10a. Quantum yield of fluorescence (Fv/Fm) response of Arundo donax sprouts at 4, 6, 8, 13, 17, 21, 25, 30, 40, 50 y 60 days after treatment.

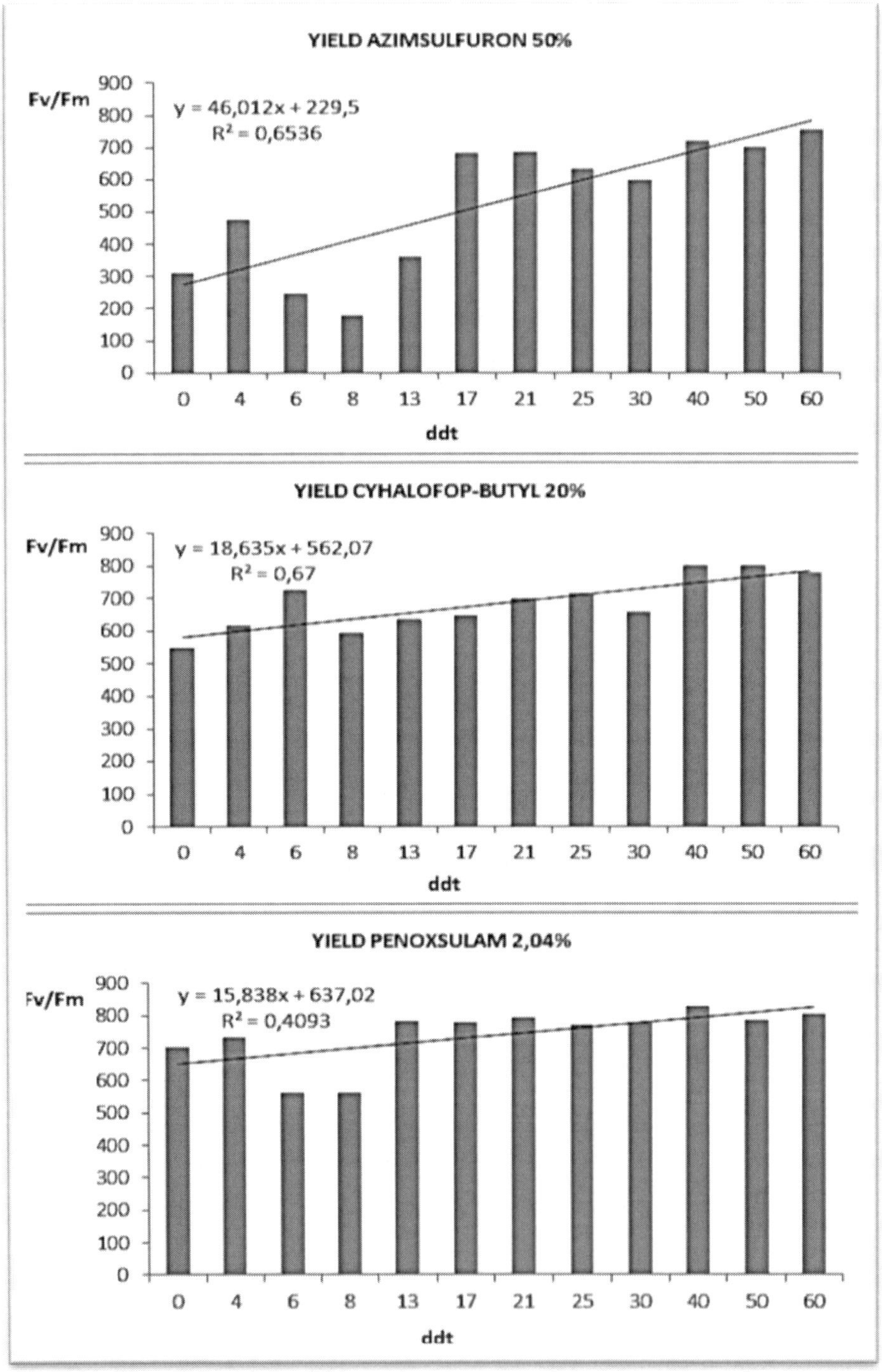

Figure 10b. Quantum yield of fluorescence (Fv/Fm) response of Arundo donax sprouts at 4, 6, 8, 13, 17, 21, 25, 30, 40, 50 y 60 days after treatment.

The cyhalofop-butyl treatment had no effect on QY at any dose. Glyphosate, from the 1/2X dose upwards, had the greatest effect, reducing QY to zero by 21 DAT.

Profoxydim, from the 1/2X dose upwards, affected photosynthetic activity between 6 and 21 DAT, but after 25 days of treatment the plants recovered. Azimsulfuron (from the 1/2X dose upwards) and penoxsulam (from the same dose upwards) reduced photosynthetic activity between 8 and 13 DAT, but at 17 DAT no effect was observed. Cyhalofop-butyl had no significant effect on photosynthetic efficiency at any dose throughout the study period.

This study covers an aspect of particular relevance in the global problem posed by alien species with invasive behavior, specially the initial control of regrowth after cutting. The cost-effectiveness has become in recent years a crucial aspect that should receive more consideration by researchers and administrations which manage the conservation of nature.

Field Study

The field study was initiated in the spring of 2010 and continued through fall of 2011 in Harnina River Basin, Almendralejo (Badajoz, Spain). The experimental site is characterized by a Mediterranean climate.

Treatments consisted of broadcast foliar application of glyphosate, azimsulfuron, cyhalofop-butyl, penoxsulam, and profoxydim. Glyphosate was applied at 10 L a.i. ha-1 over-the-top of established giant reed plants (3 to 4 m in height), after initial sprouting following cutting at 4 L a.i. L-1, and also injected into stems of established giant reed plants at 4 L a.i. L-1. Azimmsulfuron, cyhalofop-butyl, penoxsulam, and profoxydim were applied on giant reed after the initial sprouting following cutting, at label application rate, at 50 g a.i. ha-1, 1.5 L a.i. ha-1, 2 L a.i. ha-1, and 0.75 L a.i. ha-1, respectively.

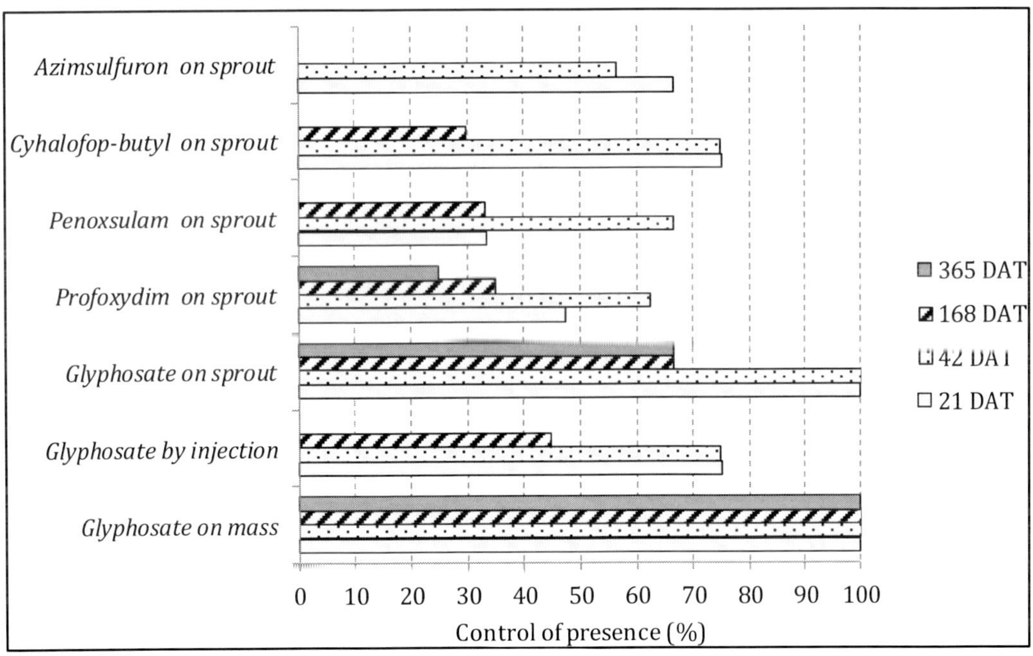

Figure 11. The control of presence by giant reed live stems or sprouts measured by visual ratings scale, and expressed as percentage (from 0%= no control to 100% = total control of presence) at 21, 42, 168 and 365 DAT.

The response to herbicides on giant reed sprout was examined using CFA techniques and showed a reduction in photosynthetic efficiency of plants treated with the herbicides. Our results showed that the effects of the different herbicides on QY at 60 day mirrored the effects visually observed on presence of new giant reed sprouts at 168 day. This suggests that the efficiency of herbicide treatment can be predicted 108 days sooner and with the CFA technique in a more objective manner (Santín-Montanyá et al., 2014).

The results also suggest that glyphosate directly applied over-the-top of established giant reed plants and glyphosate applied on new sprouts, could provide an excellent means of controlling the growth of this invasive plant species (Figure 11). The application of glyphosate by injection controlled the presence until 168 day, but eventually the plants recovered by 365 day. Profoxydim applied on sprouts of giant reed would also appear to hold promise. The effects of profoxydim on sprouts lasted until the end of the experiment (365 day). However, long-terms studies of the effects of these herbicides (glyphosate and profoxydim) on the control of this invasive species and new studies to measure the utility in combination with other products are now required.

Physical Methods

There are two main physical methods to control Giant reed, covering the infested land with plastic sheeting, and flooding of infested areas. These could both lead to 100% control of the reed if they are applied in the appropriate way. They share the advantage that they can both be used in all cases of invasive species in riparian habitats, including those zones in contact with an aquatic environment without risk of accidental contamination of other species.

The first method involves using opaque sheet – geotextile or plastic for example on previously cleared area (stump cutting). The covering should be placed in the period of rhizome growth, and must remain during two growing stages in order to obtain the total depletion of rhizome reserve. Finally, after the two growing seasons of *A. donax*, the replacement of hedges and re-vegetation of the area is recommended.

The second control method involves in flooding the reed-affected area, turning the reed's intolerance to flooding in our favour. Again, the reed must be cut previously and it is crucial to apply this method in winter, during the vegetative period of *A. donax*.

Biological Methods

The establishment of native riparian species in areas affected by *A. donax* can help to limit its growth by competing for resources. Due to the high competitive ability of giant reed, two previous rounds of sprout cutting are recommended. Several authors have shown that plantation of *Salix* species could obtain up to 100% efficacy in Mediterranean area. Taxa of the genus *Populus, Tamarix, Sambucus* and *Cornus*, could also be used, although the data on these species are not sufficient. As an additional advantage, this method is efficient for the rapid recovery of native vegetation.

Conclusion and Considerations about the Integrated Approach to Invasive Weeds in Riparian Ecosystems

Restoring native biodiversity and ecosystem function through prevention, control, and eradication is a fundamental component of management strategies. However, addressing invaders on an individual species basis might limit our ability to prevent invasions or predict future ones (Smith and Knapp 2001). Rapid response might thus be improved by considering species groups that share growth cycles, pathways of invasion, and habitat preferences (Aronson et al., 2007; Newsome and Noble 1986).

Control measures against *Arundo donax* have been widely implemented in USA and other countries, including herbicidal control, cutting and removing biomass, and prescribed fire (Bell 1997). These control measures often have short-term efficacy and can induce collateral damage on non–target species (Boose and Holt 1999). The environmental drawbacks of herbicides use are confined to some developed countries and a few regions in developing countries. It is hoped that appropriate and responsible application of herbicides will help to maintain the invasion of exotic plants and minimize contamination. At present, the integrated management of programs, along with continuous technological change, farmer participation, technology transfer, and a policy environment, are key components for attaining these objectives.

Finally, we have to bear in mind that there is no single set of recommendations on control methods of invasive plants appropriate for the diverse agricultural environments and economic conditions around the world. Rather, farmers have to be given access to, and to choose the most appropriate and cost-effective technologies for their particular circumstances.

In this regard it is important to encourage analysis and different studies of invasive behavior of species to have a better basis for determining the adequate management programs. Site-specific programs should be developed to improve efficiency of control methods and reduce potential adverse environmental impacts that could arise from a misuse of these control methodologies.

Acknowledgments

We want to show our appreciation for the technical support and interest to all staff involved in the project "*Arundo donax* control and management" of the Grupo Tragsa, as well as experts from the Ministry of Agriculture, Food and Environment for the support and funding.

References

Andreu J., Vilà M., 2007. Análisis de la gestión de las plantas exóticas en los espacios naturales españoles. *Ecosistemas*, 2007/3.

Aronson R. B., Thatje S., Clarke A., Peck L. S., Blake D. B., Wilga C. D., Seibel B. A., 2007. Climate change and invasibility of the Antarctic Benthos. Ann. *Rev. Ecol. Evol. S.*, 38:129–154.

Baker, HG 1986. Patterns of plant invasion in North America. Pages 44–57 in Mooney HA, Drake JA, eds. Ecology of Biological Invasions in North America and Hawaii. New York: Springer-Verlag.

Barney J. N., DiTomaso J. M., 2008. Nonnative species and bioenergy: are we cultivating the next invader? *Bioscience,* 58:64–70.

Bell, G. P., 1997. Ecology and management of *Arundo donax*, and approaches to riparian habitat restoration in southern California. In: Brock J, Wade M, Pysek P, Green D (eds) Plant invasions: studies from North America and Europe. Backhuys Publishers, Leiden, pp 103–113.

Boose A. B., Holt J. S., 1999. Environmental effects on asexual reproduction in *Arundo donax. Weed Research,* 39:117–127.

Bottons S. L., Webster E. P., Hensley J. B., Blouin D. C., 2011. Effects of Herbicides on Growth and Vegetative Reproduction of Creeping Rivergrass. *Weed Technology,* 25: 262-267.

Bradley, J., 1997. Bringing back the bush: The Bradley method of bush regeneration. Lansdowne Publishing Pty. Ltd. The Rocks, New South Wales, Australia.

Cadotte, M. W., Murray, B. R., Lovett-Doust, J., 2006. Ecological patterns and biological invasions: Using regional species inventories in macroecology. *Biological Invasions,* 8:809-821.

Capdevila-Argüelles, L., García, A. I., Orueta, J. F., 2006. Especies exóticas invasoras: diagnóstico y bases para la prevención y el manejo., Ed. Organismo Autónomo de Parques Nacionales, Ministerio de Medio Ambiente, Madrid.

Constán-Nava S., Boneta, A., Pastora E., Lledó M. J., 2010. Long-term control of the invasive tree *Ailanthus altissima:* Insights from Mediterranean protected forests. *Forest Ecology and Management,* 260, 1058–1064.

Colautti, R. I., Ricciardi, A., Grigorovich, I. A., Macisaac, H. J., 2004. Is invasion success explained by the enemy release hypothesis? *Ecol. Lett.*, 7: 721-733.

Craine, J. M., Tilman, D., Wedin, D., Reich, P. B., Tjoelker, M. G., Knops, J., 2002. Functional traits, productivity and effects on nitrogen cycling of 33 grassland species. *Funct Ecol.,* 16: 563-574.

Cushman J. H., Gaffney K. A., 2010. Community-level consequences of invasion: impacts of exotic clonal plants on riparian vegetation. *Biol. Inv*., 12, 2765–2776.

Davis A. S., Cousens R. D., Hill, J., Mack, R. N., Simberloff, D., Raghu, S., 2010. Screening bioenergy feedstock crops to mitigate invasion risk. *Frontiers in Ecology and the Environment*, 10, 533–539.

Decruyenaere J. G., Holt J. S., 2001. Seasonality of clonal propagation in giant reed. *Weed Sci*., 49:760–767.

Decruyenaere J. G., Holt J. S., 2005. Ramet demography of a clonal invader, *Arundo donax (Poaceae*), in southern California. *Plant and Soil,* 277: 41-52.

Deltoro Torro, V., Jiménez Ruiz, J. and Vilán Fragueiro X. M., 2012. Bases para el manejo y control de *Arundo donax* L. (Caña común). Colección Manuales Técnicos de Biodiversidad, 4. Conselleria d'Infraestructures, Territori i Medi Ambient. Generalitat Valenciana. Valencia.

DiTomaso J. M., Healy E. A., 2007. Weeds of California and Other Western States. Univ. Calif. *Agric. Nat. Res. Pub*., 3488. 1808 p.

Ellstrand, N. C., Schierenbeck, K. A., 2000. Hybridation as a stimulus for the evolution of invasiveness in plants? *Proc. Nat. Acad. Sci.,* 97: 7043-7050.

Else J., 1996. Post-flood establishment of native woody species and an exotic, *Arundo donax*, in a southern California riparian system. Master's thesis. San Diego, CA: San Diego State University.

Finn M., Minnesang D., 1990. Control of giant reed grass in a southern California riparian habitat. *Restoration and Management Notes,* 8: 53-54.

Genovesi, P. and Shine C. 2004. European Strategy on Invasive Alien Species. 67 págs. Council of Europe Publishing, Nature and Environment. 137. Strasbourg.

Godé L. X., García E., Gutierrez i Perearnau C., 2008. La gestió i recuperació de la vegetació de ribera: guia tècnica per a actuacions en riberes. Departament de Medi Ambient i Habitatge. Generalitat de Catalunya.

Grigulis, K., Lavorel, S., Davies, I. D., Dossantos, A., Lloret, F., Vilà, M., 2005. Landscape-scale positive feedbacks between fire and expansion of the large tussock grass, *Ampelodesmos mauritanica* in Catalan shrublands. *Global Change Biology,* 11: 1042-1053.

Herrera A. M., Dudley T. L., 2003. Reduction of riparian arthropod abundance and diversity as a consequence of giant reed (*Arundo donax*) invasion. *Biol. Invasions,* 5: 167-177.

Hood, W. G., R. J., Naiman. 2000. Vulnerability of riparian zones to invasion by exotic vascular plants. *Plant Ecology,* 148: 105-114.

Hooper, D. U., Vitousek, P. M., 1997. The effects of plant composition and diversity on ecosystem processes. *Science,* 277: 1302-1305.

Hulme P. E., Bremner E. T., 2006. Assessing the impacts of Impatiens glandulifera on riparian habitats: partitioning diversity components following species removal. *J. Appl. Ecol.,* 43:43–50.

Hulme, P. E., 2009. Trade, transport and trouble: managing invasive species pathways in an era of globalization. *Journal of Applied Ecology*, 46: 10-18.

Hunter J. H., 1996. Control of Canada thistle (*Cirsium arvense*) with glyphosate applied at the bud vs. rosette stage. *Weed Sci.*, 44: 934–938.

Jackson N. E. 1994. Control of *Arundo donax*; techniques and pilot project. In Jackson, N. E, p. 27-33.

Jenkins, P. 2001. "Economic Impacts of Aquatic Nuisance Species in the Great Lakes." Report prepared by Philip Jenkins and Associates. Ltd, for Environment Canada, Burlington, Ontario.

Jiménez, J, Vilán, X. M., García, J., Luquero, L. Santín, I. 2011. Estudio de la capacidad invasiva de *Arundo donax* L., en distintas regiones bioclimáticas de la Península Ibérica. XIII Congreso de la Sociedad Española de Malherbología, "Plantas Invasoras, Resistencias a Herbicidas y Detección de Malas Hierbas." La Laguna, 2011, España.

Johnson M., Dudley T., Burns C., 2006. Seed production in *Arundo donax*? Marine Science Institute, University of California. Cal-IPC News Fall.

Khudamrongsawat J., Tayyar R., Holt H. S., 2004. Genetic diversity of giant reed (*Arundo donax*) in the Santa Ana River, California. *Weed Sci.,* 52:395–405.

Kisner D. A., 2004. The Effect of Giant Reed (*Arundo donax*) on the Southern California Riparian Bird Community. M. S. thesis. San Diego, CA: San Diego State University. 90 p.

Lambert A. M., Dudley T. L., Saltonstall K., 2010. Ecology and impacts of the large-statured invasive grasses *Arundo donax* and *Phragmites australis* in North America. Invas. *Plant Sci. Manag*., 3, 489–494.

Lavorel, S., Garnier, E., 2002. Predicting changes in community composition and ecosystem functioning from plant traits: revisiting the Holy Grail. *Funct Ecol.,* 16: 545-556.

Lowe S., Browne M., Boudjelas S., De Poorter M. (2000). 100 of the World's Worst Invasive Alien Species A selection from the Global Invasive Species Database. Published by The Invasive Species Specialist Group (ISSG) a specialist group of the Species Survival Commission (SSC) of the World Conservation Union (IUCN), 12pp. First published as special lift-out in Aliens 12, December 2000. Updated and reprinted version: November 2004.

Lowrey J., Watson, J., 2004. Tamarisk and *Arundo* control on Cache Creek. In Proceedings of the California Weed Science Society; *California Weed Science Society*: San Jose, CA, 2004; Vol. 56, 82–83.

Mack R. N., 2008. Evaluating the credits and debits of a proposed biofuel species: giant reed (*Arundo donax*). *Weed Sci*., 56:883–888.

Mack, R. N., 2003. Phylogenetic constraint, absent life forms, and preadapted alien plants: a prescription for biological invasions. *Int J Plant Sci.,* 164: S183-S196.

Mack, R. N., Simberloff, D., Lonsdale, W. M., Evans, H., Clout, M. N., Bazzazz, F., 2000. Biotic invasions: causes, epidemiology, global consequences and control. *Issues in Ecology Series,* 5. Ecological Society of America, Washington.

Maskell LC, Bullock JM, Smart SM, Thompson K, Hulme PE., 2006. The distribution and habitat associations of non-native plant species in urban riparian habitats. *J Veg Sci.,* 17:499–508.

Meyerson, L. A. and Mooney H. A. 2007. Invasive alien species in an era of globalization. *Frontiers in Ecology and the Environment*, 5: 199-208.

Miller T. W., D'Auria D. E., 2011. Effects of Herbicide, Tillage, and Grass Seeding on Wild Chervil (*Anthriscus sylvestris*). *Invasive Plant Science and Management,* 4:326–331.

Moody, M. E. and R. N. Mack (1988). Controlling the spread of plant invasions: The importance of nascent foci. *Journal of Applied Ecology,* 25: 1009-1021.

Mota Freixas E., 2009. Estudi de noves tècniques per a l'eradicació de l'*Arundo donax*. Memoria del proyecto de final de carrera de Ciencias Ambientales.

Muñoz-Fuentes, V., Vilà, C., Green, A., 2007. Hybridization between white-headed ducks and introduced ruddy ducks in Spain." *Mol Ecol*., 16: 629-638.

Newsome A. E., Noble I. R., 1986. Ecological and physiological characters of invading species. Pages 1–20 in R. H. Groves and J. J. Burdon, eds. Ecology of Biological Invasions: An Australian Perspective. Cambridge, UK: Cambridge University Press.

Niienemets, U., Valladares, F., Ceulemans, R., 2003. Leaf-level phenotypic variability and plasticity of invasive Rhododendron ponticum and non-invasive Ilex aquifolium co-occurring at two contrasting European sites. *Plant Cell and Environment,* 26: 941-956.

Odero D. C., Gilbert R. A. 2012. Response of Giant Reed (*Arundo donax*) to Asulam and Trifloxysulfuron. *Weed Technol*., 26, 71–76.

Planty-Tabacchi, A. M., Tabacchi, E., Naiman, R. J., Deferrari, C., and H. Descamps., 1996. Invasibility of species rich communities in riparian zones. *Conservation Biology,* 10: 598-607.

Perrings, C., M. Williamson, E. B. Barbier, D. Delfino, S. Dalmazzone, J. Shogren, P. Simmons, and A. Watkinson. 2002. Biological invasion risks and the public good: an economic perspective. *Conservation Ecology,* 6(1): 1.

Perdue R. E., 1958. *Arundo donax*: source of musical reeds and industrial cellulose. *Econ Bot.,* 12:368–404.

Polunin O., Huxley H., 1987. Flowers of the Mediterranean. Hogarth Press, London.

Puértolas L., Damásio J., Barata C., Soares A. M. V. M., N Prat., 2010. Evaluation of side effects of glyphosate mediated control of giant reed (*Arundo donax*) on the structure and function of a nearby Mediterranean river ecosystem. *Environmental Research*, 110: 556-564.

Pysek, P., Richardson, D. M., 2007. Traits associated with Invasiveness in Alien Plants: Where Do we Stand?, Nentwig, W. (ed). *Biological Invasions*, vol. 193. Springer-Verlang Berlin Heidelberg.

Quinn L. D., Rauterkus M. A., Holt J. S., 2007. Effects of nitrogen enrichment and competition on growth and spread of giant reed (*Arundo donax*). *Weed Sci.,* 55:319–326.

Rejmanek, M., Richardson, D. M., 1996. What attributes make some plant species more invasive?. *Ecology,* 77: 1655-1661.

Renz M. J., DiTomaso J. M., 2006. Early season mowing improves the effectiveness of chlorsulfuron and glyphosate for control of perennial pepperweed (*Lepidium latifolium*). *Weed Technol.*, 20: 32–36.

Richandson, D. M., Pysek, P., 2006. Plant invasions: merging the concepts of species invasiveness and community invasibility. *Progress in Physical Geography,* 30: 409-431.

Rossa B, Tueffers A. V, Naidoo G., von Willert D. J., 1998. *Arundo donax* L. (Poaceae): a C3 species with unusually high photosynthetic capacity. *Bot Acta.,* 111:216–221.

Santín-Montanyá M. I., Jimenéz J., Ocaña L., Sánchez F. J., 2013. Effects of sprout cutting plus systemic herbicide application on the initial growth of giant reed. *Journal of Environmental Science and Health, Part B: Pesticides, Food Contaminants, and Agricultural Wastes*, 48:4, 285-290.

Santín-Montanyá M. I., Jimenéz J., Vilán X. M., Ocaña L., 2014. Effects of size and moisture of rhizome on initial invasiveness ability of giant reed. *Journal of Environmental Science and Health, Part B: Pesticides, Food Contaminants, and Agricultural Wastes,* 49, 41–44.

Sanz Elorza M., Dana Sanchez E. D., Sobrino Vesperinas E., 2004. Atlas of Invasive Alien Plants in Spain., Dirección General para la Biodiversidad. Madrid.

Seca A. M. L., Cavaleiro J. A. S., Domingues F. M. J., Silvestre A. J. D., Evtuguin D., Neto., 2000. Structural characterization of the lignin from the nodes and internodes of *Arundo donax* Reed. *J. Agric. Food Chem.*, 48:817–824.

Sharma K. P., Kushwaha S. P. S., Gopal B., 1998. A comparative study of stand structure and standing crops of two wetland species, *Arundo donax* and *Phragmites karka*, and primary production in *Arundo donax* with observations on the effect of clipping. *Trop. Ecol.*, 39:3–14.

Smith, M. D., Knapp A. K., 2001. Physiological and morphological traits of exotic, invasive exotic, and native plant species in tallgrass prairie. *Int. J Plant Sci.*, 162(4):785–792.

Spencer D. F., Tan W L. P., Ksander G. G., Whitehand L. C., 2009. Evaluation of late summer Imazapyr Treatment for Managing Giant Reed (*Arundo donax*). *Journal of Aquatic Plant Management*, 47:40- 43.

Stuart, F., Zavaleta, ES., Eviner, VT., Rosamond, N., Vitousek, PM., Reynolds, HL., Hooper, DU., Lavorel, S., Sala, OE., Hobbie, SE., Mack, MC. and Díaz, S. (2000). Consequences of changing biodiversity. *Nature,* 405, 234-242.

Thompsoon, K., Hodgson, J. G., Grime, J. P., Burke, M. J. W., 2001. Plant traits and temporal scale: evidence from a 5-year invasion experiment using native species. *Journal of Ecology,* 89: 1054-1060.

Truscott A-M., Palmer S. C., Soulsby C., Westaway S., Hulme P. E., 2008. Consequences of invasion by the alien plant *Mimulus guttatus* on the species composition and soil properties of riparian plant communities in Scotland. *Perspect Plant Ecol Evol Syst.,* 10:231–240.

USDA NRCS. United States Department of Agriculture, Natural Resources Conservation Service. 2007. http://www.nrcs.usda.gov/technical/maps. Accessed: August 2009.

Vilá, M; Valladares, F; Traveset, A; Santamaría, Luis; Castro, Pilar. (2008). Invasiones Biológicas. CSIC.

Vilá, M., Bacher, S., Hulme, P., 2006. Impactos ecológicos de las invasiones de plantas y vertebrados terrestres en Europa. *Ecosistemas*, 2.

Vilá M., Pion J., Font X., 2007. Regional assessment of plant invasions across different habitat types. *J Veg Sci.*, 18:35–42.

Yelenik, Stock, W. D., Richardson, D. M., 2007. Functional group identity does not predict invader impacts: differential effects of nitrogen-fixing exotic plants on ecosystem function. *Biol Inv.,* 9: 117-125.

Zedler, J. B. 2004. Causes and consequences of invasive plants in wetlands: opportunities, opportunists, and outcomes. *Critical Reviews in Plant Sciences,* 23(5): 431-452.

In: Riparian Zones
Editor: Oleg S. Pokrovsky

ISBN: 978-1-63484-613-4

Chapter 6

LANDSCAPE DYNAMICS AND LAKES HYDROLOGY OF KERZHENETS RIVER FLOOPLAIN

***L. E. Efimova*[1,*], *O. V. Korabliova*[2] *and D. V. Lomova*[3]**

[1]Department of Geography, Moscow State University, Moscow, Russia
[2]State Nature Biosphere Reserve "Kerzhensky," N. Novgorod, Russia
[3]Water Problems Institute Russian Academy of Sciences, Moscow, Russia

ABSTRACT

This chapter describes drivers and links between the floodplain and the river channels. We focused on the spatial patterns in floodplain topography of the Kerzhenets River (left-bank tributary of the Volga River) and flooplain evolution and lakes formation. Distribution of small lakes along the floodplain landscapes is a specific feature of the Kerzhenets River valley. Small lakes are the most sensitive to a man-caused impact and respond quickly to any changes occurring in the watershed area. We identified hydrologic conditions and chemistry of the floodplain lakes located in the territory of State Nature Biosphere Reserve Kerzhensky due to location with a special regard to distance to the river and the strength of hydraulic connection with the river, as well as the elevation level on the floodplain. It was found that the lakes were not exposed to man-caused impact and that allowed to consider the lakes as some kind of baseline conditions for the area. This information allowed to make conclusions on "normal" conditions corresponding the present state of the water bodies in the context of climatic changes. Statistical analysis of interrelations between fiven hydrochemical values and morphometric properties of the lakes was obtained. It was proved that the composition of the bottom sediments reflects the development stages of floodplain lakes.

Keywords: Kerzhenets River, channel-floodplain dynamics, lake, hydrochemistry, oxygen, conductivity, nutrients, sediments

---

* Corresponding author: L. E. Efimova. Email: ef_river@mail.ru.

INTRODUCTION

Kerzhenets River is the typical middle lowland rivers of South taiga landscapes of the Russian Plain, it flows through the territory of Nizhegorodskoe Zavolzh'ye, is a left tributary of the Volga River, and belongs to the basin of the Caspian Sea. Kerzhenets drainage area is 6140 km^2; the total length of the channel is about 300 km. Kerzhenets is a popular recreation object. In the middle stream on the left bank of Kerzhenets State Nature Biosphere Reserve Kerzhensky was established in 1993. One of the main aims of Russian nature reserves is the detailed investigation of natural processes. Scientific research of various types (geomorphological, landscaping, hydrological and hydrochemical research work as well as monitoring studies) take place in the territory of the Kerzhensky nature reserve.

River beds, floodplains, floodplain landscapes are the most dynamic and rather sensitive natural objects being formed under the influence of external geological processes. Floodplain natural systems have specific features of their development, depending on the natural zoning and regional conditions. River bed is the western border of the Kerzhensky nature reserve. As a result of the active dynamics of the channel and floodplain area of the reserve changes, as well as soil and plant cover, the new floodplain lakes are forming.

The proportion of floodplain lakes in total number of lakes of the Kerzhenets River valley exceeds 70%. Interest in studies and monitoring of water chemistry of small lakes is due to the fact these water bodies are the most sensitive to man-caused impact and respond quickly to any changes in the basin. As a rule, water bodies within conservation areas are described in terms of background hydro-ecological environmental conditions. This allows to consider baseline characteristics of water bodies for environmental assessment of protected areas.

The floodplain and riparian zone are directly formed under the influence of the processes in the course of the river and water-flooding in high water and flood seasons. However, both landscape and structure of the floodplain affect the processes in the river bed. It is important that the river bed and the floodplain should be considered as an entire natural formation and be classified as an independent subsystem which may be otherwise called as "the floodplain and the river bed riparian patterns" (Chernov, 2009).

STUDY SITES, SAMPLING AND ANALYSES

Studies of the floodplain and river bed patterns of the Kerzhenets River were based on the results of I. I. Mamy's studies (of 2005). Maps of various scales as well as satellite data collected at various time, geographical field survey, long-term monitoring in respect of the stream-banks erosion (Chernov, Korabliova, 2008), the river water level and hydrological conditions, topsoil and ground cover of the floodplain, were used in connection with the detection and analysis of development phases of natural patterns. Results of studies for several years received from separate sites, were processed within a PC database.

The information was based on hydrological and hydrochemical studies: results of field work (2012-2015) performed by officers of department of hydrology of the faculty of geography, Moscow State University, as well as bibliographic sources and library material of Kerzhensky nature reserve.

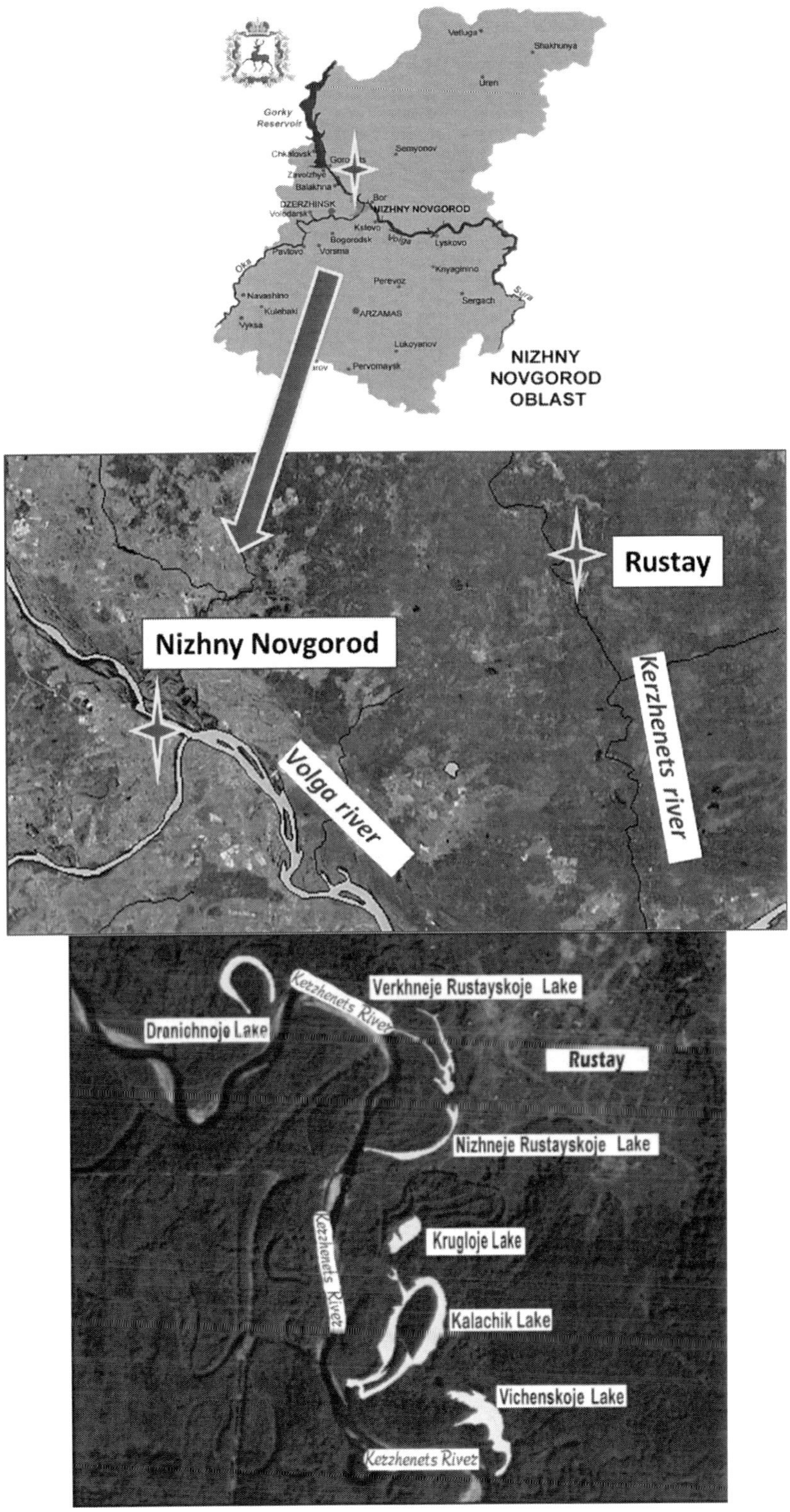

Figure 1. Schematic representation of the studied aria on the territory of Nizhny Novgorod Region.

Our research was conducted during all hydrological seasons, namely: during winter baseflow period, spring flood, summer low runoff period and autumn period. Reference heights were appointed in deep central parts of the lakes. The following measurements had been taken in the course of research work in the central parts of the lakes: water temperature, electrical conductivity, pH, dissolved oxygen concentration. The following indices had been evaluated in selected samples: basic ions content, trace elements content, organic matter content, nutrients content (Guidelines., 2003; Muraviev 1999). In the determination of trace elements content, samples had been filtered with a membrane filter (0.45 μm) and preserved with the help of nitric acid. Dissolved forms of iron and manganese had been determined by atomic absorption (Ermachenko, Ermachenko, 1999). SOD index as well as total organic matter destruction index was determined according to Kuznetsova-Romanenko's method (1985). SOD index was determined by oxygen absorption by alluvium column for a certain exposition period. Total organic matter destruction in soils was determined by CO_2 inflow from alluvium into the water. Aerobic destruction is evaluated proceeding from SOD index, anaerobic destruction is defined as the difference between total and aerobic destruction.

RESULTS

Floodplain and River Bed Patterns and Oxbowlakes: Natural Development Factors

Geological and Geomorphological Features of the Kerzhenets River Valley

The Kerzhenets River flows along low-lying territory of Nizhegorodskoje Zavolzh'ye, in the flatlands formed by the alluvial and glaciofluvial sand deposits. Mellow sand deposits are easily degradable by water flow except in some straight sections with original (pre-quaternary) clay, chalky clay, lime and dolomite deposits of Tatarian strata (cephalic Perm) on the bottom and through the banks (Fridman, Korabliova, 2001) preventing the banks from erosion. A bed, a floodplain and two terraces above the floodplain (Figure 2) can be distinguished in the Kerzhenets River valley. In the average the bed is 50 meters wide (varying from 26 meters to 70 meters).

The river channel is mainly sandy and tortuous. The floodplain is wide and two-sided in general. The first terrace above the floodplain runs along the floodplain through both banks in narrow strips. Its surface is at the height of 4.5-6 meters above the low water line of the river. Ancient river beds in the form of bedplates may be seen on the terrace (the average width of the bedplates being 500-550 meters). All the meander scars and bedplates were formed in the times past by the stream flows with water content exceeding the water content of the existing river by several times (Sidorchuk, Panin and others, 2000). The second terrace above the floodplain occupies considerable areas, covered with large bogs. The surface of the platform rises above the low-water level of the Kerzhenets River by 8-12 meters.

Figure 2. Satellite image of the western section of the Kerzhensky nature reserve with Kerzhenets River bed. 1 - the boundary of accumulative double-sided floodplain pattern, 2 - border of confined flood plain pattern.

Climate

The climatic conditions of the river basin influence the hydrological regime of the river and the formation of soil and vegetation of the floodplain. Climate is temperate continental with long cold winters and warm relatively short summers. In general, the climate is determined by the influence, first of all, the air masses of temperate latitudes, which come from the west and north-west of the Atlantic Ocean. Air masses often come from the north and north-east of the Arctic Ocean. From the south, tropical air enters; it brings a thaw in the winter and the intense heat in the summer. Most raising factors for the temperature are continental air masses from the south-east. At long as their action is set dry weather is established, drought begins with dry winds.

Seasons of the year are clearly distinguished, the most typical of the temperate climate zone. Winter in the average duration lasts 143-150 days (Annals…, 2004-2014). According to this information, the average winter temperature is from - 7.3 to - 8.0°C, the average annual rainfall is 210 mm. The amount of rainfall in the winter is of particular importance, because on the rivers of the center of the Russian Plain, as well as in most of the rivers of Russia, snow-fed is dominated. In winter, the greatest amount of rainfall falls, compared to other seasons. Spring lasts an average of 62-72 days; the average daily temperature varies from 7.9 to 8.7°C. Rainfall is from 89 to 105 mm. Summer in average has duration of 71-87 days. Average temperatures have a range depending on the year; from 17.2 to 18°C. In summer the mean value of rainfall vary from 176 to 189 mm. Autumn period lasts from 72 to 80 days. Mode average daily temperature is 7.0-7.9°C. Rainfall ranges from 134 mm to 158 mm. It has been noted there were years with a rainy autumn. Small flooding on streams is also associated with rains.

Hydrological Conditions of the River

With respect to hydrological conditions the Kerzhenets River is defined by high spring flood level, low summer water level with possible rainfall floods, low autumn water floods and low stable winter water level. Flood starts at the beginning of April. The average term of the beginning of the floods, according to monitoring observations in Kerzhensky nature reserve falls for the last 18 years on the 4th of April (Mankish, Bayanov, 2001). The morphology of the river is directly related to the hydrological regime.

The variation of the river water level produces enhanced erosion of banks. Erosion of banks enhances during high water periods due to the increase of duty of water and water flow, the banks suffer badly, corrode, trees fall as due to powerful water flow roots become eroded, there are lots of sheared trees along the hollowed-out banks. Intensive process of accumulation of fluvial sediment takes place on the hollowed-out river banks during high water periods (Korabliova, Chernov 2012).

Water level variation in small floodplain lakes is similar to that in the Kerzhenets River. However, peaks and troughs in the lakes come with some delays in comparison with water levels in the river. The sharpest rise of water in all the lakes occurs only after waters of the flooded Kerzhenets River have risen to the level of the lakes.

Peculiarities of Soil and Vegetation Cover

Soil covers in the watershed surfaces are rather monotonous; they are mainly sod-podzolic, bog-podzolic and bog. On the floodplains there are the most diverse species and their subtypes. This diversity is due primarily to the influence of high waters and erosion-accumulation activity of rivers.

There are no soils in the young floodplain (section of the floodplain located near the river bed), which has only alluvial sands. Next, soils are in the process of formation in the meander sections of the higher floodplain, so undeveloped yet or primitive alluvial soils only may be found there. Alluvial meadow stratified sandy soils are formed deep in the floodplain (central part). Signs of eluviations appear in the soil profile at the highest sections of the central floodplain which would hardly get covered with water in high water periods, and where alluvial meadow sour eluviated soils form. Alluvial meadow moors and alluvial marshy soils may be found near the oxbow depressions, in old channels, mostly near the floodplain platforms (Korabliova, 2009). Long-term surface and groundwater moistening is typical for them.

Vegetation protects against erosion of the surface of the floodplain and the coast. For the vegetation of floodplains dynamic change is marked, from pioneer species (Petasites spurius, Bromopsis inermis, Calamagrostis epigeios) to coniferous-deciduous forests. In riverine shallows herbal plants initially grow, dominated by butterbur, complemented with awnless brome, bush grass and others. Next there is a change with willow phytocenoses, which are represented by different species of willows. Deep into the floodplain willow is replaced by pine forests, in the beginning the young, then more mature. In a mature pine forest there is undergrowth of linden, oak, and spruce. Then, deep into the floodplain is a gradual complication of communities: there rejoice and linden, spruce forests. Section near the floodplain teracces is represented by dead river channels and old channels with water-resistant plants. Marshy lands of various types may be found here (Korabliova, Chernov, 2009).

Man-Caused Impact

Mass forest destruction in the Kerzhenets River valley (within the period of 1940-1980) had the strongest impact on the river bed and on all conditions of the river in general. Forest destruction had an impact on hydrological conditions of the river. Bog reclamation, peat mining also took place in this period. Considerable areas of abandoned peat mines may be found on right banks of the Kerzhenets River. Reduction of the area of bogs and dehydration of surface soil resulted in increase of amplitudes of water levels in the period between high water and low water, and further resulted in more intensive stream-bank erosion. As a consequence, amount of sand going into the river beds increased. Floating of wood which took place before the 70s of the 20th century also had an impact on the ecology of the river.

Use of floodplain natural complexes for recreation purposes results in identifiable changes of the ground cover. A tendency of substitution of rather sensitive species by the species less sensitive to recreation impact was observed in the grass stand. Erosion processes as well as washing off sand by rain and melt water down to the river bed were observed near tourists' camping sites. Green cover does not restore in the fire pit within a vegetation period.

Dynamic of Floodplain Patterns

The course of the Kerzhenets River meanders over much of the length, forming free curves (Korabliova, 2010), and the floodplain in general has a segmented and ridged structure. As a result of erosion and deposition floodplain and channel is formed in the friable rock. This pattern dominates along a confined floodplain (Figure 2.1). In turn, every floodplain and channel pattern type includes parts (natural pattern with typical specific features, namely: geography and rock material of which river beds and floodplains are built, soils, green cover). The latter will be the main diagnostic properties for the definition of floodplain and the river bed pattern and natural pattern. Composition of the rocks composing the floodplain by morphological structure of floodplain and the river patterns can be divided into two types: the accumulative and the confined (Korabliova, Chernov, 2012).

The accumulative floodplains and the river course patterns have a normal or increased width of own alluvium, thus the floodplain may freely change there. Drifting sands which are forming the river banks, do not prevent erosion. Water flow in such rocks intensively erodes the coast (Figure 3) and accumulates a solid (in this case sand) material; then the channel becomes tortuous, a broad bilateral floodplain is formed.

The confined floodplains (Figure 2) are built of thin alluvium layers deposited on the bedding rock so that the low-water bed of the river is cut into the rock. Hard and dense rocks (loam, clay, marl, limestone, etc.) are resistant to water erosion and hinder the development of channel deformation (erosion of the coast).

Figure 3. The erosion of the banks and the fall of trees on the Kerzhenets River.

There are three floodplain zones (development phases) in accumulative floodplain: pre-floodplain (A), young floodplain (B), mature floodplain (C). All development phases of the floodplain, with the inclusion of a fossiliferous layer (D), are well seen at all levels from the young stage to mature. Here, the process of accumulation of sediments and process of formation and development of floodplains are the most intensive on raised banks of the meanders. A sand ridge is formed first (A) (Figure 4) constituting a pre-floodplain phase. Later primitive plants appear as a prove of the process of formation of a young (primitive) section of the floodplain (B) distinct from the mature floodplain. The duration of this phase, the origin and formation of the floodplain, is up to 100, sometimes up to 150 years.

The young floodplain is formed step by step following the increase of the curve degree of the meander. Sub-phases B1, B2, B3 (Figure 5) differing from each other by age, height, type of soil, type of plants, the floodplain conditions, are formed within the phase B.

At the initial sub-phase (B1) it is just sand deposits, without soil and with individual pioneer plant species, in this case, butterbur (Petasites spurius). Then willow, grasses, sedges settle appear, although soils still are not formed. The height above the river's edge is small, averages 1.5 m, flooded every year during the spring floods. Then, in the middle sub-phase (B2) willow and pine trees are growing, the average height of 3 m, flooded during the spring flood 5-6 times in 10 years. Soils are not yet formed, on the surface of the sand humus and moss spots appear. The process of formation of the young floodplain finishes by sub-phase of formation (B3), which is characterized by the formation of a mature pine forest; its age is about 100 years old. The soils are poorly developed, and the height and the flooding regime is the same as at sub-phase B2. The last sub-phase B3 usually stands on a sharp bends. The first two are on the flat bends of the river (Figure 5). Young floodplain as a whole has a stepped surface.

In the accumulative floodeplaine-chanel pattern fragments of rectilinear channels of bilateral floodplain are found; here an accumulation of sandy alluvium and formation of young floodplain occurs on both sides (Figure 6). Peculiarities of such dynamics are due to the influence of anthropogenic factor, this is primarily the construction of bridges and the strengthening (concreting) the least stable areas of the coast, where intense erosion could occur with the formation of the new bend.

Mature floodplain is in the second phase (C) that's the phase of stable existence and slow development, its duration being several hundred or even several thousand years. This phase is characterized by mature stratified soils. The following floodplain zones may be distinguished within a mature phase: the meander (C1), central zone (C2) and teracce near the floodplain (C3). These zones are located at different distances from the channel. Riverine floodplain (Figure 7) is adjacent to the channel. Here is the erosion actively manifested and shores washouts occur. The altitude over the edge is on the average 3.7 m and above, with ridged surface and riverine shafts where pine, oak, linden wood grow. At high flood, the riverine floodplain area is under water. Typically it occurs 2 times in 10 years.

The central zone (C2) is located at some distance from the river bed, so water flow rates are much lower in high water periods and no active erosion or accumulative processes take place. Surface is wavy with an average height of 3.5 m. The central floodplain consists mainly of coniferous-deciduous forests (Figure 7) in various combinations of pine, spruce, oak, linden; most often on the turf layered sandy and loamy soils. Dead river channels (floodplain lakes, former river bed sections) are situated in central zone. Old channels (the oldest sections of the river bed) are separated from the main course as a result of the lining process some hundreds years ago, has become filled with alder, birch and willow trees. The teracce near the floodplain (C3) is located at a considerable distance from the river bed. In the average it is 2.5 meters above the encroachment line. The teracce near the floodplain is located lower than other floodplain parts, and water flow is slow. Meltwater coming from the teracce slopes or original banks is accumulated here, the water is often stagnant, so the teracce near the floodplain is swamped to different extents, and muddy soils similar to low-land bogs, can be formed here (Korabliova, 2009).

Figure 4. Formation of the young floodplain on flat bends of Kerzhenets River.

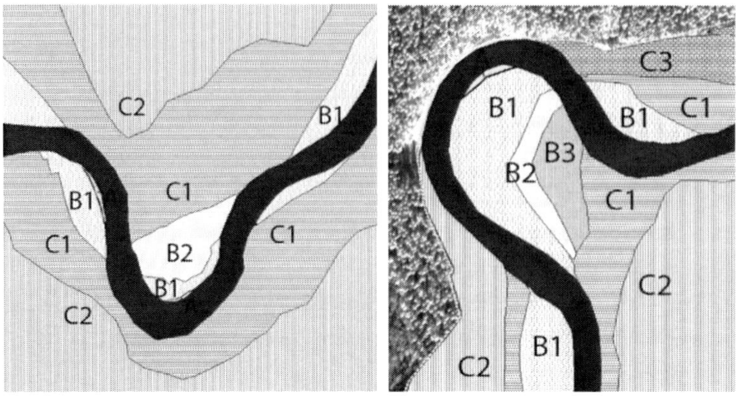

I II

Figure 5. Schemes with phases and sub-phases of development (A, B1, B2, B3, C1, C2, C3) on the flat (I) and steep (II) bends of Kerzhenets River.

Figure 6. The rectilinear channel in accumulative floodeplaine-chanel pattern at the bridge over the Kerzhenets River.

Photo by E. N. Korshunov.

Figure 7. Kerzhenets River: mature (C) riverine flood plain on a high erosion bank, sandbar (A) on a low accumulative bank and the young floodplain (B).

The final development phase (D), namely, the change of the old natural pattern with a new one, taking place when high surfaces which have been rarely covered with water, get to be short of water. Signs of eluviations appear in the soil profile; formation of soddy-podzolic soils with pine and spruce forests mainly including lime or oak trees, or pine forests only, takes place. The floodplain pattern turns gradually into a teracce above the floodplain.

Accumulative floodplain and the river course pattern undergoes all the dynamic processes typical to the meandering river, namely: stream bank erosion, formation of shallows and young floodplain sections, changes in the river course and in addition, formation of floodplain lakes, old channels and new natural patterns then. Accumulative floodplain and the river course patterns are more sensitive to the impact of natural and anthropogenic factors and respond to any changes in the environment more quickly than confined floodplains.

Confined floodplain has got some differences with respect to properties and morphology of the floodplain. No side erosion or sand accumulation take place in such areas as the banks are built of rather erosion resistant rock. Incomplete floodplain and the river bed complex with one-sided floodplain (without a young floodplain) is formed here (Figure 8). Part of the mature river bed near the bed has got richer light clay-loam soils and is often represented by oak-woods or spruce forests.

In confined floodplain pattern in which the bed and banks are built of more solid rock materials (clay, chalky clay, lime) the floodplain dynamics is as follows: **C (C1–C2–C3)→D**. Confined floodplain with a straight line course and one-sided floodplain is the most stable. The rock of which the floodplain and the river bed are built limits the development of erosion and accumulative activity of the river.

Figure 8. The central floodplain of Kerzhenets River - a phase of sustainable existence and slow development.

Hydrological and Hydrochemical Specific Features of the Floodplain Lakes in the Kerzhenets River Valley

Typically, floodplain lakes have a horseshoe form, with width varying from 15 m to 40 m depending on the initial size of the river streambed in the dead arm of the river, and intensity of further vegetal invasion of waters. Lakes sizes do not exceed 1 кm^2, the water volume varies from 7-10 to 50-60 thousand m^3, that allows to classify them as “small lakes.” The following factors have an impact on the lakes conditions: plain low-lying ground consisting of well washed sand and clay sand, widely wooded and inundated land. These factors provide for the balancing of annual water flow and the forming of water with low salt content.

First of all, stream and hydrochemical conditions of floodplain lakes are determined by their location within the floodplain and streambed complex, taking the following aspects onto account: the distance to the river, the hydraulic connection with the river, and the elevation level on the floodplain. The lakes concerned are located in the mature floodplain: in the area near the river bed (Nizhneje Rustayskoje Lake, Verkhneje Rustayskoje Lake), in central floodplain area (Kalachik Lake, Krugloje Lake) and on the teracce near the floodplain (Chernozerje Lake). The above areas are at various distances from the river bed.

It is known that upon formation lakes mostly develop under the influence of the processes taking place in their watersheds. The intensity of such influence is determined by specific water inflow (water catch). The more significant is the influence, the more intensive is terrigenic sedimentation leading to the filling of a lake's watersink with sediments and to changing of the initial shape. In the Kerzhenets River valley, the influence of the water catch is especially intensive in respect of the lakes with the smallest water areas.

We have compared variability of fluctuations of water level in the lakes. Such variability depends on the morphometry of the lake's watersink, the size of the lake basin and the flowage. N. Rustayskoje Lake had shown the best flowage among the lakes examined. It's volume of water is the smallest (by 6.5 times less water than in Kalachik Lake). In addition, this lake is connected with the river throughout the year.

Specific limnological features were determined depending on their location, frequency and duration of contacts with the river waters during high water period. V. Rustayskoje Lake and N. Rustayskoje Lake have got a continuous contact with the river. These lakes are located in the area near the floodplain which takes the water flow coming during high water period, first. This forces the biggest solid particles (including coarse sand) to deposit in lakes, which is proved by bottom sediments of N. Rustayskoje Lake. The rates of water flow running to the floodplain, reaches 1.5 m/s, however the rates go down quickly, which results in accumulation of sand deposits brought by the river. The hydrochemical properties of water in the lakes are similar to those of the river, which proves faster water exchange than in other floodplain lakes. Kalachik Lake is located in central area of the floodplain and has contacts with the river in 85% of cases. The river water flow reaches the lakes free of sand deposits, flow rates decrease and composition of deposits change. In the absence of hydraulic connection with the river, the water level in the lake decreases, and snow, rain and underground waters become the basic source of water with an impact on the lakes water chemistry. The Krugloje Lake is located 1 meter higher than the river bed level, so this lake has got more rare hydraulic connection with the river than Kalachik Lake.

Seasonally recurring curve of surface water temperature of the lakes is defined by variation of air temperature. Principles of distribution of temperature through the water layer mostly depend on morphometric properties of the lake's watersink. The inverse temperature stratification may be seen in winter in the studied lakes. In average, surface water temperature of the lakes differ insignificantly. Differences in temperature become more obvious below the depth of 2 meters. Data analysis in respect of the distribution of temperature in small lakes of the Kerzhenets River valley showed that at a depth of 2-4 meters waters in Kalachik Lake and N. Rustayskoje Lake are warmer than water in Chernoje Lake (by 3°C almost), located at a significant distance from the river in the watershed.

In summer, the constant direct temperature stratification with the surface maximum is formed in the lakes. The temperature of epilimnion of Kalachik Lake and Chernozerje Lake is by 2.5°C colder than that of Krugloje Lake. Most likely, this was due to different sunlight intensity in water areas of the lakes and different warming up of surface water layers. Temperature jump layer (thermocline layer) is at the depth of 1-2 meters in all lakes practically (sometimes 0.5 m higher or deeper). During the survey, thermal gradient was 5-7°C (9°C in Kalachik Lake). Similar situation is registered in the lakes every year (Annals…, 1997-2014). Specific morphometric properties, location at places protected from winds, and water plants in water area preventing water mixing, provide for the formation of a steady thermocline layer.

Formation of spring homothermal conditions is due to the level of flood water of the Kerzhenets River and due to high intensity of river water flow into the floodplain lakes. At flood decay, Nizhneje and Verkhneje Rustayskoje Lakes are filled with the river waters and are practically arms of the river, that's why water temperature in these lakes does not vary with the depth, thus is as that of the river. As for other floodplain lakes, spring homothermal conditions are not registered every year and if any, than for a short period of time. Short autumn homothermal conditions are formed in the lakes in the end of September-beginning of October, what is typical for shallow-water ponds within this geographical zone. Homothermal conditions are formed faster in the lakes which are more open for wind circulation (Krugloje Lake).

Distribution of specific conductivity of the water (æ) (its value is settled to 18°C) in the lakes concerned is characterized by dimensional and time-dependent variability. Differences in oscillating amplitude of electrical conductivity of water are marked depending on the flowage. Specific conductivity values increase in the lakes in low-water periods when amount of ground and underground waters feeding the lakes, increases. Ground waters feeding the lakes are mostly perched ground water or downstream underground waters, their chemistry being very similar to that of the Kerzhenets River. Seasonally recurring differences in specific conductivity of water in the floodplain lakes may be rather substantial, increasing in low water periods by more than two times as compared to high water periods. The most significant seasonally recurring differences in specific conductivity of water are registered in N. Rustayskoje Lake, Chernozerje Lake, are the least in Krugloje Lake (Figure 9).

Analysis of seasonally recurring changes in specific conductivity of water allowed to divide the lakes in two groups. Watershed lakes of organogenic origin belong to one group. A deep floodplain Krugloje Lake located rather far from the river-bed, having hydraulic connection with the river in 60% of cases during the period of flood, may be ascribed to the same group. Floodplain lakes with max depth of 3 meters located not far from the river, belong to the second group. Upon decay of flood, specific conductivity of water in dead lakes grows as a result of increase of volume of ground waters with higher percentage of salt in total volume of feeding water. The lower lake is located in the flood plain, the greater is its electrical conductivity of water. The highest gradients of specific conductivity are registered in summer in thermocline layer in Kalachik Lake (60 μS/cm per 1 m), Chernozerje Lake (90 μS/cm per 1 m), N. Rustayskoje Lake (70 μS/cm per 1 m). Thus, stability of water in this lakes depend on both temperature and mineralization level.

Officers of the Kerzhensky nature reserve (Bayanov, Krivdina, 2013) analyzed the ionic composition ofwaters in various conditions in 1998-2004. Results of the this analysis showed that water of the lakes may be classified as hydrocarbonate category water (calcium group).

However, the authors state that "composition of waters of the Kerzhenets river and floodplain lakes is subject to changes in all seasons, so that the category of waters may change." Salts content in the surface waters (river and lake) in the territory of the Kerzhensky nature reserve during low-water period does not exceed 100 mg/l in the average, the largest values (120-150 mg/l) are in the bottom epipedon of the lakes), due to the ground and underground water surge (N. Rustayskoje, V. Rustayskoje lakes). Results of our research work also showed that waters of small lakes in the Kerzhenets River valley belong to hydrocarbonate and sulphate water (calcium and magnesium group). Proportion of sulfates (20%) in the water of the lakes is rather big. Anionic composition of water in Kalachik Lake and Krugloje Lake during high water periods may be specified as mixed because percentage of hydrocarbonate and sulfate ions are practically equal. Proportion of sulfates increases especially in the end of high water period, as a result of chemical degradation of the products of decaying of plant and animal residue in the leaf mold and upper level of soil.

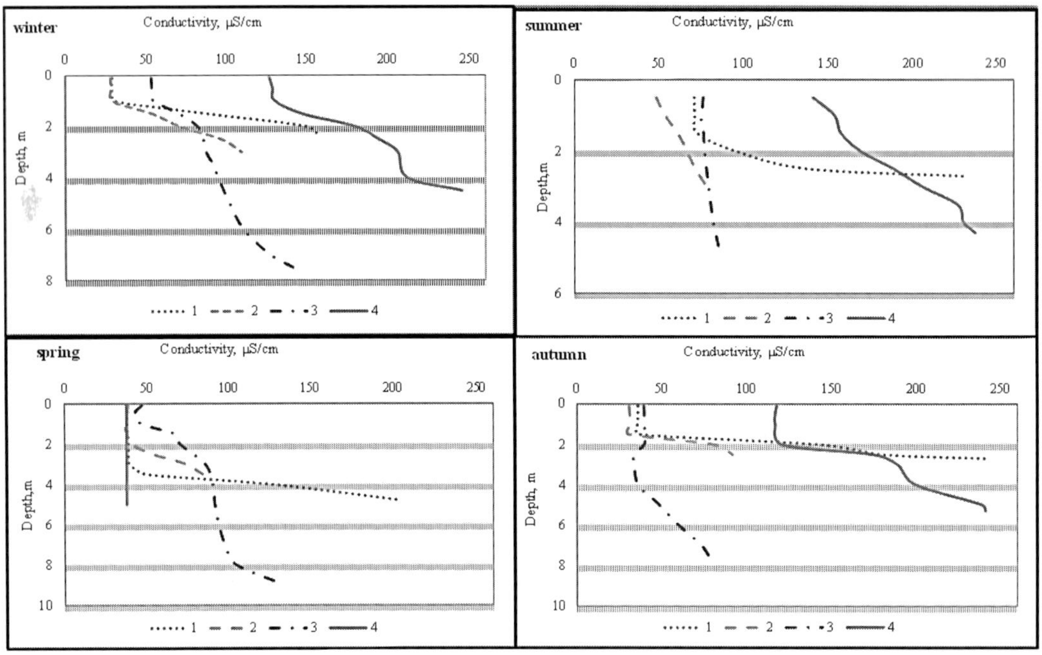

Figure 9. Vertical distribution of electrical conductivity of water in lakes in various hydrological seasons: 1 – Chernozarje Lake, 2 – Kalachik Lake, 3 – Krugloje Lake, 4 – N. Rustayskoje Lake.

The existence of oxygen deficiency and anoxia zones in the floodplain lakes practically throughout the year is typical for the oxygen expansion drive. For example, in February-March, 2013 and 2014, anoxia zone in Krugloje Lake and Kalachik Lake extended over the whole water layer (from bottom up to the surface). Water oxygenation in the surface layer of N. Rustayskoje Lake did not exceed 4%. Similar situation is typical for low-water periods, however, may be seen in the end of high water periods, as the process of stirring of water in the lakes in spring does not always lead to the oxygenation of water (Annals…, 1997-2014).

During a high water period in 2015 water layer in Verkhneje Rustayskoje and Nizhneje Rustayskoje lakes was fully aerated, however, water oxygenation level did not exceed 72%. In other floodplain lakes thickness of the oxygenated water layer (by 50% and more) varied.

Thickness of the layer in Chernozerje Lake was 2 meters, when in Krugloje Lake it was 0.7 m only. Water-dissolved oxygen content decreased from spring to summer. In summer, there are anoxia zones in all the lakes practically at the depth of 1.5-2.0 meters, as a result of stable stratification (Figure 10). Significant amount of oxygen is in the upper water layer (1 meter wide) only, thus creating good conditions for aquatic organisms living in dead stream branches (Bayanov, 2012). In autumn, along with the temperature and density destratification and formation of homothermal conditions, as a rule, oxygen content increases, however, there are still oxygen deficiency zones in bottom water layers of all the lakes concerned.

The process of photosynthesis in the epilimnion of N. Rustayskoje Lake may develop rather intensively then in other lakes. This leads to increase of pH value up to 7.2-7.8 and of water-dissolved oxygen content. pH of surface water layers in Kalachik Lake and Krugloje Lake ranges from 6.40 to 6.75. Processes of destruction in the bottom layers of these lakes causing the decrease of water-dissolved oxygen proportion to analytic zero and pH values to 5.7-6.05.

Location in the flood plain provides for the active development of productional processes in the epilimnion of the floodplain lakes. The lakes concerned are located at various distances from the river, amidst various plants and with various sunlight intensity. N. Rustayskoje Lake is located in the flood plain covered with brushwood thus being open to sunlight, while Krugloje Lake is located in the middle of the wood and is strongly shadowed. Registered water transparency value of N. Rustayskoje Lake in summer is 1.5-1.9 m, water in Krugloje Lake is less transparent (0.7-0.8 m) (Figure 11, 12). In addition to sunlight intensity conditions in the water shed, color of water (stipulated, in turn, by correlation between proportions of various water types in the feeding water of a certain lake) also affects the transparency value of water.

The pH in the water of floodplain lakes is characterized by dimensional and time-dependent variability. The following may be specified as basic factors determining the variability of this index: supply of organic matter from the watershed territory, intensity of internal processes (first of all, photosynthesis and destruction), as well as CO_2 content in the water. Seasonally recurring changes in CO_2 content in the water of the floodplain lakes are well defined. The lowest CO_2 concentrations were observed in the end of spring-beginning of summer period and the highest, in the second half of winter. Extended CO_2 content in the water of the lakes in winter (up to 30 mg/l) may be due to underground supply of water, and accumulation of CO_2 - due to the oxidation of organic matter and aquatic organism's breathing, if water is covered with ice.

Practically all waters of reserve are subacidic (5-6.5); pH increases in the surface water layers in summer only, due to photosynthesis. Monitoring data analysis for a long-term period (Annals…., 1998-2014) showed tendency for the decrease of pH value in all the water basins in the territory of Kerzhensky nature reserve (excluding bogs) commenced in 2006-2007. Decrease of the average annual values and increase of the average annual electrical conductivity caused by total decrease of water content in the basin of the Upstream Volga River, were also registered in the period. Beginning from the 70-s decrease of stream flow in spring and increase of stream flow in low-water periods have been registered, what was evidenced by the increased degree of natural flow control (Frolova and other 2013). Changes in annual distribution of stream flow of the Kerzhenets River affect the percentage of salt and chemistry of water in the floodplain lakes.

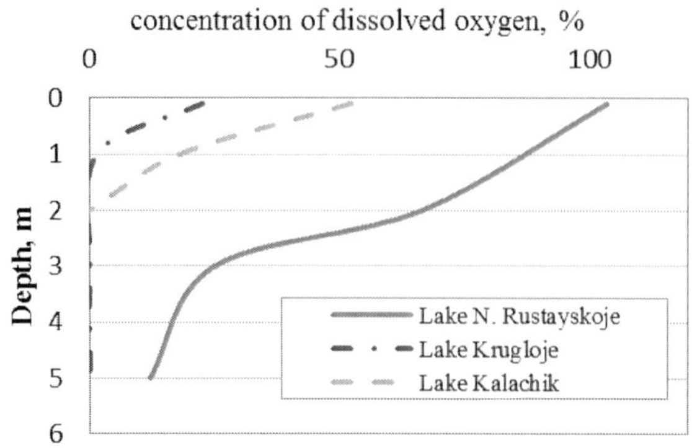

Figure 10. Vertical distribution of oxygen proportion in floodplain lakes.

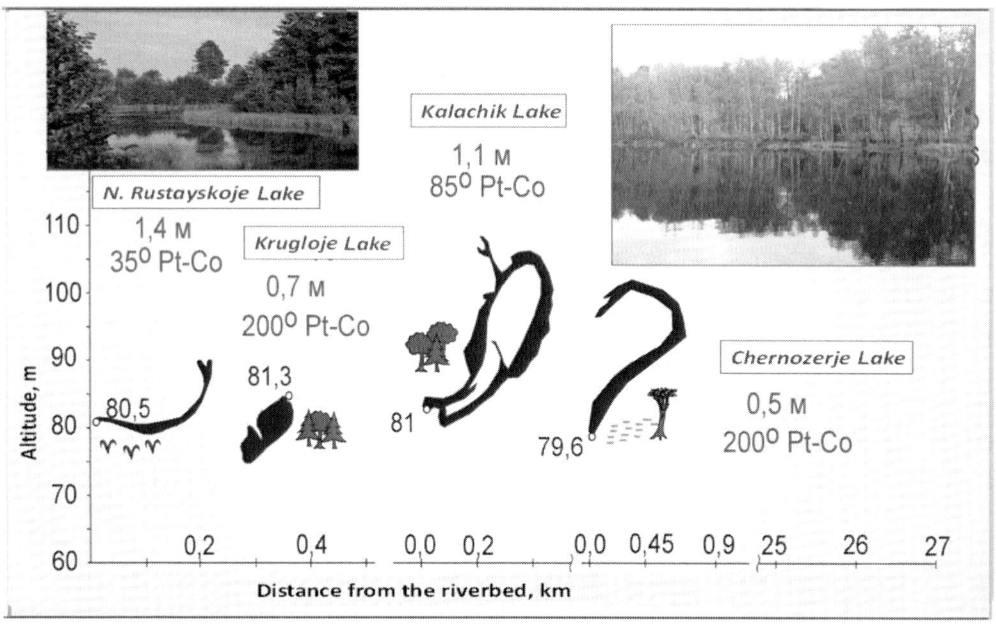

Figure 11. The location of the lakes within the floodplain, transparency of survey color of water.

A B

Figure 12. Floodplain lakes in the Kerzhenets River valley (A) N. Rustayskoje Lake and (B) Krugloje Lake.

Increased amount of organic matter in water of studied water bodies is due to a great amount of bogs in the territory. Landscaping conditions in the watershed significantly stipulate intensity of removal of soluble humus substances, thus affecting the color of water, concentration and chemistry of water humus in rivers and lakes. High correlation factor between the values (color and permanganate value) shows that the water mainly contains organic matter resistant to biochemical oxidation of allochthonic origin (Skopintsev, Goncharova, 1987). Relationship between the permanganate value and the color was very stable practically throughout the year. For the most water bodies within this geographical zone color of water in the lakes depends, as a rule, on the existence of not only organic matter but of dissolved iron compounds also. In the average, the permanganate oxidation value for the water of N. Rustayskoje Lake was 5-8 mgO/l, corresponding to the DOC of 15 to 22 mg/l, for Krugloje Lake it was 31-32 mgO/l corresponding to 56-58 mg/l. Vertical distribution of dissolved organic matter in the water of studied lakes is rather equal. Both concentration of organic matter resistant to biochemical oxidation and total organic matter content increase from the surface to the bottom.

During the survey, the color of water varied from 35° (N. Rustayskoje Lake, the surface water layer) to 350° (Kalachik Lake, the bottom water layers) in various water bodies increasing up to 700° Cr-Co-scale (water in bogs) as shown in Figure 13. The above mentioned differences between lakes persist through all hydrological seasons being connected with specific nature of feeding process of the lakes, reaching maximal values in the end of high-water periods due to enhanced supply of organic matter of humus origin.

Figure 13. The confluent of Kerzhenets River.

The phosphorus in water of the floodplain lakes is mainly represented by inorganic form with concentration of 70-90% of total phosphorus near the bottom. Relatively low concentration P_{min} was registered in the water of some lakes in winter 2014, as compared to winter 2013. Thus, concentration of mineral phosphorus in the bottom layers in Kalachik Lake was twice as low as registered in winter 2013 (0.33 and 0.68 mg/l correspondingly). The highest concentration of mineral phosphorus in the bottom layer (almost 1 mg/l) was registered in Chernozerje Lake; the same situation was registered in all the seasons of 2013. The highest concentration of mineral phosphorus in the bottom layer was registered in Kalachik Lake (by 4-5 times as high as P_{min} registered in the water of all other lakes concerned). Concentration P_{min} in the surface layers of the lakes is considerably lower than that in the bottom layers (Figure 14).

As compared to winter, in summer concentration of mineral phosphorus in the surface layers decreases considerably. Organic form of phosphorus prevails in the surface water layers of the lakes during a vegetal period, proportion of P_{min} being 6.5-25% P_{tot}. In contrast, P_{min} had prevailed in the bottom layers throughout the year (50-70% P_{tot}). Accumulation of mineral phosphorus in the bottom layers is due to stable stratification (in summer), underground water surge. Higher concentration of phosphorus may be determined by its income from bottom sediments in anoxic zone (Martynova 1977, 2010). Essential part of phosphorus flow in bottom water layers of the floodplain lakes may be formed by compounds, released during organic matter destruction process occurring in the surface of bottom sediments (Martynova 2010). High content of dissolved manganese and iron regularly registered in the most water bodies in the territory of Kerzhensky nature reserve, is typical for this geographical zone (Efimova, Frolova 2013; Nedogarko, 2007; Shvareva and other, 2014). Landscape conditions in the watershed of the lakes provide the delivery of manganese. Tree biomass contains a considerable amount of manganese, being the basic factor of high concentration of its mobile forms. Here, manganese enter the watershed landscape in the process of mineralization of wood residue and top humus, then being delivered to the lakes along with the surface water and sedimentating in the bottom of water bodies thus enriching bottom sediments (Nedogarko, 2007).

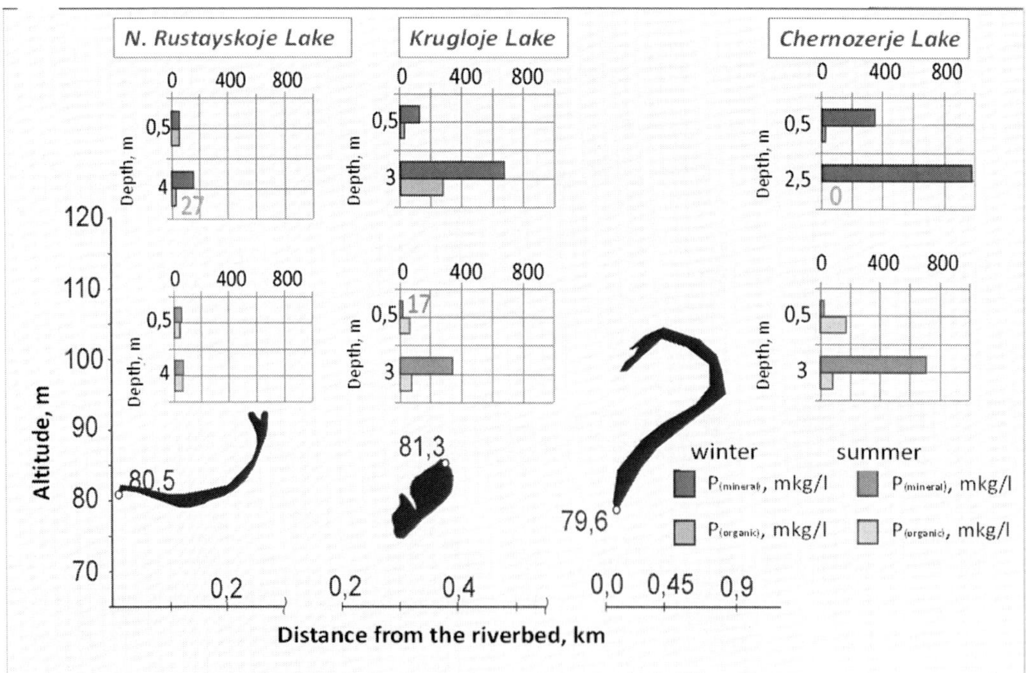

Figure 14. The concentration of mineral and organic phosphorus in floodplain lakes.

Bottom Sediments and Interreactions in the Shear Layer "Water – Bottom Sediments"

Chemistry and properties of bottom sediments reflect the entire complex of biological, chemical and physical processes occurring in the water bodies. Accumulation of chemical matter on the bottom as well as its removal from bottom sediments are one of the key mechanisms controlling the concentration of chemical substances in the water column having impact on the productivity of water ecosystems and water quality. Gaseous interchange conditions, temperature and chemical composition of the near bottom water layers, properties of bottom sediments, concentration of organic matter, sediment thickness and number of benthic organisms are basic factors defining the nature and intensity of processes occurring in the water-bottom subsurface horizon.

The composition of bottom sediments reflects various development stages of the floodplain lakes. Bottom sediments of Chernozerje Lake and Krugloje Lake are represented by black and brown alluvium, very unconsolidated, fine-dispersed (almost colloid) and non-stratified. Alluvium in Kalachik Lake is lighter, more airated and contains a lot of fallen leaves. Bottom sediments of N. Rustayskoje Lake are similar to those of the Kerzhenets River being sandy alluvial sediments. The chemical compostion of bottom sediments of the lakes is mainly represented by silicon, alluminium, and iron.

Concentration of considered components decreases in bottom sediments of the floodplain lakes in the following sequence: Chernozerje Lake → Krugloje Lake → Kalachik Lake → N. Rustayskoje Lake. The silicon content in bottom sediments of the lakes increases in the same sequence. SiO_2 concentration in the ground increases along with the increase of concentration of sand and sandy loam containing much silica. According to the results of our studies soil in Chernozerje Lake demonstrates great concentration of organic matter (50% as compared to 70% in Chernoje Lake sediment, 20% in Krugloje Lake sediment and 10% in Kalachik Lake sediment). Sandy sediments of N. Rustayskoje Lake contains trace amount of organic matter (Table 1).

Table 1. The composition of bottom sediments offlood plain lakes

The content (%)	Chernozerje lake	Krugloje lake	Kalachik lake	N. Rustayskoje lake
OB	50.0	20.0	10.0	-
Na_2O	0.25	0.93	0.70	0.47
MgO	0.54	1.29	1.14	0.33
Al_2O_3	3.77	13.05	12.19	3.56
SiO_2	39.8	70.8	70.8	93.5
P_2O_5	0.75	0.36	0.22	0.045
K_2O	0.50	1.51	1.43	0.36
CaO	3.39	0.68	0.59	0.083
TiO_2	0.27	0.65	0.59	0.062
SO_3	2.27	2.12	1.80	0.11
MnO	0.201	0.055	0.038	0.004
Fe_2O_3	5.38	3.93	3.19	0.391

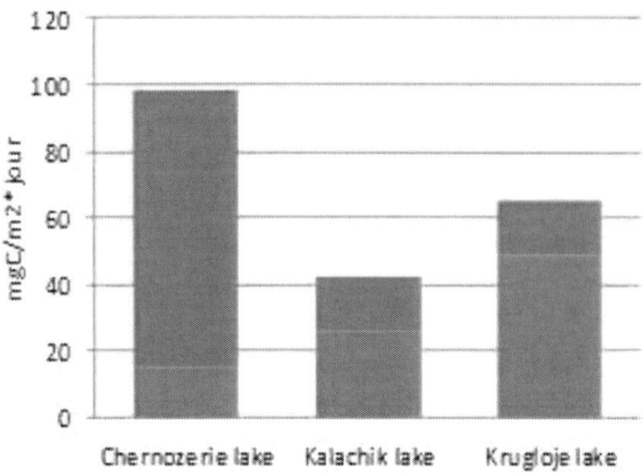

Figure 15. Correlation between aerobic and anaerobic destruction of organic matter in bottom sediments in autumn. Red column - anaerobic degradation and blue column - aerobic degradation.

Accumulation of organic and mineral matter in Chernozerje Lake alluviums is due to the location of the lake in the back marsh and connected with the bogging process providing for the revivification conditions in the near bottom layers of the lake throughout a year. Mobile forms of manganese and iron linked to organic matter (humic and fulvic acids), are present in soils and bottom sediments of marshy lakes. The highest concentration of phosphorus compounds (1-2%), manganese and iron (0.1-0.2% and 6-10% correspondingly) is found in the Chernozerje Lake alluviums.

It is known that not only mineral phosphorus but manganese and iron also come into the water from the bottom sediments in anoxic conditions (Martynova 2012, 2014). Increase of Mn concentration in the bottom layers in Kalachik Lake, Krugloje Lake, Chernozerje Lake confirms that these process takes place. Increase of concentration may result from the diffusion of Mn from interstitial waters with Mn, as well as in connection with the underground waters surge to the bed and river banks. Among the lakes considered in this work, maximal manganese content was registered in the bottom water layers in Kalachik Lake in winter. High concentration of dissolved forms of iron and manganese in the near bottom water layer is due to anoxia. Diffusion flows of these elements from bottom sediments into the water column provide high concentration of the above elements in the near bottom water layers, are formed.

Continuous processes of organic matter destruction take place in the water and bottom sediments. Intensity of organic matter destruction processes in alluviums is determined not by gross content but by the amount of digestible compounds (Alexandrova, 1973) and by SOD index. Existence of anoxic zones throughout a year is a specific condition of dissolved oxygen in the floodplain lakes of the Kerzhenets River valley. For this reason, the organic matter destruction process occurs mostly under the influence of anaerobic bacterial processes. In winter, total mineralization did not exceed 45 $mgS/m^2/day$ in the lakes. Massive sedimentation of dead phytoplankton and input of fresh labial organic matter onto the bottom are observed in autumn. This provides nutrients for the growth of bacterial inhabitants, whose peak of growth occurs in autumn and results in the increased total organic matter destruction in soils. Maximal amount (98 $mgS/m^2/day$) is observed in Chernozerje Lake alluviums, assuming that anaerobic destruction of organic matter prevails here. On the contrary, organic matter destruction in other floodplain lakes mainly occurs at the expense of aerobic processes (Figure 15).

Macrozoobenthos has a direct impact on the intensity of destructive processes in sediments. Via digging a network of holes, the burrowing organisms change the nature of masses interchange in water-bottom sediments subsurface horizon from diffusive to turbulent. In general, the diversity of macrozoobenthos of the flood plain lakes of the Kerzhenetz River is not large. The following organisms were found in macrobenthos: Chironomids larvae (four species), Chaoborus larvae (Chaoborus flavicans), Oligochaetes (Tubificidae family) and very little amount of biting midges (Ceratopogonidae family). Invertebrate predaceous organisms (Chaoborus flavicans) significantly determine species composition, size, structure and seasonally recurring variability of macrozoobenthos. In autumn and winter periods Chaoborus larvae (Chaoborus flavicans) concentrated in the deepest part of a water body. The maximal density of bottom species (6335 pieces/m^2) was registered in autumn in Chernozerhe Lake; population of Chaoborus larvae (Chaoborus flavicans) in Kalachik Lake and Krugloje lake amounts to 1000-2000 pieces/m^2.

CONCLUSION

Kerzhenets River is a typical medium-sized river in the southern boreal forest zone of the Russian plane. Meandering channel pattern with wide floodplain and straight confined channel with relatively narrow floodplain existing on just one bank of the river dominates valley morphology.

The changes of long-term states of the floodplain patterns occur in accumulative floodplain and streambed. These changes are expressed in the transition of pre-floodplain development phase (sand banks) to the phase of initiation and formation of a young floodplain, later – to the phase of stable existence and slow development of mature floodplain with floodplain areas (near streambed, central area and terrace near the floodplain). The final development phase is the substitution of the floodplain pattern by a terrace. Young floodplain is formed stepwise following the increase of degree of curve of the ancon and has got various sub-phase development

Hydrological conditions of the Kerzhenets River and duration of hydraulic connection between floodplain lakes and the river are the most important drivers of lakes water chemestry. We characterized the water chemistry in small lakes of the Kerzhenets River valley went from open lakes located in the area near river channel, to bog lakes located in the terrace near floodplain.

Temperature and concentration of dissolved oxygen mainly define the intensity of productive processes occurring in the lakes. The temperature conditions of the lakes depend on duration of ice period, on the formation of temperature jump layer (thermocline layer) and the photic layer depth in summer, as well as, on specific morphometric properties of the lakes' watersink.

Water oxygenating during a short-term flood period is a specific condition of the oxbowlakes located in the Kerzhenets River valley. However the stirring of waters in spring will not always result in water oxygenation. Anoxia zones will appear in the lakes already at the depth of 1-1,5 meters and will persist throughout the year. Existence of anoxia zones is supported by higher concentration of organic matter (resistant to biochemical oxidation of allochthonic origin mostly) coming from the watershed. Accumulation of inorganic phosphorus in the near bottom layers of the lakes throughout the year is a result of stable stratification, release of ground waters and at the expense of bottom sediments in anoxia conditions.

Composition of bottom sediments provides a historical data on the floodplain lakes. Location and the age of the old channels (recent lakes) within the floodplain and flooding regime govens organic and mineral matter concentration in bottom sediments.

Accumulative floodplain and streambed patterns are more sensitive to natural and man-caused impact and respond to any environmental changes more quickly than confined river valleys. Meander zones are the most sensitive to this impact. However, at present water chemistry of the oxbowlakes within the floodplain and streambed patterns may be classified as apristine, and the lakes water parameters are linked to baseline conditions of the area which could be useful for environmental monitoring of the water bodies within the territory surrounding the protected area.

ACKNOWLEDGMENTS

The financial support from Russian Science Foundation (project 14-17-00155) is acknowledged.

REFERENCES

Alexandrova, D. N., 1973. Microbiology and primary production of Onega Lake. Moscow, "Nauka" Publishing House, 5–83 (in Russian).

Annals of nature of Kerzhensky Nature Reserve for 1997-2014 years, 1998-2015. N. Novgorod, State Nature Biosphere Reserve Kerzhensky Publishing (in Russian).

Bayanov, N. G., 2012. Intensity of productive and destructive processes in old channels of the Kerzhenets River. Izvestiya KGU, ser. *Biology, Agricultural sciences*, No 24, 36–47 (in Russian).

Bayanov, N. G., Krivdina, T. V., 2013. Off-season dynamics of hydrological and hydrochemical properties of the Kerzhenets River and old channels. Proceedings of Russian Academy of Sciences, ser. *Geography*, No 2, 52–68 (in Russian).

Chernov, A. V., 2009. Geography and geo-ecological conditions of riverbeds and floodplains of North Eurasia. Moscow, 684 pp (in Russian).

Chernov, A. V., Korabliova, O. V., 2008. Experience of monitoring channel deformations on wide-floodplain rivers (Kerzhenets River). *Geography and natural resources*, No 2, 158–165 (in Russian).

Guidelines for chemical analysis of the sea and fresh waters in connection with ecological monitoring of fishery water bodies and zones of the World ocean perspective for industry, 2003. Moscow, 202 pp (in Russian).

Ermachenko, L. A., Ermachenko, V. M., 1999. Nuclear and absorption analysis with a graphite furnace. Moscow, 219 pp (in Russian).

Efimova, L. E., Frolova, N. L., 2013. Hydrological monitoring within the limits of conservation areas. *Water: chemistry and ecology*, No 5, 20–28 (in Russian).

Freedman, B. I., Korabliova, O. V., 2001. Geology and topography of Kerzhensky Nature Reserve. Proceedings of the State Nature Reserve Kerzhensky, vol. 1. N. Novgorod, 45–61 (in Russian).

Frolova, N. L., Agaphonova, S. A., Nesterenko, D. P., Povalishnikova, E. S., 2013. Natural flow control in respect of the Volga River water basin in changing climate conditions. *Water industry of Russia*, Moscow, 32–49 (in Russian).

Komarova, N. V., Kamentsev, Y. S. 2006. Practice guidelines for the use of capillary electrophoresis systems "Kapel." Saint-Petersburg, 212 pp (in Russian).

Korabliova, O. V., 2009. Organization of monitoring the dynamics of floodplain natural complexes in Kerzhensky nature reserve. In: digest of international science conference "Modern problems of ecology and environmental education," Orekhovo-Zuevo, 39–40 (in Russian).

Korabliova, O. V., 2010. The morphology of the valley and Kerzhenets River bed deformation. *Geomorphology*, No 2, 69–78 (in Russian).

Korabliova, O. V., 2011. The dynamics of floodplain- channel complexes of southern taiga landscapes of Nizhegorodskoe Zavolzh'ye. *Problems of regional ecology*, No 3, 13–21 (in Russian).

Korabliova, O. V., Chernov, A. V., 2012. The dynamics of floodplain-channel complexes of Nizhegorodskoe Zavolzh'ye (river Kerzhenets). Proceedings of the State Nature Biosphere Reserve "Kerzhensky," vol. 5. N. Novgorod, 196 pp (in Russian).

Mamai, I. I., 2005. Dynamics and functioning of landscapes. MGU Publishing House, 138 pp (in Russian).

Mankish, V. D., Bayanov, N. G., 2001. Hydrological and hydrochemical regime of Kerzhenets River and its tributaries in the middle and lower reaches, In: Proceedings of the National Nature Reserve "Kerzhenskiy," vol. 1. N. Novgorod, 79–108 (in Russian).

Martynova, M. V., 1977. Contribution of bottom sediments to the phosphorus cycle in a water body, In: *Hydrochemical studies in respect of surface and underground waters in Mozhaisk water storage area*. Moscow, 52–61 (in Russian).

Martynova, M. V., 2010. Bottom sediments as components of limnic ecosystems. Moscow, 240 p. (in Russian).

Martynova, M. V., 2012. Concentration of manganese in the alluviums of Mozhaisk water storage. *Water resources*, vol. 39, No 2, 212–217 (in Russian).

Muraviev, A. G., 1999. *Manual for the determination of water quality using field methods*. Saint-Petersburg, 204 pp (in Russian).

Nedogarko, I. V., 2007. Chemistry of bottom sediments in Valday lakes. Reporter of Tver state university, *Ser.s Geography and geoecology*, No 19 [47], 74–85 (in Russian).

Romanenko, V. I., 1985. Microbiological processes of the production and destruction of organic matter in inland water bodies. Leningrad, 294 pp (in Russian).

Shvareva, I. S., Triphonov, K. I., Nikiphorov, A. F., 2014. Hydrochemical monitoring of water ecosystems of the national park "Meschera," *Water industry of Russia*, No 1, 58–74 (in Russian).

Skopintsev, B. A., Goncharova, I. A., 1987. Use of values concerning relations between various properties of organic matter in natural waters for the quality assessment. In: *"Contemporary problems of regional and applied hydrochemistry,"* Leningrad, 95–117 (in Russian).

Sydorchuk, A. B., Panin, A. V., Chernov, A. V., 2000. Water flow and morphology of river beds of the Russian Plain rivers in Holocene and Late Valday time. *Soil erosion and channel processes*, vol. 12. Moscow, MGU Publishing House, 196–230 (in Russian).

In: Riparian Zones
Editor: Oleg S. Pokrovsky
ISBN: 978-1-63484-613-4

Chapter 7

Colloidal Speciation and Size Fractionation of Dissolved Organic Matter and Trace Elements in Small Subarctic Watershed and Its Riparian Zone

S. M. Ilina*[1,2,*]*, S. A. Lapitskiy*[2]*, Yu. V. Alekhin*[2]*,
***O. Yu. Drozdova*[1,2]*, J. Viers*[1] *and O. S. Pokrovsky*[1]**
[1]GET UMR 5563 CNRS, University of Toulouse, France
[2]Moscow State University, Russia

Abstract

Size distribution and speciation of organic matter (OM), major and trace elements (TE) linked to organo-mineral colloids along the landscape continuum "soil water – peat bog – river and its riparian zone – terminal lake" have been investigated during several years in the basin of a stream and river located in the subarctic region (Karelia, Russia) in the summer base-flow period, when the effect of riparian zone is most pronounced. A large volume of natural waters was filtered in the field using cascade filtration and ultrafiltration (UF) with a progressively decreasing pore size (100, 20, 10, 5, 0.8, 0.4, 0.22, 0.1, 0.046, 0.0066 (100 kDa), 0.0031 (10 kDa), and 0.0014 µm (1 kDa)) followed by multi-elemental ICP-MS, dissolved organic carbon (DOC) analysis, UV-vis and size exclusion chromatography measurements. Surrogate parameter SUVA (specific UV absorbance: absorbance at 254 nm normalized for DOC concentration ($L \cdot mg^{-1} \cdot m^{-1}$)) and C/N ratio were also applied for the characterization of OM in filtered and ultrafiltered water from soil solution, bog, river and lake.

In the < 0.22 µm filtrates, there was a systematic decrease in DOC concentration, C/N ratio, SUVA (hydrophobicity and aromaticity) and proportion of colloidal (1 kDa – 0.22 µm) OC along the watershed profile from peat bog soil solution and feeding humic lake which forms the large, 10 to 20 m wide, riparian zone of the upper reaches of the stream. This decrease further followed from the middle course of the stream towards the terminal oligotrophic lake. Within the filtrates and ultrafiltrates of soil solution and

* Email: svetik-msu@ya.ru; s.ilina@brgm.fr.

terminal lake, C/N increased from 100 to 140 and from 7 to 25 for 0.22-10 μm and < 1 kDa fractions, respectively. SUVA, degree of humification, hydrophobicity and aromaticity generally increased from high molecular weight (HMW) to low molecular weight (LMW) fractions, being highest in < 1 kDa fraction.

According to the TE distribution among different size fractions in the continuum soil (bog) – feeding humic lakes – river – terminal (oligotrophic) lake, the following groups of elements could be distinguished: (*i*) elements significantly bound to the colloidal (1 kDa – 0.22 μm) fraction and decreasing the proportion of this fraction from the feeding lake and stream to the terminal oligotrophic lake (Fe, Ti, and U); (*ii*) colloidally dominated (>80%) elements exhibiting similar size fractionation in all studied settings (Y, REEs, Zr, Hf and Th); (*iii*) elements appreciably bound to colloids (20 to 40%) in organic-rich soil and bog waters and decreasing their colloidal fraction to approximately 10% in the oligotrophic lake (Mg, Ca, Sr, Rb, and Mo); and (*iv*) elements linked to the colloidal fraction (40 to 80%) and not demonstrating any systematic variation between different landscape units (Ni, Co, Cu, Cd, Cr, Mn, Zn, and Pb).

Thermodynamic modeling of trace element speciation using available codes demonstrated that complexation with dissolved organic matter (DOM) and adsorption at the surface of ferric colloids can adequately model the observed colloidal speciation of divalent metals. However, the modeling could not describe the distribution of trivalent and tetravalent hydrolysates ($TE^{3+,4+}$) among different size fractions and cannot reproduce the experimentally observed proportion of their colloidal forms. Therefore, coprecipitation with organo-ferric colloids should be considered to account for the partitioning of $TE^{3+,4+}$ between truly dissolved (< 1 kDa) and various colloidal fractions.

Significant enrichement by organo-mineral colloids of the stream water occurs within the riparian zone of the stream where anoxic underground Fe^{2+}-rich waters interacting with basic rocks of the basement meet surface, well oxygenetaed waters rich in dissolved organic matter from the upper reaches of the river, located within the bog zone rich in vegetation leachates.

It follows from the results of this study that autochthonous processes of organic matter fractionation, such as 1) transformation of initially allochthonous soil-derived colloids via photo- and bio-degradation or 2) new organic ligand production by plankton and peryphyton, cannot appreciably affect the distribution of trace elements among various size fractions of colloids and particles along the landscape gradient from soil water to terminal lake. The main features of colloidal chemical composition and size fractionation are therefore acquired during fe-DOC interaction within the riparian zone.

Keywords: speciation, colloids, chemical elements, organic matter, cascade filtration, boreal waters

INTRODUCTION

Boreal regions of the Russian Arctic play a crucial role in transport of elements from continents to the ocean at high latitudes. In view of the importance of these circumpolar zones for our understanding of ecosystem response to global warming, it is very timely to carry out detailed regional studies of TE geochemistry in boreal landscapes. Arctic and subarctic regions are among the most fragile zones in the world due to their low resistance to industrial impact, low productivity of terrestrial biota and limited biological activity over the year (Reuss et al., 1987). Changing climatic conditions for the boreal and arctic regions means that a longer growth period for terrestrial plants will enhance organic production and cycling and possibly increase the reservoir of organic carbon in soils, which is a likely source for organic colloids in surface waters (Dahlqvist et al., 2007). These zones represent one of the most important organic carbon reservoirs in the form of peat bogs, wetlands and soils very rich in organic matter and iron. As a consequence, trace elements in water are likely to be transferred in the form of organic colloids (humic and fulvic acids, microbial exudates, polysaccharides) and organo-mineral entities (Fe and Al hydroxide stabilized by organic matter) (Allard, 2006; Allard and Derenne, 2007; Dahlqvist et al., 2004, 2007; Viers et al., 1997; Ingri and Widerlund, 1994; Gustafsson and Gschwend, 1997; Gustafsson et al., 2000; Andersson et al., 2001; Pokrovsky et al., 2005, 2006, Filella, 2008). The pathways of formation of these colloids are not well understood. It is certain however, that orgaano-ferric colloids can form within the main river bed (hyporheic zone) whereas the main source of DOM which stabilizes Fe hydroxide is the riparian zone of the stream.

Metals speciation in natural waters is of increasing interest and importance because toxicity, bioavailability, environmental mobility, biogeochemical behavior, and potential risk in general are strongly dependent on the chemical species of metals (Fytianos, 2001). To allow an explanation of the diverging degrees of bioavailability and toxicity of different elements, enhanced knowledge is needed about the chemical forms in which the trace elements are present in water. Mobility of the organic carbon (OC) and associated TE during permafrost thawing caused by the climate warming is the main change happening in boreal zones and one of the principal environmental and scientific challenges nowadays. Continuous increase of the runoff of Russian arctic rivers during the last several decades (Serreze et al., 2003) together with the liberation of carbon and metals scavenged to present day by this permafrost (Guo et al., 2004) can modify fluxes of elements exported to the oceans (Hölemann et al., 2005), as well as their speciation in river water and soil solutions.

A detailed chemical and physical description of aqueous colloids and particles in terrestrial surface and ground waters may be of great relevance for future studies in the light of global climate change and warming. Changing climatic conditions are likely to significantly affect annual mean temperatures in the boreal and arctic regions and modify both type and amount of precipitation, thus directly affecting hydro-geological pathways. A longer growth period for terrestrial plants will enhance organic production and cycling and possibly increase the reservoir of organic carbon in soils, which is a likely source for organic colloids in surface waters. At the same time thawing permafrost will release large quantities of organic carbon to stream waters. It is difficult to assess how global climatic changes will affect regional hydro-geochemistry. It can be argued that ground waters with long residence times are sources for surface waters with high concentrations of Fe-rich colloids. In contrast to this are soil waters rapidly washing the uppermost organic-rich layers, e.g., during episodes of intense snowmelt or rainstorms, acting as a source for organic-rich colloids. However, it is clear that the colloidal fraction will continue to be a significant component in surface waters and a carrier for other trace elements. It is therefore important to understand current processes and conditions under which certain types of colloids are present and of significance for the total transport of major and trace elements from the terrestrial environment to the ocean (Dahlqvist et al., 2007).

Study Sites, Sampling and Analyses

Study Site

The study site is located in the Northern Karelia (N 66°, E 30°), ca. 40-60 km south of the Arctic Circle. The site is pristine in terms of local industrial or agricultural pollution and receives only long-range atmospheric impact of metals. The basin of the Vostochniy stream was chosen as the object of study (Figure 1). The stream flows from west to east and empties into lake Tsipringa. It is ca. 1 km long and at a relative altitude of 50 m; the catchment area is 0.95 $km^{2.}$ The bedrock of the catchment comprises amphibolitic gabbroids of the low Proterozoic intruzive (Ilina et al., 2013b).

The climate of the region is mild-cold, transitional between oceanic and continental, with a determinant influence of the Arctic and Northern Atlantics air masses. Average temperature in January is -13°C, and +15°C in July, but extremes can reach -45° to +35°C in the winter and summer periods, respectively. Average annual precipitation ranges between 450 and 550 mm/yr. The snow period lasts from October to April-May with an average thickness of cover of 70 to 80 cm. Our study area is in the most elevated part of Karelia, within a landscape of tectonic denudation hills, plateaus and ridges with an average altitude of 300-400 m, with separate insulated massifs (Maksimova, 1967; Vasyukova et al., 2010).

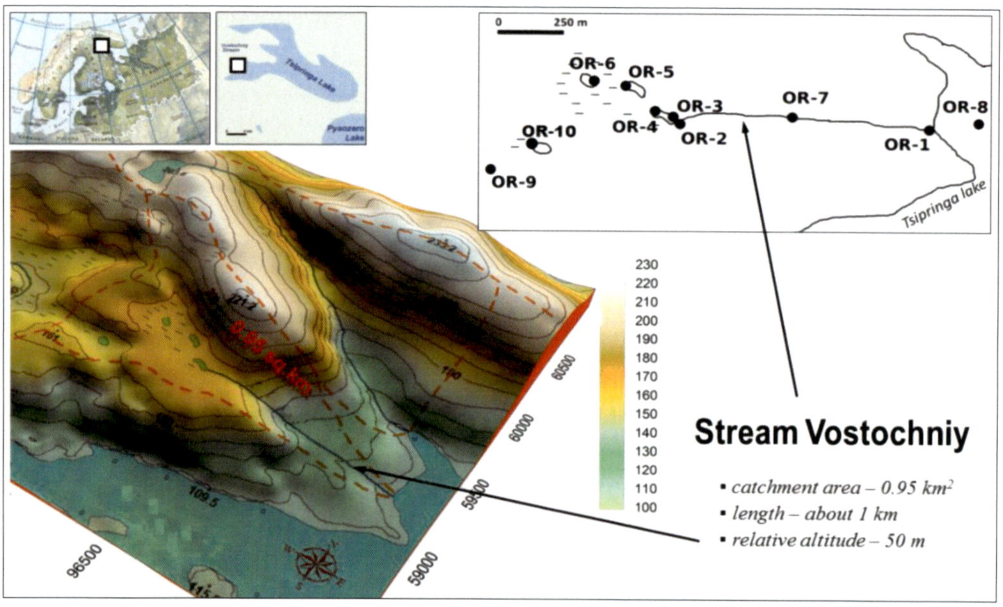

Figure 1. Sampling scheme of a subarctic watershed, its sources and riparian zone.

The composition of the river water in Karelia is determined by the weathering of silicate bedrocks of the Baltic crystalline shield and Quaternary deposits, and the presence of numerous peatlands. Typical values for total dissolved solid (TDS) for the rivers of the region are 15-30 mg/L (Maksimova, 1967; Zakharova et al., 2007); the concentration of river suspended matter is very low. The soil cover of the region is very young and is often absent from ledges of bedrock and steep slopes. Low temperature, in combination with high humidity, is responsible for the slow humification and mineralization of plant residues. Therefore, much OM has accumulated in the form of peat deposits, and on better drained sites, in the form of coarse humus. Predominant soils are illuvial-humic and illuvial-ferruginous-humic podzols. All the types of podzol exhibit a highly acidic reaction and low base saturation of the upper layers. Coniferous forest dominates the vegetation of the region. The main conifers are pine and spruce. The common deciduous trees are birch, aspen and alder. Sparse understory consists of mountain ash and juniper, being dominated by blueberries and cranberries in the shrub layer, and moss in the lower layer. The rocks are usually covered with patches of black, gray, yellow, red, brown crustose lichens.

Sampling, Filtration, Dyalisis

Figure 1 shows a simplified scheme of the Vostochniy stream watershed sites along with the sampling points, with the list of collected water samples in Table 1. The feeding humic lakes of the bog zone (OR-6, OR-5, OR-4), waterlogged shores of the feeding lake (OR-3, OR-2), middle course of the stream (OR-7), its mouth reach (OR-1), interstitial soil solution (OR-9), feeding bog (OR-10) and large clear water terminal lake (OR-8) were sampled in 2008-2013 field seasons during the base flow period. In this landscape continuum, the feeding lake and the bog zone locate dupland the watershed belong to the riparian zone of the Vostochniy stream.

Gravitational soil solution (OR-9) of the peat bog zone feeding the watershed was collected from a depth of 5-10 cm with a piezometer. Large volumes (20–30 L) were collected in pre-cleaned, light-protected PVC bottles for the size fractionation procedure employing 100, 20, 10, 5, 0.8, 0.4, 0.22, 0.1, 0.046, 0.0066 (100 kDa), 0.0031 (10 kDa) and 0.0014 μm (1 kDa) cascade filtration and ultrafiltration conducted directly in the field using a specially prepared polyethylene-covered clean space. The main filtration characteristics are listed in Table 2 and the scheme for the size fractionation procedure is given in Figure 2. The sampling period of this study was always July corresponding to summer baseflow. The most complete data series were collected in 2009 but some additional series of the Vostochniy stream and adjacent surface streams were performed in 2008, 2010, 2011 and 2013. The terminal oligotrophic lake (OR-8) was sampled in 2008, 2009, 2010, 2011, and 2013. The sampled years were different in mean summer months temperature and precipitation as shown in Table 3.

Pre-filtration through 100 μm was performed using nylon net (Fisherbrand). Cascade frontal filtration with a decreasing pore size from 20 to 0.1 μm was performed using a 250 ml vacuum polycarbonate cell (Nalgene) and nylon membranes (Osmonics). Frontal cascade ultrafiltration (UF) in the series 100 → 10 → 1 kDa was performed using a 400 ml polycarbonate cell (Amicon 8400) equipped with a suspended magnet stirring bar located above the filter to prevent clogging during filtration. Vacuum filtration was performed using a portable hand pump and the ultrafiltration was performed at 2-3 bars using a portable automobile pump with a 0.22 μm Teflon filter installed before the Amicon cell. Large volumes of samples were passed through the Lavsan (polyethylene tereph-thalate, PETP) filters of 0.4 μm pore size and 500 cm^2 surface area. Filtration occurred via gravitational flow (0.3-0.5 kPa).

Table 1. Water samples from Vostochniy stream watershed

Sample	Description
OR-9	Soil solution near top feeding lake
OR-10	Feeding bog
OR-6	Top feeding lake surface ca. 150 m^2, depth 2.6 m
OR-5	Middle feeding lake surface ca. 210 m^2, depth 3 m
OR-4	Low feeding lake, surface ca. 200 m^2, depth 2.5 m
OR-3	Waterlogged shore of another low feeding lake, surface area ca. 50 m^2
OR-2	Waterlogged shore of low feeding lake, surface ca. 50 m^2
OR-7	Middle course, 600 m from the mouth
OR-1	Stream, mouth reach
OR-8	Tsipringa lake, 50 m from the mouth reach of the stream

It is known that reproducible and accurate results for size fractionation of DOC require rigorous cleaning and strict sampling protocols (Guo and Santschi, 1996). To this end, before each filtration, the system was cleaned by flushing with EasyPure water, then 3% ultrapure HNO_3, and finally, abundant EasyPure water. Each filter was soaked in EasyPure water for at least 1 day before the experiment and used only once. Preliminary experiments demonstrated that flushing 100 ml of MilliQ water (after 1 day's soaking) through Amicon UF and Nalgen filtration cell with a membrane is sufficient to decrease the OC blank to as low as 0.2-0.5 mg/L, at least an order of magnitude lower than the typical concentration in filtrates and ultrafiltrates.

Table 2. Main filtration characteristics

Pore size	Filer size	Filter material	Filtration pressure	Filtration unit
100 μm	300*300 mm	Nylon	gravity flow	-
20 μm	Ø 37 mm	Nylon	-80 - 0 kPa	Nalgen, 250 ml
10 μm	Ø 37 mm	Nylon	-80 - 0 kPa	Nalgen, 250 ml
5 μm	Ø 37 mm	Nylon	-80 - 0 kPa	Nalgen, 250 ml
0.8 μm	Ø 37 mm	Nylon	-80 - 0 kPa	Nalgen, 250 ml
0.4 μm	2×100*250 mm	Lavsan	130 - 150 kPa	-
0.22 μm	Ø 37 mm	Nylon	-80 - 0 kPa	Nalgen, 250 ml
0.1 μm	Ø 37 mm	Nylon	-80 - 0 kPa	Nalgen, 250 ml
0.046 μm	Ø 37 mm	Lavsan	-80 - 0 kPa	Nalgen, 250 ml
100 kDa	Ø 76 mm	Regenerated cellulose	0 - 100 kPa	Amicon, 8400
10 kDa	Ø 76 mm	Regenerated cellulose	0 - 350 kPa	Amicon, 8400
1 kDa	Ø 76 mm	Regenerated cellulose	0 - 350 kPa	Amicon, 8400

During filtration, the first 50 ml of sample solution were discarded, thereby allowing saturation of the filter surface and collecting vessel prior to filtrate recovery. This greatly decreased the probability of cross contamination during sample filtration, while improving the OC blank. It also provided identical conditions of filtration for all samples and allowed good recovery of colloidal particles. Discussion of the technique and precautions against possible filtration artifacts is given by Viers et al. (1997), Dupré et al. (1999), Pokrovsky and Schott (2002), Pokrovsky et al. (2005, 2006, 2010), Pokrovsky and Shirokova, 2013; Alekhin et al. (2010) and Ilina et al. (2013a).

Dialysis experiments were performed using 20- to 50-ml pre-cleaned dialysis bags placed directly in the river or bog water (in situ dialysis) as described elsewhere (Vasyukova et al. 2010; Pokrovsky et al. 2011, 2012a).

Table 3. The weather characteristics of the samplings periods by the data of meteostation 22217 KANDALAKSHA (25m – 67°09N – 32°21E), www.mundomanz.com

Precipitations, mm	2007	2008	2009	2010	2011	2013
June	42.3	99.0	68.4	72.8	17.4	44.8
July	102.8	69.7	79.7	37.6	90.3	2.3
August	38.8	99.3	105.5	65.0	54.1	-
Year	602.2	631.8	562.2	506.4	494.5	-
Temperature, °C	2007	2008	2009	2010	2011	2013
June	11.1	11.2	10.6	10.4	14.0	13.5
July	14.4	13.7	13.5	16.3	16.0	14.8
August	14.0	10.7	13.4	11.9	11.8	-
Year	1.7	1.2	0.9	0.1	1.9	-

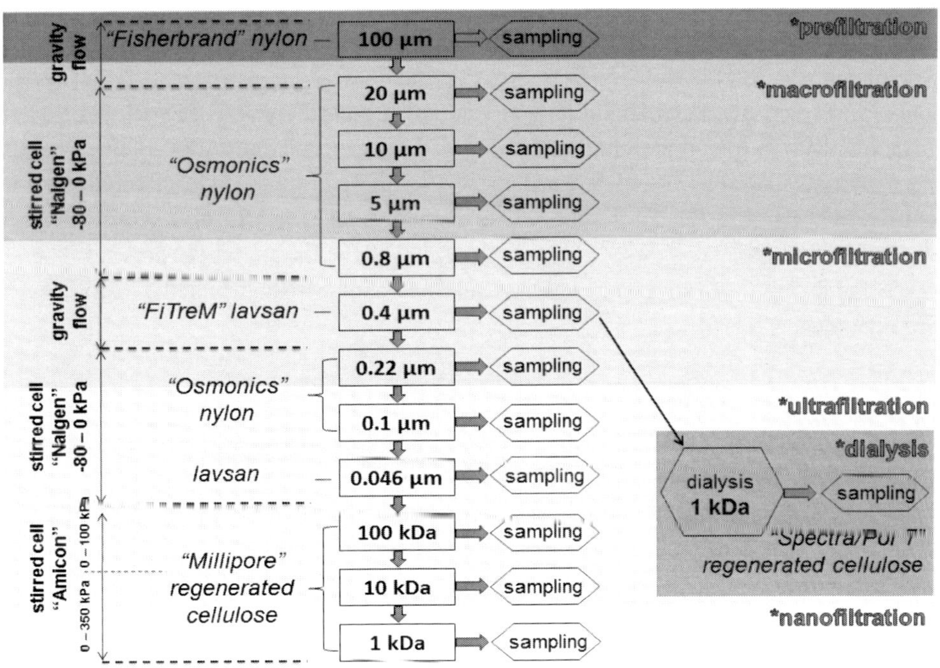

Figure 2. The scheme of cascade filtration or ultrafiltration /dialysis size fractionation procedure used in this study.

Analytical Techniques

Water temperature, pH and conductivity were measured in the field. The pH was measured with an uncertainty of 0.02 using a combination glass electrode calibrated against NIST buffer solutions. Major anion concentrations (Cl^-, SO_4^{2-}) were measured using ion chromatography (Dionex 2000i) with an uncertainty of 2%. Alkalinity was measured in situ by Gran titration with HCl using phenolphthalein as indicator. DOC and N concentration was determined in Toulouse using a Shimadzu CNS Analyzer and in Moscow using Elementar TOC analyzer with an uncertainty of 3% and a detection limit of 0.05 mg/L (Ilina et al., 2014).

Specific UV-absorbance [SUVA, $L \cdot mg^{-1} \cdot m^{-1}$] is absorbance of a given sample at 254 nm divided by the DOC concentration of the sample. The ratio describes the nature of the DOM in terms of hydrophobicity and hydrophilicity; a value > 4 indicates mainly hydrophobic and especially aromatic material, whilst a value < 3 corresponds to the presence of mainly hydrophilic material (Edzwald and Tobiason, 1999; Minor and Stephens, 2008; Matilainen et al., 2011). Several studies have emphasised that good agreement may exist between the ability for OM removal by coagulation and a high SUVA value (Archer and Singer, 2006; Bose and Reckhow, 2007). Absorption of the filtrates at 254 nm was measured in the laboratory within 1 month of sampling, using Specord 50 instrument.

Weight average molecular weight (WAMW) was measured via size exclusion chromatography (SEC) (Hagel, 2001) using an Agilent 1100 chromatographic system (Agilent Technologies, USA) with diode array detector and Ultropac column at 280 nm wavelength (TSK G2000SW; 7.5 x 300 mm; LKB, Sweden). A solution of 0.1 M Na_2HPO_4 buffer (pH 7) and 0.1% sodium dodecyl sulfate was used as effluent. All samples were purified from low molecular weight (LMW) contaminants by elution through a Sephadex G-10 column. The calibration was performed on certified globular protein solutions (Pharmacia Fine Chemicals, Sweden).

Major and trace elements, including Fe, were measured by ICP-MS (7500ce, Agilent Technologies) without preconcentration. Indium and rhenium were used as internal standards. The international geostandard SLRS-4 (Riverine Water Reference Material for Trace Metals certified by the National Research Council of Canada) was used to check the validity and reproducibility of each analysis (see Vasyukova et al. (2010) for analytical details). There was good agreement between our replicated measurements of SLRS-4 and the certified values (relative difference < 15%).

Results and Discussions

General Hydrochemical Parameters

The measured dissolved organic carbon, trace and major elements concentrations, alkalinity, pH and specific conductivity values in the water samples are reported in Table 4.

The waters of the Vostochniy Stream, its riparian zone and the terminal Tsipringa Lake were neutral, with pH values ranging from 6.3 to 7.5, while the waters of waterlogged humic lake and soil solution were acidic, with pH values of 5.8 and 4.3, respectively. The pH decreased by 0.5 units during filtration from 20 μm to 1 kDa. All waters contained a low concentration of total dissolved solids (TDS ≤ 30 mg/L) with the dominance of Ca^{2+} and HCO_3^- ions in rivers and lakes or Ca^{2+}, Cl^- and Na^+ ions in the soil solution (Ilina et al., 2015). The inorganic ion charge balance $((\sum^+-\sum^-)/\sum^+)$ was below 0.1 for all samples except soil solution with a deficit of anions of 0.4-0.5. This deficit was correlated with the DOC concentration in filtrates and ultrafiltrates and was similar to that reported in other surface waters of North Karelia (Vasyukova et al. 2010) that are typical of organic-rich boreal surface waters; the DOC concentrations were 18-22, 20-25 and 140 mg/L in stream, lakes, and soil solutions, respectively (Ilina et al., 2014). The evolution of the DOC concentration along the landscape profile of the Vostochniy Stream (Figure 3) had a systematic decrease of [DOC] down the catchment, from soil solution (OR-9) through the feeding bog (OR-6) and small feeding lakes (OR-5, OR-4, OR-3, OR-2), along the stream itself (OR-7, OR-1) and finally to the terminal Tsipringa Lake (OR-8). The largest decrease was observed between the soil solution (OR-9) and the bog lake (OR-6) and between the mouth of the spring and the terminal clear water lake; the variations in the [DOC] within the upper humic lakes and within the stream were rather small (< 15%).

Table 4. Composition and DOM parameters for 0.22 μm fractions from surface waters of the Vostochniy stream watershed (nd, not determined)

	Soil water	Riparian Zone					Main Stream		Lake
Sample	OR-9	OR-6	OR-5	OR-4	OR-3	OR-2	OR-7	OR-1	OR-8
pH	4.3	6.3	6.6	5.8	6.6	6.3	6.6	6.7	7.5
t, °C	17.2	21.7	20.1	19.9	18.3	18.9	15.7	12.8	17.8
O_2, mg/L	1.9	4.1	nd	nd	4.2	nd	nd	4.3	4.8
R[a], μSm/cm	57.4	20.8	19.1	15.7	16.6	18.1	14.3	14.4	42.5
TDS[b]; mg/L	31.9	10.2	9.7	8.1	7.5	8.2	7.9	8.5	23.3
Na^+, mg/L	1.23	0.98	0.88	0.84	0.72	0.75	0.88	0.96	1.2
Mg^{2+}, mg/L	0.49	0.59	0.61	0.55	0.55	0.51	0.58	0.56	1.6
K^+, mg/L	0.38	0.04	0.04	0.04	0.07	0.01	0.06	0.04	0.80
Ca^{2+}, mg/L	1.3	3.3	2.5	2.0	2.5	2.3	2.0	2.2	5.9
SO_4^{2-}, mg/L	0.06	0.32	1.15	0.97	0.29	1.1	0.84	0.89	0.04
NO_3^-, mg/L	0.12	0.03	0.64	0.61	0.14	0.50	0.64	0.39	0.08
Cl^-, mg/L	0.69	0.33	0.35	0.28	0.38	0.40	0.24	0.42	0.64
HCO_3^-, mg/L	nd	13.4	11.0	9.0	9.7	9.2	9.5	17.4	33.1
$(\sum^+-\sum^-)/\sum^+$), %	65	3.8	3.3	-3.2	5.2	-3.3	-3.1	-3.6	1.3
N, mg/L	0.49	0.33	nd	nd	nd	nd	nd	0.34	0.18
DOC, mg/L	144	18.0	18.5	19.0	19.0	18.0	16.5	16.0	7.0
C/N	104	58	nd	nd	nd	nd	nd	48	24
WAMW[c]	1260	1020	nd	nd	nd	nd	nd	1010	960
SUVA, $L\cdot mg^{-1}\cdot m^{-1}$	4.9	4.2	nd	nd	nd	3.2	nd	4.1	1.1

[a] Specific conductivity;

[b] Total dissolved solid;

[c] Weight average molecular weight.

The concentration of dissolved organic N and C/N decreased systematically along the watershed profile, from soil solution through stream and to the terminal lake (Table 4), and the higher was the DOC concentration, the higher was the C/N ratio. The C/N value for the soil solution was equal to 104, similar to the biomass of coniferous trees (Onstad et al., 2000; Twichell et al., 2002; Tremblay and Benner, 2006), confirming the dominant role of the ligno-cellulose complex of pine and birch litter in the formation of the aqueous OM of peat bog soil water (Guggenberg et al., 1994; See and Bronk, 2005; Tremblay and Benner, 2006). The samples from the upper lakes (OR-6, OR-5, OR-4, OR-3) and the stream (OR-2, OR-7, OR-1) were very similar to each other but drastically different from the bog soil waters. As such, the dominant source of OM in the stream should be bog lakes rather than interstitial peat soil solution. The water of terminal Lake Tsipringa had the lowest C/N value (24), which is typical for aquatic phytoplankton and macrophytes and their humification products (Wolfe et al., 2002). Therefore, the contribution from allochthonous river water and bog water to the DOM pool of this large oligotrophic lake was rather small.

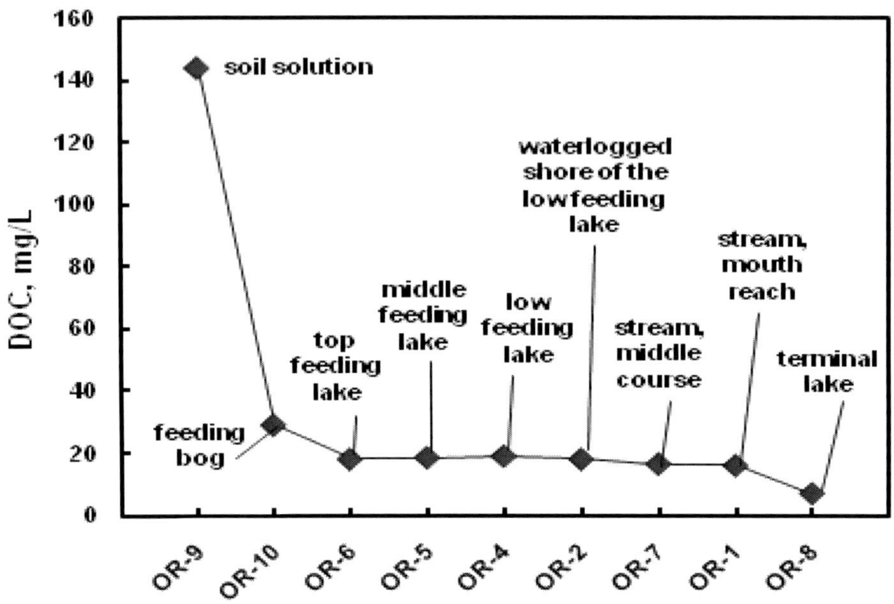

Figure 3. DOC concentration (mg/L) for samples from Vostochniy stream. The solid line is for guiding.

Size Fractionation of DOM

The relative proportion of various size fraction of DOC, from 100 μm to 1 kDa for 5 representative samples (OR-9: soil solution, OR-6: humic feeding lake at the top of the watershed, OR-2: waterlogged shore of another lake, OR-1: mouth reach of the stream, OR-8: terminal clear water lake) is presented at the Figure 4.

In all samples, the mass fraction of low molecular weight (LMW) DOM (<1-10 kDa) dominated the DOC with a significant proportion of high molecular weight (HMW) (10 – 100 kDa) colloids. It is important to note that the molar fraction distribution of different size organic components was dramatically different from that of the mass fraction. Since the molar weight is proportional to the third degree of molecular diameter, the difference in molecular mass between the association of molecules of 0.2 μm diameter and a molecule of 10 kDa diameter (ca. 3 nm) reaches 6 orders of magnitude, being equal to 8×10^{-3} and 8×10^{-9} μm^3, respectively. As a result, the molar concentration of LMW fulvic acids was several orders of magnitude higher that of the HMW OM. The molar fraction and molar concentration of LMW (<1-10 kDa) fulvic acids therefore dominated the DOC, regardless of its allochthonous or autochthonous nature.

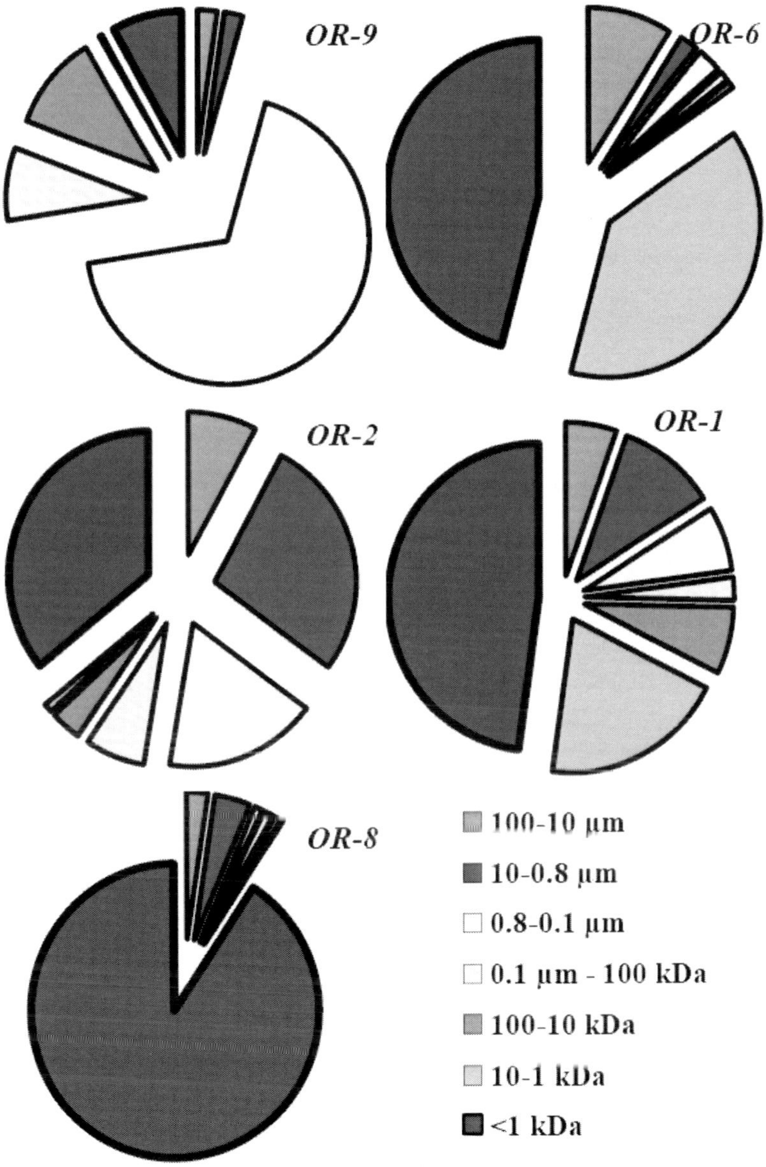

Figure 4. OC mass fractionation in the samples of Vostochniy stream.

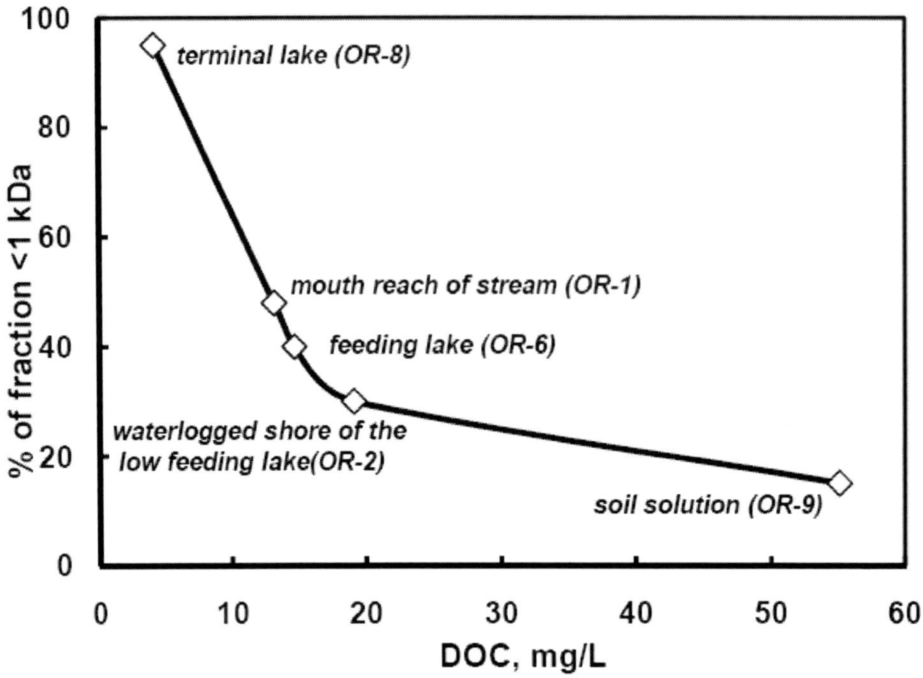

Figure 5. Plot the % of < 1 kDa form as a function of DOC in 0.22 μm fraction. The line is for guiding purposes.

It can be seen from Figure 5 that the fraction LMW DOM increased with downstream transit in parallel with a decrease in specific UV absorbance (Table 4). This would suggest that LMW molecules are linked to relatively low SUVA values, presumably because of low aromatic contents. In contrast, the SUVA measurements that were performed on filtrates representing different molecular size fractions (Table 4) consistently showed that LMW DOM fractions were linked to high SUVA, e.g., high aromaticity. In fact, SUVA often increased sharply with decreasing filter pore size. This paradox stems from the huge difference in SUVA (and aromatic OM content) between soil solution OR-9 and oligotrophic lake OR-8. In fact, the lowest SUVA observed in the oligotrophic lake (sample OR-8) was independent on the poresize fraction. This strongly suggests that the main transformation of LMW, high SUVA fraction occurs between soil solution and first surface water reservoir, an intermediate small lake feeding the stream or the mouth reach of the stream. In large oligotrophic lake, the majority of the soil (humic) aromatic fraction of LMW is removed via autochthonous process in the water column, presumably photo- and bio-degradation. It follows that similar and dramatic transformation of DOM occurs precisely in the riparian zone along the main stream course, where the DOM from interstitial soil solution meet the oxygenated river waters.

Along the landscape profile of the watershed, with the decrease of the width of riparian zone, the relative proportion of LMW (< 1 kDa) OC significantly increased from soil solution to stream and finally, to the terminal clear water lake, following the decrease in concentration of conventionally dissolved $DOC_{<0.22\ \mu m}$ (Figure 5). The mouth reach of the stream (OR-1) exhibited around 50% of DOC in < 1 kDa form, a typical value for other boreal landscapes (Prokushkin et al., 2011; Guo et al., 2004). The proportion of LMW < 1 kDa OC was also measured by equilibrium dialysis procedure in 2008 and 2009; the obtained results (25 to 55% of LMW fraction for the intermediate samples of the Vostochniy stream) were in agreement (± 5 to 10%) with the ultrafiltration results of 2009.

The DOM size fraction evolution in Figure 5 corresponds to progressive depolymerization of HMW soil humic acids whether via heterotrophic aerobic bacterioplankton activity, as known for other boreal landscapes (Tranvik, 1988; Pokrovsky et al., 2011). Alternatively, it could be photodegradation in the feedstock lakes and stream channel (De Haan, 1993; Zuo and Jones, 1997; Wang et al., 2001; Albinet et al., 2010; Thorn at al., 2010). The increase in the proportion of bio-mineralized or photodegraded products of allochthonous HMW organic components may be responsible for the increase in the proportion of LMW ligands in the continuum soil solution – feeding humic lake – stream – terminal oligotrophic lake. It can be hypothesized therefore that, after leaving the soil, the DOM is subjected to progressive degradation in stagnant water reservoirs. It is known from other boreal, permafrost-bearing aquatic systems that the longer the residence time of allochthonous DOM in the system, the higher the proportion of LMW products of photodegradation and bacterioplankton DOM transformation (Pokrovsky et al.,2011, 2013; Shirokova et al., 2013). An additional factor responsible for the trend in Figure 5 may be the summer phytoplankton activity, which produces LMW exometabolites dominating the speciation of OC in open water systems, especially in the large clearwater lake.

Therefore, we may tentatively attribute the increase of $LMW_{<1\ kDa}$ fraction to the appearance of small-size autochthonous OC in the form of phytoplankton exometabolites accompanied by the consumption of allochthonous soil-derived OM by heterotrophic bacterioplankton. Similar to Siberian thaw lakes (Pokrovsky et al., 2011), in addition to exometabolites production, the higher proportion of LMW carbon in the largest lake can be a result of the decreasing input of soil and bog-derived organic matter to this lake, mostly due to a large water-body in relation to the length of the shoreline or the watershed area. Therefore, the average residence time of the allochthonous organic macromolecules in these lakes is longer, exposing them to degradation by the bacterioplankton for a longer time (e.g., Amon and Benner, 1996a, b). As a result, dominant organic macromolecules are smaller in size as compared to mainly allochthonous DOM of small lakes located within the bog or the DOM of the forest stream. In a similar manner, the photo-oxidation of DOM is most pronounced in clear water (oligotrophic) lake, notably due to much longer residence time of organic ligands in this lake. Given that photochemical processes degrade a part of the refractory pool of DOM that is not readily available to bacteria (Amon and Benner, 1996b), the importance of photodegradation in boreal lakes deserve further investigation.

Chromatograms of the soil solution and stream water samples had a unimodal distribution (Figure 6). In accord with the data on molecular size distribution, the SEC results demonstrated the absolute molar dominance of 1 kDa size compounds (Figure 6, Table 4). Within the range 0.2 µm to 10 kDa (2.8 nm) and a concentration of OC 5.2 mg l^{-1}, 8 molecules of mass 5,200,000 Da are equivalent to the presence of 4.4×10^6 molecules of mass 10 kDa. This strongly suggests a potentially high importance of the LMW fraction in metal complexation reactions in boreal aquatic environments. The dominance of 1000 Da nominal molecular mass in the stream water is in agreement with results on most rivers of the Arctic Ocean basin (cf. Dittmar and Kattner, 2003).

Despite the dominant consensus that the bioavailability of organic carbon decreases as its size decreases (Amon and Benner, 1996a, b), with the LMW fraction in Arctic rivers being more refractory than the colloidal (1 kDa–0.45 μm) fraction (Guo and MacDonald, 2006), potential bioavailability of LMW organic carbon may be still quite high. Indeed, the LMW complexes (<1 kDa) or conventionally dissolved species are bioavailable in the case of passive diffusion through the biological membranes as the pore sizes of the cell walls transport channels (10–30 Å in bacteria, 35–50 Å in plant cells (Carpita et al., 1979; Trias et al., 1992; Colombini, 1980) and that of the 1 kDa dialysis membrane (1–3 nm) are comparable. However, in-situ biodegradation experiments of various size fraction organic matter are necessary to constrain the bioavailability in boreal subarctic settings.

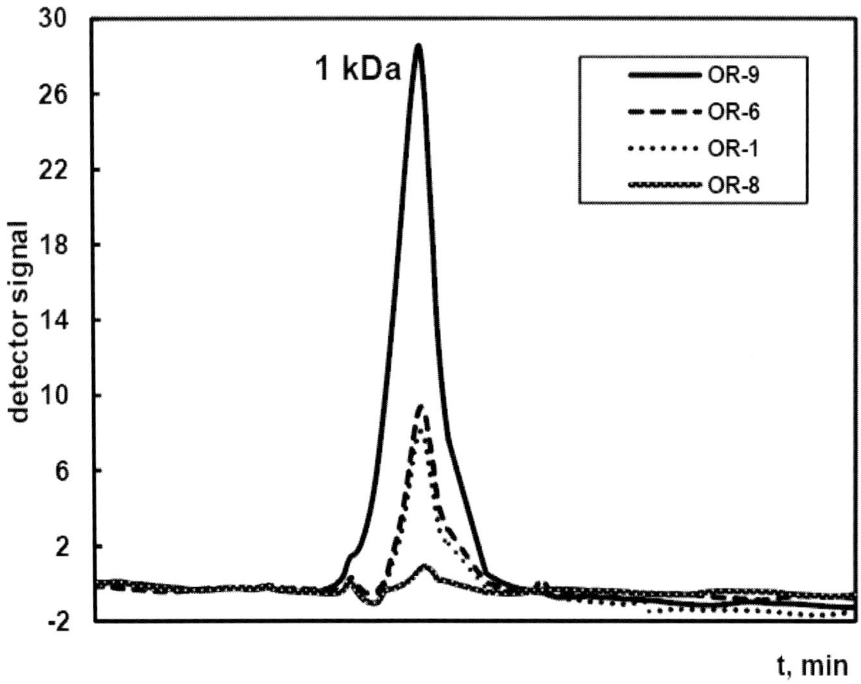

Figure 6. Weight average molecular weight (WAMW) distribution in filtrates (0.22 μm) of samples (OR-9 – soil solution, OR-6 – top feeding lake, OR-1 – mouth reach of the stream, OR-8 – terminal lake), determined from size exclusion chromatography.

Size Fractionation of Chemical Elements

Major and Trace Element Distribution within the Catchment Profile in the Main Filtrates

Iron

There was a two-orders-of-magnitude decrease in the Fe concentration (< 0.22 μm) along the Vostonychiy Stream profile, from the soil solution (2600 μg/L, OR-9) through the feeding lake (125 μg/L, OR-6) and downstream (170 μg/L, OR-2, OR-1) towards the terminal lake (10 μg/L, OR-8). An extremely high concentration of total dissolved Fe in the soil solution probably reflects the presence of soluble Fe(II) in this partially anoxic solution (40% O_2 saturation). Therefore, the most pronounced concentration decreases were observed between the soil solution and an adjacent small lake and between the stream mouth and an oligotrophic lake. The variations in the Fe concentration between the top feeding lake (OR-6) and the stream (OR-1) were insignificant ($p > 0.05$). It could be hypothesized that the largest change in Fe concentration and speciation occurs precisely within the riparian zone of the stream, located at the border of two highly contrasting aquatic environment: suboxic soil solution and oxygenated stream water.

Alkali and Alkaline-Earth Metals, Neutral Molecules and Oxyanions

The behavior of all alkali metals (Na, K, Rb) along the profile of the Vostochniy Stream was uniform, with maximal concentrations observed in the large terminal lake (OR-8) and soil solution (OR-9) and minimal concentrations observed in the feeding bog lake (OR-2). Down the landscape profile, there was a decrease in the concentrations from OR-9 to OR-2, which was followed by an increase from OR-2 to OR-8 along the stream flow. Unlike other alkali metals, the Cs concentration decreased from OR-1 to OR-8. Likewise, the alkali, Ca, Mg, Sr and Ba concentrations decreased from the soil solution and bogs towards the terminal lake; there was a pronounced increase in the concentrations of these elements in the river mouth compared to the bog and lakes of the upper reaches. Such a pattern of highly soluble elements probably reflects the progressive increase in the input of the groundwater; as the depth of the water body increased, the impact of the water – rock interaction at a depth that enriched the fluids in highly mobile alkali and alkaline earths also increased.

The evolution of the concentration of the elements present as neutral molecules and oxy-anions (Si, V, Mo, Ge, W, As, and Sb) in both studied watershed profiles did not demonstrate any systematic trend within the watershed continuum; the concentrations varied by factors of 3 to 10 without any detectable influence of the landscape context, DOC or Fe concentration. Apparently, these elements are not strongly influenced by the water-rock interaction or by the atmospheric deposition via accumulation in the mosses of the bog zone and subsequent release. Therefore, it can be concluded that alkali and alkaline-earths are least affected by the processes occurring in the riparian zone.

Insoluble Trivalent and Tetravalent Elements

Within the Vostochniy watershed, the Al concentration steadily decreased by a factor of 2 downstream from the soil solution (OR-9) towards the river mouth (OR-1) and further by a factor of 6 in the terminal lake (OR-8). The Ga concentration varied non-systematically between 0.004 and 0.02 µg/, with minimal values observed in the humic lakes (OR-2 and OR-6). Likewise, the REE concentrations were similar among all samples of the Vostochniy watershed, with values in the terminal lake, OR-8, being a factor of 4 to 5 smaller than the others.

The concentration of tetravalent elements (Ti, Zr, Hf, and Th) evolved in a manner similar to that of REE in the Vostochniy Stream. The concentrations of the tetravalent elements (Ti, Zr, Hf, and Th) remained fairly constant (±10%) between OR-9 and OR-1, decreasing by 30-50% in the terminal lake, OR-8.

Other Trace Elements: V, Mn, Ni, Cr, Co, Cu, Zn, Pb, Cd, and U

The changes in the V, Ni, Cr, Co, Cu, and Mn concentrations along the watershed of the Vostochniy Stream were rather similar; the concentrations decreased by 30-80% from OR-9 to OR-6 and then remained constant until the stream mouth (OR-1), strongly decreasing towards the terminal lakes OR-8, similar to the behavior of Fe and organic carbon. The concentrations of Cd, Zn, Co, Mn and Pb do not exhibit any systematic variations along the Vostochniy watershed profile. Other metal micronutrients or biologically relevant elements (V, Ni, Cr, Cu, and Zn) exhibited a small evolution of the total dissolved concentration. The uranium concentration systematically increased by a factor of 8 along the profile of the Vostochniy Stream, from the soil solution to the terminal lake.

Element Distribution among Different Size Fractions

Iron

The removal of Fe during cascade filtration and ultrafiltration occurred in two steps, and the maximal decrease occurred between 1 and 0.1 μm and below 100 kDa, which may correspond to the size of Fe-rich coarse colloids and LMW organic complexes, respectively (Figure 7). The correlation of Fe with DOC in ultrafiltrates series was not significant at $p < 0.05$; R^2 ranged from < 0.60 to 0.83 (Table 5).

Typically, a significant drop in the Fe concentration, notably in the region of coarse colloids, was not accompanied by any significant DOC concentration change, and even in the region of LMW complexes, the DOC concentration only decreased by a factor of 1.5-2 with up to a 10× Fe concentration decrease.

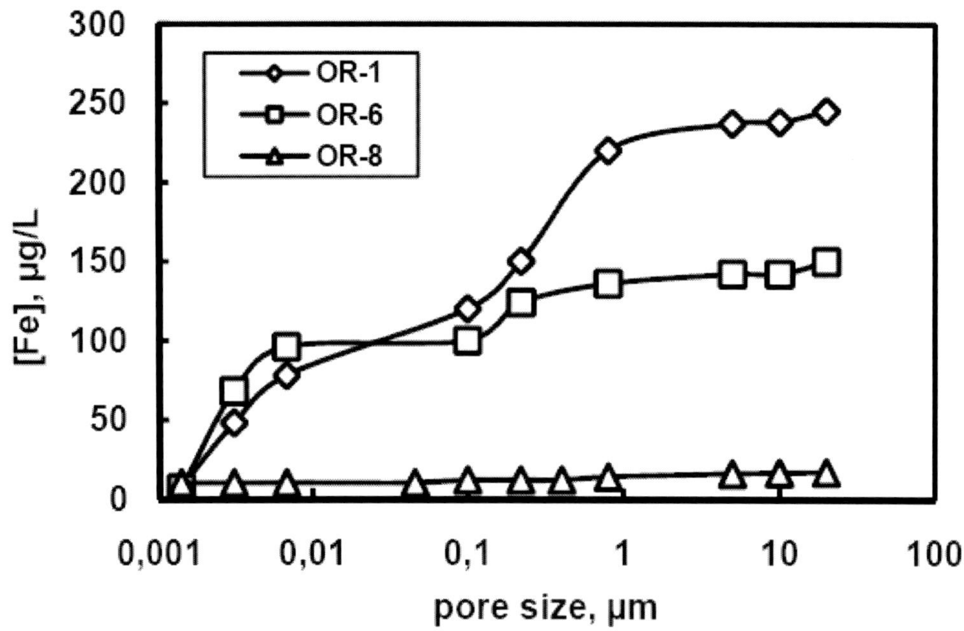

Figure 7. Iron distribution in the filtrates of of Vostochniy stream (OR-1 – middle course of the stream, OR-6 – upper feeding lake, OR-8 – terminal lake) watershed. The waterlogged shore can be considered as most affected by the processes in the riparian zone.

Table 5. Correlation coefficients (R^2) of TE with Fe and OC in filtrates and ultrafiltrates

	TE-Fe (R^2)				TE-DOC (R^2)			
	OR-9	OR-6	OR-1	OR-8	OR-9	OR-6	OR-1	OR-8
Al	0.99	0.76	0.99	0.81	<0.60	0.91	0.51	<0.60
Ti	0.97	0.82	0.95	<0.60	<0.60	0.87	0.6	<0.60
V	0.97	0.82	0.95	<0.60	<0.60	0.68	<0.60	<0.60
Cr	0.63	<0.60	0.71	<0.60	<0.60	0.79	<0.60	<0.60
Mn	0.69	0.88	0.98	<0.60	<0.60	0.87	0.64	<0.60
Fe	-	-	-	-	<0.60	0.83	<0.60	0.77
Co	0.87	<0.60	0.94	<0.60	<0.60	0.7	<0.60	0.82
Ni	0.88	0.6	<0.60	<0.60	<0.60	0.88	<0.60	<0.60
Cu	0.72	<0.60	0.72	<0.60	<0.60	0.88	<0.60	<0.60
Zn	<0.60	<0.60	<0.60	<0.60	<0.60	<0.60	<0.60	<0.60
Ga	0.8	<0.60	<0.60	<0.60	<0.60	<0.60	<0.60	<0.60
Ge	<0.60	<0.60	<0.60	<0.60	<0.60	<0.60	<0.60	<0.60
As	<0.60	<0.60	0.73	<0.60	<0.60	0.75	0.84	<0.60
Rb	<0.60	<0.60	<0.60	<0.60	<0.60	<0.60	<0.60	<0.60
Sr	0.88	0.69	<0.60	0.6	<0.60	0.89	<0.60	<0.60
Y	0.97	0.83	0.89	0.71	<0.60	0.95	<0.60	<0.60
Zr	0.81	<0.60	0.78	<0.60	<0.60	0.84	0.68	<0.60
Mo	<0.60	<0.60	<0.60	<0.60	<0.60	<0.60	<0.60	<0.60
Cd	<0.60	0.91	<0.60	<0.60	<0.60	0.74	<0.60	<0.60
Sn	0.76	<0.60	<0.60	<0.60	<0.60	<0.60	<0.60	<0.60
Cs	<0.60	<0.60	<0.60	<0.60	<0.60	<0.60	<0.60	<0.60
Ba	0.96	0.75	<0.60	<0.60	<0.60	0.88	<0.60	<0.60
La	0.97	0.88	0.94	0.81	<0.60	0.9	0.67	0.65
Ce	0.99	0.87	0.96	0.85	<0.60	0.89	0.72	0.69
Pr	0.94	0.85	0.96	0.8	<0.60	0.87	0.73	0.63
Nd	0.97	0.83	0.91	0.76	<0.60	0.9	0.71	0.64
Sm	0.82	0.78	0.93	0.6	<0.60	0.86	0.83	<0.60
Eu	0.68	0.76	0.71	<0.60	<0.60	0.81	0.7	<0.60
Gd	0.66	0.6	0.69	<0.60	<0.60	<0.60	0.75	<0.60
Tb	0.76	0.75	0.62	<0.60	<0.60	0.85	<0.60	<0.60
Dy	0.75	0.84	0.72	<0.60	<0.60	0.88	<0.60	<0.60
Ho	0.66	0.75	<0.60	<0.60	<0.60	0.8	0.6	<0.60
Er	0.8	0.81	0.64	<0.60	<0.60	0.94	<0.60	<0.60
Tm	<0.60	<0.60	0.7	<0.60	<0.60	0.86	<0.60	<0.60
Yb	<0.60	<0.60	<0.60	<0.60	<0.60	0.89	<0.60	<0.60
Lu	<0.60	<0.60	<0.60	<0.60	<0.60	0.78	<0.60	<0.60
Hf	<0.60	<0.60	<0.60	<0.60	<0.60	<0.60	<0.60	<0.60
W	<0.60	<0.60	<0.60	<0.60	<0.60	<0.60	<0.60	<0.60
Pb	0.79	<0.60	0.95	0.97	<0.60	0.72	0.75	0.7
Th	<0.60	<0.60	<0.60	<0.60	<0.60	0.91	<0.60	<0.60
U	<0.60	<0.60	0.68	<0.60	<0.60	0.92	<0.60	<0.60

Elements Not Affected by the Filtration and Ultrafiltration Procedure (<10-20% in Colloids)

The alkaline metals (Na, K, Rb, and Cs) were not affected by the ultrafiltration and did not exhibit any correlation ($p > 0.05$) with Fe or DOC because they were largely present as ionic, non-complexed species. Additionally, the alkaline-earth metals did not change the concentration during filtration or decrease in concentration by 10-20% after filtration through 1-10 kDa filters. The latter decrease often coincided with that of organic carbon rather than Fe, suggesting the presence of weak colloidal organic complexes of these elements.

Neutral molecules and oxyanions (Si, Ge, As, Mo, Sb, and W) demonstrated the lack of concentration variation in the full series of filtrates and ultrafiltrates, over 5 orders of magnitude in the size fraction, from 100 μm to 1 kDa. There was no correlation ($R^2 < 0.6$) between the concentrations of these elements with Fe or OC in the ultrafiltrates ($p > 0.05$). This strongly suggests the absence not only of colloids and LMW complexes but also of large size silicate particles, suspended mineral material or phytoliths capable transporting these elements in surface waters.

Elements Affected by the Size Separation Procedure: Trivalent and Tetravalent Hydrolysates

In the course of the filtration and ultrafiltration procedure, the aluminum concentration was highly correlated with the Fe concentration in the filtrates and ultrafiltrates ($R^2 = 0.75$-0.99), whereas the correlation with OC is only detectable in fractions < 0.22 μm ($R^2 = 0.50$-0.91). It is fairly possible that this reflects the presence of both inorganic Fe-Al-rich oxyhydroxides (stabilized by organic matter) and Al-organic matter colloids and LMW complexes; the latter was also demonstrated by Dia et al. (2000). The correlation between Ga and Fe was rather poor in the filtrates of river waters ($R^2 = 0.21$-0.73), but it was much better in the feeding bog (KAR-2, $R^2 = 0.99$); the correlation of this element with OC was rather weak ($R^2 < 0.6$).

Among the insoluble tetravalent elements, Ti was well correlated with Fe in the filtrates and ultrafiltrates ($R^2 = 0.98$), especially in soil solutions; Th was also correlated with OC ($R^2 = 0.7$ to 0.9), although this dependence is less pronounced than that with Fe. A good correlation of the Th with DOC in different size fraction filtrates in the groundwaters was also demonstrated by Dia et al. (2000).

Other Trace Elements: V, Mn, Ni, Cr, Co, Cu, Zn, Pb, Cd, and U

In filtrates and ultrafiltrates, Mn, Co and V were much better correlated with Fe compared to DOC (Table 5). Cr, Ni, Cu, Zn, Cu, and Cd did not exhibit any significant correlation with Fe or DOC. A link between Fe and Pb is observed in the results from the soil solution and the Vostochniy stream and lake (OR-1, OR-8, OR-9). Vanadium was also linked to Fe in samples OR-1, OR-6 and OR-9, which did not exhibit any connection to DOC concentration in the filtrates and ultrafiltrates. Uranium was was not correlated with Fe or OC in the samples from the Vostochniy Stream watershed.

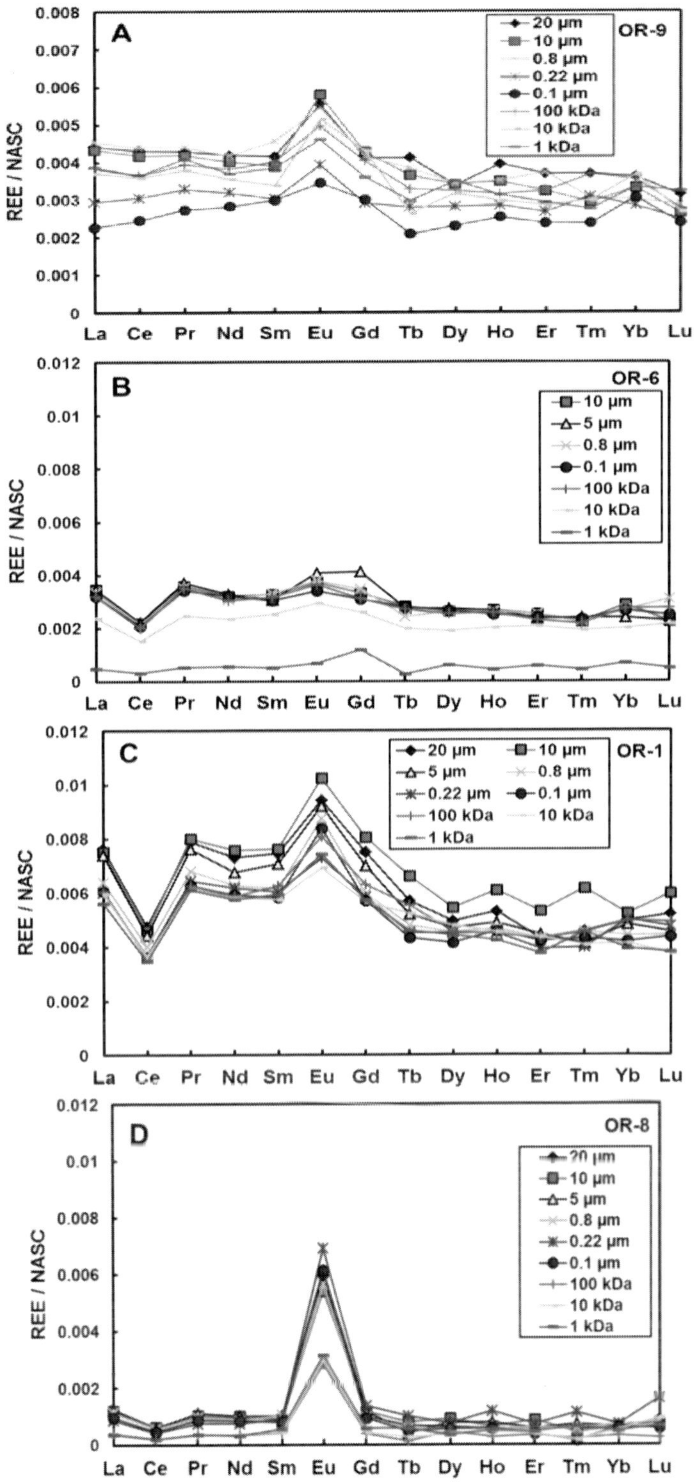

Figure 8. REE patterns normalized to NASC for filtrates and ultrafiltrates of Vostochniy stream watershed (A – soil solution, B – upper feeding lake, C – mouth reach of the stream, D – terminal lake Tsipringa).

Geochemistry and Size Fractionation of Rare Earth Elements (REE)

Further insights on size fractionation of REE can be assessed from REE spectra (concentrations normalized to NASC) for various filtrates and ultrafiltrates of both studied watersheds (Figure 8 A-D). The following typical features can be noted. The REE spectra of soil solution (OR-9, Figure 8 A) is flat without Ce minimum, typical for anoxic or suboxic aquatic environments (Sholkovitz et al., 1992; Sholkovitz, 1992, 1993, 1995; German et al., 1995; Sultan and Shazili, 2009). Oxygenated humic lake (OR-6) also exhibits flat spectrum but the Ce minimum becomes visible in HMW colloids > 10 kDa (Figure 8 B). Vostochniy stream filtration series (OR-1) demonstrates strongly pronounced Ce minimum detectable down to < 1 kDa ultrafiltrates and Eu maximum, the latter is also visible in all filtrates of the terminal Tsipringa lake (OR-8, see Figure 8 D).

It is known that the presence of Ce minimum is tightly linked to its oxidation on Fe oxyhydroxide colloids and particles as follows from numerous previous studies (Sholkovitz, 1992, 1993, 1995; Ingri et al., 2000; Pourret et al., 2008). For this minimum to be pronounced in all filtration and ultrafiltration series, high dissolved Fe concentrations and O_2-saturated environments are necessary. These conditions are met for samples OR-1, OR-6 and OR-8. The Ce minima clearly persists in 1 kDa ultrafiltrates of Vostochniy stream (OR-1, Figure 8 C). This confirms the importance of Fe-rich colloidal and dissolved carriers in creating Ce anomaly in oxygenated surface waters. Note that the fractionation between HREE and LREE disappears for all $LMW_{< 1\ kDa}$ fractions having Fe concentration lower than 100 μg/L.

Compared to bog soil solution (OR-9), the humic lake feeding the stream Vostochniy and receiving its water from adjacent bogs (OR-6) exhibit much lower REE concentrations in LMW fraction and the absence of Ce anomaly in 1 kDa ultrafiltrates (Figure 8 B). The most likely cause for such a transformation of REE spectrum between bog water and lake waters is the deposition of Fe oxy(hydr)oxides in this oxygenated lake via coagulation induced by gravitational sedimentation, photo- and bio-degradation of organo-mineral colloids. Because colloidal Fe oxy(hydr)oxides are the main carriers of REE in studied waters, Fe-poor LMW organic fraction does not have Ce anomaly and exhibits an order of magnitude lower concentration of all REE.

Strong positive Eu anomaly in all filtrates and ultrafiltrates of large terminal clear water lake (OR-8, Figure 8 D) may be linked to its feeding by groundwaters bearing the signature of basic rocks. Note that this Eu anomaly is also visible in soil waters located within the basic rocks (OR-9, Figure 8 A).

Thermodynamic Modeling of NOM and Metal Speciation in Colloids

Element speciation in the presence of major colloidal constituents was assessed using the Visual MINTEQ computer code (Gustafsson 2011, version 3.0 for Windows) in conjunction with a database and the NICA-Donnan humic ion binding model (Benedetti et al. 1995; Kinniburgh et al. 1999; Milne et al. 2001, 2003) as well as hydrous ferric oxide with the diffuse layer model (HFO) (Dzombak and Morel 1990). The input parameters of the model were pH, alkalinity, anions, cations, total TE, DOC concentrations and the respective levels in the suspended particulate matter (particulate organic carbon (POC) and hydrous ferric oxide (HFO)). For this model, we assumed that the $DOM_{< 1\ kDa}$ fraction consists of fulvic acid (FA) and $DOM_{0.2\ \mu m\ -\ 1\ kDa}$ fraction consists of humic acid (HA). The suspended organic matter

(SOM) consists of 100% of HA, and its concentration was calculated as the difference between the total OM (< 100 μm) and $DOM_{<0.2\mu m}$. We considered a ratio of the active OM to DOC and POC of 1.65 (Sjöstedt et al., 2010).

The results of cascade filtration show that in the studied samples, the inorganic Fe colloids were mostly present in the fraction of > 0.1 μm; therefore, the concentrations of HFO were calculated as a difference between the total Fe (< 100 μm) and the Fe in the < 0.1 μm filtrate. Speciation calculations were performed for the samples (OR-8, OR-1, OR-6 and OR-9) of the Vostochniy watershed (Table 1, Figure 1). The results of the speciation calculation are presented in Figure 9 and described below. The model allowed for prediction of Na, K, Cs, Mg, Ca, Sr, Ba, Al, Fe, REE (Y, La, Ce, Nd, Dy, and Yb), Th, Mn, Co, Ni, Cu, Cd, Pb, U, Cl^-, SO_4^{2-}, and NO_3^- fractions complexed with DOM, adsorbed onto Fe oxy(hydr)oxide, adsorbed on particulate OM (> 0.22 μm) and present as free (inorganic) species.

In all samples, the major anions exist as free species, which is in agreement with the UF results. Alkali metals (Na, K, and Cs) are weakly (electrostatically) bound to DOM and have very few (≤ 5%) organic complexes. This is also in agreement with the UF results of all samples of the Vostochniy watershed. Contrasting speciation is observed in the soil solution for which the model predicts that Na, K and Cs are electrostatically bound to DOM (7-8%) and electrostatically bound to solid organic matter (SOM, >0.2 μm) (2-3%), while the remaining 90% is in the form of free ions.

All divalent cations (Mg, Ca, Sr, and Ba) have similar behavior and are complexed, predominantly by electrostatic bonds, with DOM at 55-65, 45-50, 62-66, and 6%, and with SOM at 15-30, 2-4, 5-9, and 0% for OR-9, OR-6, OR-1 and OR-8, respectively. The remaining parts are in the form of free ions.

Aluminum is 100, 95, 73 and 94% complexed with OM in the soil solution (OR-9), small feeding lake (OR-6), stream mouth (OR-1) and terminal lake (OR-8), respectively. The bonds are predominantly due to the phenolic functional groups. Dissolved iron is completely complexed by DOM in the water samples (OR-6, OR-1, and OR-8), which is mainly with the phenolic functional groups bonds, and 88% complexed by OM (DOM and SOM) in the soil solution (OR-9). The rare-earth elements are 30-99% complexed with OM; furthermore, they are 30-80% bound with DOM and 10-65% bound with SOM; the nature of the bonds is predominantly electrostatic. Contrasting speciation is observed in the terminal lake (OR-8), where REE are mainly in the form of free ions (up to 90%), and at 10-45% is absorbed onto HFO. The adsorption of Yb onto Fe oxy(hydro)oxide varies from 4 (for OR-6) to 45% (OR-8).

For the samples from the Vostochniy watershed (OR-9, OR-6, and OR-1), the divalent metals Mn, Co, Ni and Cd exhibited similar behavior and were present as OM complexes at 12-20, 9-12, 2-3 and 0-1%, respectively; furthermore, in the soil solution sample, part of the SOM complexes predominates. Different speciation is observed for the terminal lake (OR-8), where 78, 83, 44 and 4% of the Mn, Co, Ni and Cd are in the free ion form, respectively, and the remaining portion of these metals are in the form of OM complexes. The nature of the OM complexes for all samples is multiple and includes both electrostatic interactions and carboxylic and phenolic functional groups bonding. All samples exhibit similar speciation of Cu, Pb, Th and U with complete (> 99%) binding with OM (as adsorbed on SOM or complexed with DOM), which is principally due to the carboxylic and phenolic functional group bonds. Overall, the soil solution (OR-9) is the richest in organic complexes (including complexes with SOM), and the waters of Tsipringa Lake (OR-8) are significantly depleted in OM complexes compared to the samples of the Vostochniy Stream.

There was reasonable agreement between the predicted % of TE bound to colloids in the form of both organic complexes and adsorbed on Fe hydroxides, and the experimentally assessed fraction of colloids using dialysis or ultrafiltration. Al, Fe, Mn, Co, Ni, Cu, Cd, Pb and U exhibit reasonable agreement between the experimental and theoretical colloidal content (within ± 20-30% difference). In contrast, REE and Th demonstrate significant discordance between the theoretical and experimental colloidal content (> 30%) in all samples except OR-6. This sample contained the lowest Fe concentration. We hypothesize that the reason for the observed disagreement is underestimation by the model of the amount of TE bound inside organo-ferric colloids (coprecipitated with Fe oxyhydroxide).

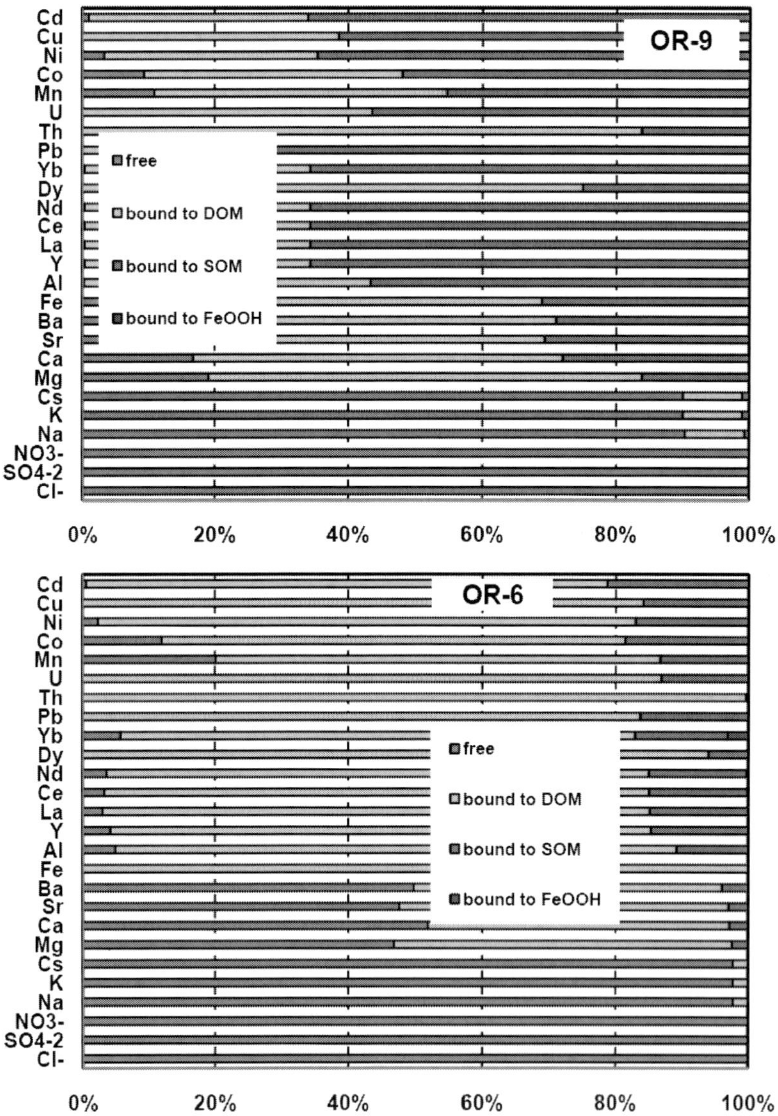

Figure 9. (Continued).

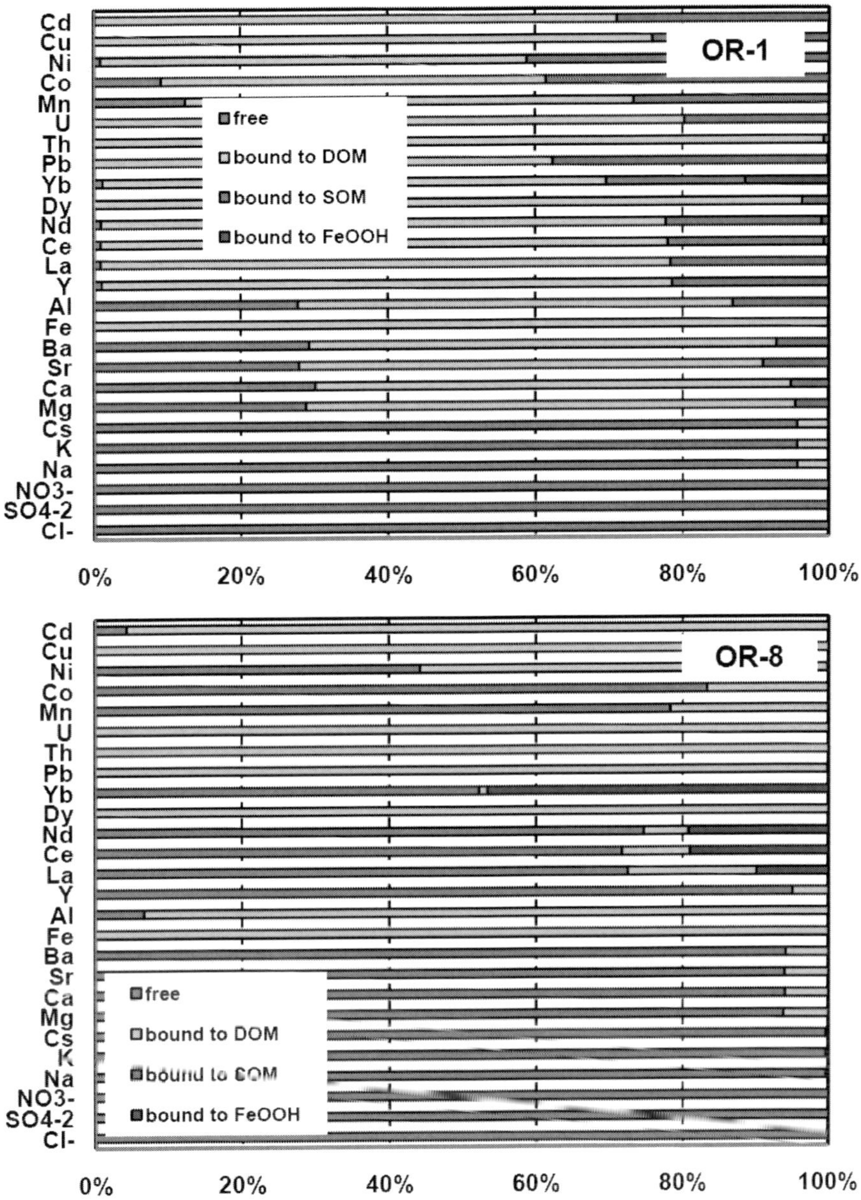

Figure 9. Results of Visual MINTEQ speciation calculation for soil solution (OR-9), upper feeding lake (OR-6), mouth reach of the stream (OR-1), and terminal lake Tsipringa (OR-8) of Vostochniy stream watershed.

Although usually considered trivalent and tetravalent hydrolysates, Pb and U are predicted to be bound to organic complexes; according to cascade filtration, they are 1) strongly correlated with Fe and 2) removed from solution during UF together with Fe rather than with OM. It follows from this comparison that the TE speciation in the studied waters may be more complex than a combination of adsorption on Fe oxy(hydr)oxide and OM and complexation with OM colloids. Presumably, ternary complexes TE-Fe-OM and TE coprecipitated within the bulk of Fe oxy(hydr)oxides globules and stabilization by organic matter, as hypothesized in previous works (Vasyukova et al., 2010), can be evoked. A very poorly studied aspect is TE complexation with OM in the low-molecular-weight (< 1 kDa) fraction that still contains up to 99 molar % DOC and is thus capable of complexing the metals.

A number of physico-chemical and methodological factors may also be responsible for the observed difference between the model and the experiment. First, our assumption that all DOM in the < 1 kDa fraction is represented by FA and the colloidal DOM (0.2 μm to 1 kDa) is humic acid may not be true because of the limited knowledge of the proportion of fulvic acids (e.g., Ren et al. 2015). If there are aggregates of FA larger than 1 kDa forming complexes with TE, there will be an underestimation of the theoretical value of the colloidal fraction. Second, in addition to TE-OM colloids, other colloidal forms, such as TE-OM-Fe, may exist, which could not be adequately modeled. Third, the complexation constants of the TE-soil DOM reaction obtained by extrapolation in the vMinteq database may be inappropriate for water in the boreal zone. Fourth, various experimental artifacts may increase the measured % of the colloidal fraction, including the decrease in the nominal pore size due to filter clogging, streaming potential, other electrochemical ultrafiltration artifacts and diffusional layer formation on the dialysis membranes.

Relatively low transformation of organic and organo-mineral colloids along the watershed profile of stream Vostochniy strongly suggests the weak degree of impact of riparian processes on colloidal chemistry in the zone along the stream. Rather, the main transformation of size and chemical nature of colloids occurs between the soil solution and upland humic lakes. The water residence time in surface reservoirs determined the degree of colloidal transformation via photo- and biodegradation in the water column. In contrast, the role of riparian processes may be more important at the scale of larger watershed such as river Palajoiki of Northern Karelia (Ilina et al., 2013a, b, 2015). In that case, more significant input of the groundwater produces massive oxidation of Fe(II) and coagulation of Fe(II)oxy(hydr)oxides stabilized by dissolved organic matter.

CONCLUSION

The results of this study demonstrated a significant and systematic change in hydrochemical (DOC, TE concentrations, C/N ratio, SUVA) and molecular size (100 μm – 1 kDa filtrates and ultrafiltrates) parameters of DOM in the continuum soil solution → feeding bog and humic lakes → stream → terminal oligotrophic lake of a representative boreal subarctic watershed. These crucial changes in basic DOM parameters occurred over very short distance, < 2 km and included the riparian zone of the stream, notably within its upper reaches and feeding bogs. For any part of the landscape continuum, either a lake or a stream, the key parameter controlling the transformation of DOM in the water column is the water residence time. From the one hand, it determines the relative effect of DOM input to the water body from the surface flow or DOM in-situ production by aquatic phytoplankton, periphyton and macrophytes. From the other hand, it controls the intensity of the DOM removal from in this body, and DOM transformation via heterotrophic aerobic consumption or the photo-degradation. As a result, the DOM evolution encountered in the Vostochniy watershed components reflects the complex process of the interplay between two main sources of DOC

– soil humic and fulvic acids from the plant litter and bog water and aquagenic exometabolites of phytoplankton and macrophytes. Both sources are subject to biodegradation and photodegradation, whose extent depends on the residence time of DOM in a given aquatic reservoir.

The on-site size fractionation study of boreal surface water colloids using a unified cascade filtration/ultrafiltration conducted within the soil solution – bogs – humic lakes – streams and their riparian zones – terminal clear water lake continuum allowed for an assessment of the degree of major and TE binding to various colloidal pools in the dominant surface water bodies of the subarctic. Possible transformation of the allochthonous humic–rich soil colloids and progressive enrichment of waters by autochthonous DOM within the sequence from soil solution to the stream and then further to the terminal lakes does not significantly change the TE size fractionation. At the same time, we observed a progressive decrease in the degree of TE association with ferric colloids from soil solution towards the terminal lake, which was also reflected in the decrease in the colloidal Fe concentration. This suggests that there is significant transformation of colloids formed by *i)* photo- and bio-degradation of allochthonous (soil) humic and fulvic acids, leading to sedimentation of Fe-rich phases, and *ii)* progressive enrichment of dissolved pool by autochthonous organic-rich, Fe-poor dissolved organic matter of periphyton and plankton exometabolites. The main role of riparian zone in colloid formation in the main stream consists in providing the site of Fe(II) input from groundwater reservoirs and its mixing within DOM-rich surface waters.

We observed reasonable agreement between the proportion of trace elements (divalent metals and U) bound to organic material and adsorbed onto Fe-rich colloids, calculated using available speciation codes and experimentally measured fractions of colloids. The theoretical model underestimates the proportion of trivalent and tetravalent hydrolysates bound to colloids, which is presumably a result of not accounting for the coprecipitation of TE with Fe hydroxides (1 kDa – 0.22 μm) stabilized by DOM.

As such, small streams may deliver to the ocean the DOC and organo-ferric colloids that is more significantly chemically fractionated among different size fractions and transformed by biodegradation and photodegradation compared with the DOM load of large rivers. This may be especially true for the most labile $LMW_{<1\ kDa}$ fraction, by far dominant in molar concentration of boreal DOM and susceptible to travel through the freshwater – seawater mixing zone without significant coagulation (cf. Dittmar and Kattner, 2003; Amon and Meon, 2004; Krachler et al., 2010; Pokrovsky et al., 2014).

To place this work in the context of permafrost thawing, one has to consider the difference in DOM transformation within small watersheds of the coastal zone and within the large arctic rivers. It is possible that climate warming at high latitude will change the flux and the speciation of carbon in large rivers to a smaller degree than in small watersheds along the Arctic Ocean coast. Due to the relatively small discharge and baseflow regime during the Arctic summer, sufficient DOM residence time in these small water bodies, together with elevated surface temperature will stimulate production of phytoplankton, heterotrophic mineralization and photooxidation of allochthonous DOC (cf. Porcal et al., 2009; Jansson et al., 2008). In contrast, large subarctic rivers will be mostly affected by the increase in allochthonous soil OC input, due to the change in river discharge and the increase in the active layer thickness (Prokushkin et al., 2011; Bagard et al., 2011), as well as the increase in the winter discharge of terrestrial DOM (Stedmon et al., 2011). As such, further studies of small subarctic watersheds with high seasonal resolution are equally important as for large rivers.

Acknowledgments

The work was supported by Grant "BIO-GEO-CLIM" of Russian Ministry of Science and Education (No 14.B25.31.0001 of 24.06.2013) and RSF grant No 15-17-10009.

References

Albinet, A., Minero, C., Vione, D., 2010. Photochemical generation of reactive species upon irradiation of rainwater: negligible photoactivity of dissolved organic matter. *Sci. Total Environ.*, 408, 3367-3373.

Alekhin, Yu. V., Ilina, S. M., Lapitsky, S. A., Sitnikova, M. V., 2010. Results of a study of co-migration of trace elements and organic matter in a river flow in a boreal zone. *Moscow Univ. Geol. Bull.*, 65(6), 380-386, doi: 10.3103/S0145875210060050.

Allard, B., 2006. A comparative study on the chemical composition of humic acids from forest soil, agricultural soil and lignite deposit: bound lipid, carbohydrate and amino acid distributions. *Geoderma*, 130(1-2), 77-96.

Allard, B., Derenne, S., 2007. Oxidation of humic acids from an agricultural soil and a lignite deposit: Analysis of lipophilic and hydrophilic products. *Org. Geochem.*, 38(12), 2036-2057.

Amon, R. M. W., Benner, R., 1996a. Bacterial utilization of different size classes of dissolved organic matter. *Limnol. Oceanogr.*, 41, 41-51.

Amon, R. M. W., Benner, R., 1996b. Photochemical and microbial consumption of dissolved organic carbon and dissolved oxygen in the Amazon River System. *Geochim. Cosmochim. Acta*, 60, 1783-1792.

Amon, R. M. W., Meon, B., 2004. The biogeochemistry of dissolved organic matter and nutrients in two large Arctic estuaries and potential implications for our understanding of the Arctic Ocean system. *Mar. Chem.*, 92, 311-330.

Andersson, P. S., Dahlqvist, R., Ingri, J., Gustafsson, O., 2001. The isotopic composition of Nd in a boreal river: a reflection of selective weathering and colloidal transport. *Geochim. Cosmochim. Acta*, 65(4), 521-527.

Archer, A. D., Singer, P. C., 2006. An evaluation of the relationship between SUVA and NOM coagulation using the ICR database. *J. Am. Water Works Ass.*, 98, 110-123.

Bagard, M.-L., Chabaux, F., Pokrovsky, O. S., Viers, J., Prokushkin, A. S., Stille, P., Rihs, S., Schmitt, A. D., Dupre, B., 2011. Seasonal variability of element fluxes in two Central Siberian rivers draining high latitude permafrost dominated areas. *Geochim. Cosmochim. Acta*, 75, 3335-3357.

Benedetti, M., Milne, C., Kinniburgh, D., van Riemsdijk, W., Koopal, L., 1995. Metal ion binding to humic substances: Application of the non ideal competitive adsorption model. *Environ. Sci. Technol.*, 29, 446-457.

Bose, P., Reckhow, D. A., 2007. The effect of ozonation on natural organic matter removal by alum coagulation. *Water Res.*, 41, 1516-1524.

Carpita, N., Sabularse, D., Montezinos, D., Delmer, D., 1979. Determination of the pore size of cell walls of living plant cells. *Science,* 205, 1144-1147.

Colombini, M. 1980. Pore size and properties of channels from mitochondria isolated from Neurospora crassa. *J. Membrane Biol.,* 53, 1432-1424.

Dahlqvist, R., Andersson, K., Ingri, J., Larsson, T., Stolpe, B., Turner, D., 2007. Temporal variations of colloidal carrier phases and associated trace elements in a boreal river. *Geochim. Cosmochim. Acta*, 71(22), 5339-5354.

Dahlqvist, R., Benedetti, M. F., Andersson, K., Turner, D., Larsson, T., Stolpe, B., Ingri, J., 2004. Association of calcium with colloidal particles and speciation of calcium in the Kalix and Amazon rivers. *Geochim. Cosmochim. Acta,* 68(20), 4059-4075.

De Haan, H., 1993. Solar UV-light penetration and photodegradation of humic substances in peaty lake water. *Limnol. Oceanog.,* 38, 1072-1076.

Dittmar, Th., Kattner, G., 2003. The biogeochemistry of the river and shelf ecosystem of the Arctic Ocean: a review. *Mar. Chem.,* 83, 103-120.

Dupré, B., Viers, J., Dandurand, J.-L., Polve, M., Bénézeth, P., Vervier, Ph., Braun J.-J., 1999. Major and trace elements associated with colloids in organic-rich river waters: ultrafiltration of natural and spiked solutions. *Chem. Geol.,* 160, 63-80.

Dzombak, D. A., Morel, F. M. M., 1990. Surface complexation modelling: hydrous ferric oxide. *Wiley-Interscience,* New York, pp 393.

Edzwald, J. K., Tobiason, J. E., 1999. Enhanced coagulation: US requirements and a broader view. *Water Sci. Technol.,* 40, 63-70.

Filella, M., Rellstab, C., Chanudet, V., Spaak, P., 2008. Effect of the filter feeder Daphnia on the particle size distribution of inorganic colloids in freshwaters. *Water Res,* 42(8-9), 1919-1924.

Fytianos, K., 2001. Speciation analysis of heavy metals in natural waters: a review. *J. AOAC Int.,* 84, 1763-1769.

German, C. R., Masuzawa, T., Greaves, M. J., Elderfield, H., Edmond, J. M., 1995. Dissolved rare earth elements in the Southern Ocean: Cerium oxidation and the influence of hydrography. *Geochim. Cosmochim. Acta,* 59(8), 1551-1558.

Guggenberger, G., Christensen, B. T., Zech, W., 1994. Land-use effects on the composition of organic matter in particle-size separates of soil: I. Lignin and carbohydrate signature. *Eur. J. Soil Sci.,* 45, 449-458.

Guo, L., MacDonald, R. W., 2006. Source and transport of terrigenous organic matter in the upper Yukon River: Evidence from isotope ($\delta^{13}C$, $\Delta^{14}C$, and $\delta^{15}N$) composition of dissolved, colloidal, and particulate phases. *Global Biogeochem. Cy.*, 20, GB2011, doi:10.1029/2005GB002593.

Guo, L., Santschi, P. H., 1996. A critical evaluation of the cross-flow ultrafiltration technique for sampling colloidal organic carbon in seawater. *Mar. Chem.,* 55, 113-127

Guo, L., Semiletov, I., Gustafsson, Ö., Ingri, J., Andersson, P., Dudarev, O., White, D., 2004. Characterization of Siberian Arctic coastal sediments: Implications for terrestrial organic carbon export. *Global Biogeochem. Cy.,* 18, GB1036, doi:10.1029/2003GB002087.

Gustafsson, J., 2011. Visual MINTEQ ver. 3.0. http://www2.lwr.kth.se/English/OurSoftware/Vminteq.

Gustafsson, O., Gschwend, P. M., 1997. Aquatic colloids: Concepts, definitions, and current challenges. *Limnol. Oceanogr.,* 42(3), 519-528.

Gustafsson, O., Widerlund, A., Andersson, P. S., Ingri, J., Roos, P., Ledin, A., 2000. Colloid dynamics and transport of major elements through a boreal river - brackish bay mixing zone. *Mar. Chem.,* 71, 1-21.

Hagel, L., 2001. Gel-Filtration Chromatography. Current Protocols in *Mol. Biol.*, 10.9, Supp. 44, 1-32, doi: 10.1002/0471142727.mb1009s44.

Hölemann, J. A., Schirmacher, M., Prange, A., 2005. Seasonal variability of trace metals in the Lena River and the southeastern Laptev Sea: Impact of the spring freshet. *Global Planet. Change,* 48 (1-3), 112-125.

Ilina, S. M., Drozdova, O. Yu., Lapitskiy, S. A., Alekhin, Yu. V., Demin, V. V., Zavgorodnyaya, Yu. A., Shirokova, L. S., Viers, J., Pokrovsky, O. S., 2014. Size fractionation and optical properties of dissolved organic matter in the continuum soil solution-bog-river and terminal lake of a boreal watershed (North Karelia, Russia). *Org. Geochem.,* 66, 14-24.

Ilina, S. M., Lapitskiy, S. A., Alekhin, Viers, J., Benedetti, M., Pokrovsky, O. S., 2015. Speciation, size fractionation and transport of trace elements in the continuum soil water – bog lake - river – terminal oligotrophic lake of a subarctic watershed. *Aquat. Geochem.,* in press.

Ilina, S. M., Poitrasson, F., Lapitskiy, S. A., Alekhin, Yu. V., Viers, J., Pokrovsky, O. S., 2013a. Extreme iron isotope fractionation between colloids and particles of boreal and temperate organic-rich waters. *Geochim. Cosmochim. Acta,* 101, 96-111.

Ilina, S. M., Viers, J., Lapitsky, S. A., Mialle, S., Mavromatis, V., Chmeleff, J., Brunet, P., Alekhin, Y. V., Isnard, H., Pokrovsky, O. S., 2013b. Stable (Cu, Mg) and radiogenic (Sr, Nd) isotope fractionation in colloids of boreal organic-rich waters. *Chem. Geol.,* 342, 63-75.

Ingri, J., Widerlund, A., 1994. Uptake of alkali and alkaline-earth elements on suspended iron and manganese in the Kalix River, northern Sweden. *Geochim. Cosmochim. Acta,* 58(24), 5433-5442.

Ingri, J., Widerlund, A., Land, M., Gustafsson, O., Andersson, P., Ohlander, B., 2000. Temporal variations in the fractionation of the rare earth elements in a boreal river; the role of colloidal particles. *Chem. Geol.,* 166, 23-45.

Jansson, M., Hickler, Th., Jonsson, A., Karlsson, J., 2008. Links between terrestrial primary production and bacterial production and respiration in lakes in a climate gradient in subarctic Sweden. *Ecosystems,* 11, 367-376.

Kinniburgh, D. G., van Riemsdijk, W. Hv., Koopal, L. K., Borkovec, M., Benedetti, M. F., Avena, M. J., 1999. Ion binding to natural organic matter: competition, heterogeneity, stoichiometry and thermodynamic consistency. *Colloids Surf. A Physicochem. Eng. Asp.,* 151, 147-166.

Krachler, R., Krachler, R. F., von Kammer, F., Suphandag, A., Jirsa, F., Ayromlou, S., Hofmann, Th., Keppler, B. K., 2010. Relevance of peat-draining rivers for the riverine input of dissolved iron into the ocean. *Sci. Total Env.,* 408, 2402-2408.

Maksimova, M., 1967. Inorganic and organic composition of major ions in rivers of Karelian coast of the White Sea (in Russian). Gidrobiologicheskie issledovaniya na Karelskom poberezhie Belogo moray. *Nauka,* Leningrad, 9-20.

Matilainen, A., Gjessing, E. T., Lahtinen, T., Hed, L., Bhatnagar, A., Sillanpää, M., 2011. An overview of the methods used in the characterisation of natural organic matter (NOM) in relation to drinking water treatment. *Chemosphere,* 83, 1431-1442.

Milne, C. J., Kinniburgh, D. G., Tipping, E., 2001. Generic NICADonnan model parameters for proton binding by humic substances. *Environ. Sci. Technol.,* 35, 2049-2059.

Milne, C. J., Kinniburgh, D. G., van Riemsdijk, W. H., Tipping, E., 2003. Generic NICA-donnan model parameters for metal-ion binding by humic substances. *Environ. Sci. Technol.,* 37(5), 958-971.

Minor, E., Stephens, B, 2008. Dissolved organic matter characteristics within the Lake Superior watershed. *Org. Geochem.,* 39, 1489-1501.

Onstad, G. D., Canfield, D. E., Quay, P. D., Hedges, J. I., 2000. Sources of particulate organic matter in rivers from the continental USA: Lignin phenol and stable carbon isotope compositions. *Geochim. Cosmochim. Acta,* 64, 3539-3546.

Pokrovsky, O. S., Dupré, B., Schott, J., 2005. Fe–Al–organic Colloids Control of Trace Elements in Peat Soil Solutions: Results of Ultrafiltration and Dialysis. *Aquat. Geochem.,* 11(3), 241-278.

Pokrovsky, O. S., Schott, J., 2002. Iron colloids/organic matter associated transport of major and trace elements in small boreal rivers and their estuaries (NW Russia). *Chem. Geol.,* 190(1-4), 141-179.

Pokrovsky, O. S., Schott, J., Dupre, B., 2006. Trace element fractionation and transport in boreal rivers and soil porewaters of permafrost-dominated basaltic terrain in Central Siberia. *Geochim. Cosmochim. Acta,* 70, 3239-3260.

Pokrovsky, O. S., Shirokova, L. S., 2013. Diurnal variations of dissolved and colloidal organic carbon and trace metals in a boreal lake during summer bloom. *Water Res.,* 47(2), 922–932.

Pokrovsky, O. S., Shirokova, L. S., Gordeev, V. V., Sevchenko, V. P., Viers J., et al. (2014) Fate of colloids in the Arctic estuary. Ocean Science 10, 107–125.

Pokrovsky, O. S., Shirokova, L. S., Kirpotin, S. N., Audry, S., Viers, J., Dupré, B., 2011. Effect of permafrost thawing on organic carbon and trace element colloidal speciation in the thermokarst lakes of western Siberia. *Biogeosciences,* 8, 565-583.

Pokrovsky, O. S., Shirokova, L. S., Kirpotin, S. N., Kulizhsky, S. P., Vorobiev, S. N. 2013. Effects of anomalous high temperatures on carbon dioxide, methane, dissolved organic carbon and trace element concentrations in thaw lakes in Western Siberia in 2012. *Biogeosciences,* 10, 5349-5365.

Pokrovsky, O. S., Shirokova, L. S., Zabelina, S. A., Vorobieva, T. Y., Moreva, O. Yu., Chupakov, A., Audry, S., Viers, J., 2012. Size fractionation of trace elements in a seasonally stratified boreal lake: control of organic matter and iron colloids. *Aquat. Geochem.,* 18, 115-139.

Pokrovsky, O. S., Viers, J., Shirokova, L. S., Shevchenko, V. P., Filipov, A. S., Dupré, B., 2010. Dissolved, suspended, and colloidal fluxes of organic carbon, major and trace elements in the Severnaya Dvina River and its tributary. *Chem. Geol.,* 273(1-2), 136-149.

Porcal, P., Koprivnjak, J.-F., Molot, L. A., Dillon, P. J., 2009. Humic substances-part 7: the biogeochemistry of dissolved organic carbon and its interaction with climate change. *Environ. Sci. Pollut. R.,* 16, 714-726.

Pourret, O., Davranche, M., Gruau, G., Dia, A., 2008. New insights into cerium anomalies in organic-rich alkaline waters. *Chem. Geol.,* 251(1-4), 120-127.

Prokushkin, A. S., Pokrovsky, O. S., Shirokova, L. S., Korets, M. A., Viers, J., Prokushkin, S. G., Amon, R. M. W., Guggenberger, G., McDowell, W. H., 2011. Sources and the flux pattern of dissolved carbon in rivers of the Yenisey basin draining the Central Siberian Plateau. *Environ. Res. Lett.*, 6, 045212, doi:10.1088/1748-9326/6/4/045212.

Ren, Z. L., Tella, M., Bravin, M. N., Comans, R. N. J., Dai, J., Garnier, J. M. et al., 2015. Effect of dissolved organic matter composition on metal speciation in soil solutions. *Chem. Geol.,* 398, 61-69.

Reuss, J., Cosby, B., Wright, R., 1987. Chemical processes governing soil and water acidification. *Nature*, 329, 27-32.

See, J. H., Bronk, D. A., 2005. Changes in C:N ratios and chemical structures of estuarine humic substances during aging. *Mar. Chem.,* 97, 334-346.

Serreze, M. C., Bromwich, D. H., Clark, M. P., Etringer, A. J., Zhang, T. and Lammers, R. 2003. The large-scale hydroclimatology of the terrestrial Arctic drainage system. *J. Geophys. Res.,* 108(D2), 8160, doi:10.1029/2001JD000919.

Shirokova, L. S., Pokrovsky, O. S., Kirpotin, S. N., Desmukh, C., Pokrovsky, B. G., Audry, S., Viers, J., 2013. Biogeochemistry of organic carbon, CO_2, CH_4, and trace elements in thermokarst water bodies in discontinuous permafrost zones of Western Siberia. *Biogeochemistry,* 113, 573-593.

Sholkovitz, E. R., 1992. Chemical evolution of rare earth elements – fractionation between colloidal and solution phase of filtrated river water. *Earth Planet. Sci. Lett.,* 114(1), 77-84.

Sholkovitz, E. R., 1993. The geochemistry of rare earth elements in the Amazon River estuary. *Geochim. Cosmochim. Acta,* 57(10), 2181-2190.

Sholkovitz, E. R., 1995. The aquatic chemistry of rare earth elements in rivers and estuaries. *Aquat. Geochem.,* 1, 1-34.

Sholkovitz, E. R., Shaw, T. J., Schneider, D. L., 1992. The geochemistry of rare earth elements in the seasonally anoxic water column and porewaters of Chesapeake Bay. *Geochim. Cosmochim. Acta,* 56(9), 3389-3402.

Sjöstedt, C. S., Gustafsson, J. P., Köhler, S. J., 2010. Chemical equilibrium modeling of organic acids, pH, aluminum and iron in Swedish surface waters. *Environ. Sci. Technol.,* 44, 8587-8593.

Stedmon, C. A., Amon, R. M. W., Rinehart, A. J., Walker, S. A., 2011. The supply and characteristics of colored dissolved organic matter (CDOM) in the Arctic Ocean: Pan Arctic trends and differences. *Mar. Chem.,* 124, 108-118.

Sultan, K., Shazili, N. A., 2009. Rare earth elements in tropical surface water, soil and sediments of the Terengganu River Basin, Malaysia. *J. Rare Earth,* 27(6), 1072-1078.

Thorn, K. A., Younger, S. J., Cox, L. G., 2010. Order of functionality loss during photodegradation of aquatic humic substances. *J. Environ. Qual.,* 39, 1416-1428.

Tranvik, L. J., 1988. Availability of dissolved organic carbon for planktonic bacteria in oligotrophic lakes of differing humic content. *Microbial Ecol.,* 16, 311-322.

Tremblay, L., Benner, R., 2006. Microbial contributions to N-immobilization and organic matter preservation in decaying plant detritus. *Geochim. Cosmochim. Acta,* 70, 133-146.

Trias, J., Jarlier, V., Benz, R., 1992. Porins in the cell wall of mycobacteria. *Science,* 258 (5087), 1479-1481.

Twichell, S. C., Meyers, P. A., Diester-Haass, L., 2002. Significance of high C/N ratios in organic-carbon-rich Neogene sediments under the Benguela Current upwelling system. *Org. Geochem.,* 33, 715–722.

Vasyukova, E., Pokrovsky, O., Viers, J., Oliva, P., Dupré, B., Martin, F., Candaudap, F., 2010. Trace elements in organic- and iron-rich surficial fluids of boreal zone: Assessing colloidal forms via dialysis and ultrafiltration. *Geochim Cosmochim Acta,* 74, 449-468.

Viers, J., Dupre, B., Polve, M., Dandurand, J., Braun, J., 1997. Chemical weathering in the drainage basin of a tropical watershed, Nsimi-Zoetele site (Cameroon): comparison between organic-poor and organic-rich waters. *Chem. Geol.,* 140, 181-206.

Wang, G. S., Liao, C. H., Wu, F. J., 2001. Photodegradation of humic acids in the presence of hydrogen peroxide. *Chemosphere,* 42, 379-387.

Wolfe, A. P., Kaushal, S. S., Fulton, J. R., McKnight, D. M., 2002. Spectrofluorescence of sediment humic substances and historical changes of lacustrine organic matter provenance in response to atmospheric nutrient enrichment. *Environ. Sci. Technol.,* 36, 3217-3223.

Zakharova, E., Pokrovsky, O. S., Dupré, B., Gaillardet, J., Efimova, L., 2007. Chemical weathering of silicate rocks in Karelia region and Kola peninsula, NW Russia: Assessing the effect of rock composition, wetlands and vegetation. *Chem. Geol.,* 242, 255-277.

Zuo, Y., Jones, R. D., 1997. Photochemistry of natural dissolved organic matter in lake and wetland waters—production of carbon monoxide. *Water Res.,* 31, 850-858.

In: Riparian Zones
Editor: Oleg S. Pokrovsky ISBN: 978-1-63484-613-4

Chapter 8

Barrier Function of Floodplain and Riparian Landscapes in River Runoff Formation

I. A. Avessalomova, A. V. Khoroshev and A. V. Savenko*
Moscow M. V. Lomonosov State University, Moscow, Russia

Abstract

Agricultural pollution of water bodies is a crucial problem that impose limitations for both land use in floodplains and water supply. The solution can be found either in improvement of agricultural technologies or in better adaptation of land use to landscape spatial pattern and matter flows. The paper focuses on possible natural mechanisms that can perform protective functions in relation to small rivers in the intensively cultivated agro-landscape. The typical attributes of the floodplains landscape structure were studied on the example of the East-European middle taiga (the Severnaya Dvina River basin). We examined factors of the floodplain development in connection with landscape-geochemical structure of elementary catchments with particular focus on system-forming lateral flows. We identified various flowpaths of the dissolved matter washed out from eroded slopes in the agro-landscapes. In the lowest sections of the catenas specific landscape neighborhoods induce geochemical contrasts resulting in the development of the barrier zones at the floodplains. Retention capacity of biogeochemical barriers was evaluated quantitatively as well as involvement of nutrients in water migration. We detected differences in retention functions of biogeochemical barriers dependent on bio-availability of trace elements in acid and alkaline soils. The loss of nutrients with ion discharge in the agro-landscapes requires preservation of floodplains barrier functions to ensure effective ecological network.

Keywords: floodplain, taiga, agro-landscape, catena, geochemical barrier, trace elements, nutrients, flowpath, retention capacity

* Corresponding author: A. V. Savenko. E-mail: Alla_Savenko@rambler.ru.

INTRODUCTION

Floodplains and riparian zones are commonly treated as natural objects of extreme importance due to their high landscape and biological diversity and regulating role in matter transport. Connectivity in hydrological systems is a key attribute (Ahern, 2004). High interest in study of linkages between the river and surrounding landscape indicates the important new stage in science of river ecology (Likens and Bormann, 1995; Townsend and Riley, 1999). Longitudinal links along the length of the river system are the focus of the river continuum concept (Vannote et al., 1980). Lateral links with the adjacent terrestrial systems, on the one hand, depend to a great extent on flood events which integrate flows within the entire river basin (Junk et al., 1989; Khromykh, 2007). Widely known spiralling concept is based on the recurrent use of matter in ecosystems along the river and describes strong interactions between the stream and the riparian systems (Pinay et al., 1990). On the other hand, matter can reach floodplains and water bodies directly from the valley slopes and terraces independently of floods. In that case floodplains and adjacent spatial units can function as a barriers or filters which control matter delivery to the river. In landscapes with immense anthropogenic activity (e.g., agro-landscapes) this function is crucially needed since the rivers and floodplains are multifunctional ecological systems both for nature and man. In this paper we focus on buffer functions of floodplain and riparian landscapes and on natural mechanisms that protect these landscapes from man-induced contamination.

Buffer zones often attract attention of landscape ecologists. They provide an example of small spatial units that demonstrate powerful contribute much to the development of ecological processes (Turner et al., 2001). Most studies deal with contribution of riparian phytocoenoses to interception of pollutants. Problem of quantitative assessment of nutrients interception is still unresolved (Baker et al., 2006). Nowadays, we have no conceptual model of buffer potential evaluation; in most studies proportion of forests at the fixed distance from the water body is taken as a proxy (Weller et al., 2007). Researchers concentrate their efforts on working out metrics that afford to relate pollutants flows to the needed spatial parameters of buffer zones at the basin or catena scale (Baker et al., 2006). Important priorities of research are seen in the influence of landscape pattern on matter transport with surface and subsurface flows (Wickham et al., 2003), in spatio-temporal dynamics of hydrologically controlled physical and chemical properties of a landscape (Malard et al., 2000). These studies reflect switch from focus on geometrical attributes only to the analysis of processes generating spatial patterns. This requires choosing appropriate methodology. In this paper we base our findings on methodology of landscape geochemistry which is traditionally being applied by Central and Eastern European landscape ecologists in the studies of radial and lateral interactions between the natural bodies (e.g., Opp, 1991; Klimo et al., 1996; Sugier and Czarnecka, 1998; Banaszuk et al., 2000; Roder and Syrbe, 2000; Wicik, 2001; Ratas et al., 2003).

Landscape-geochemical approach proved to be appropriate for the studies of matter migration and buffer function of ecosystems. The approach was based initially on the concept of soil catena (Milne, 1935) and integrates knowledge about interactions between soil, vegetation, water, rocks depending on topographic position (Polynov, 1956; Perel'man, 1977; Fortescue, 1980; Glazovskaya, 2007). The basic idea involves determination of dispersion, transit and accumulation functions of spatial units. Landscape-geochemical approach assumes application of several concepts that are relevant for describing various aspects of floodplain ecosystems structure and functioning as well as their ecological significance. The widely used concept of cascade landscape-geochemical system based on general system theory (Glazovskaya, 2007) is the particular case of cascade systems determined by lateral matter and energy flows between dynamically connected subsystems with special focus on input/output ratio (Chorley and Kennedy, 1971). Holistic principle is in the core of the concept as well as hierarchical diversity. In regional studies two hierarchical orders of the cascade landscape-geochemical systems are distinguished: (1) landscape-geochemical catena, and (2) combination of catenas within the catchment (Glazovskaya, 2007).

Landscape-geochemical catena commonly includes elementary landscapes connected by one-way lateral flows. Geochemical classification of elementary landscapes is commonly based on position in catena (so-called autonomous, trans-eluvial, trans-accumulative, superaqual, and aqual elementary landscapes) and dominant chemical elements that determine conditions of water migration (H, H-Ca, Ca, H-Fe classes). Geochemical fields within a catena differ in chemical elements concentrations. Occurrence of monotonous or step-like gradients shows environment diversity (Fortescue, 1980). Floodplains are superaqual geochemically subordinate units that receive matter both from the river and from the catenas of neighbouring catchments. Basin approach is applied to unite catenas into cascade landscape-geochemical system. The river network serves as a means of connection. Floodplains are subject to downstream migration that contributes to matter export from the basin. Thus, to explain landscape structure one needs to consider the diversity of catenas in a river basin with particular focus on factors that determine migration conditions, transfer distance and possibility of accumulation in the lowest sections of a catena.

The functional role of floodplains as accumulators of chemical elements is based on the geochemical barrier concept elaborated by A.I. Perel'man (1977). The core idea of the concept is the existence of sites where rate of matter migration changes abruptly. The location of geochemical barriers is closely related to landscape heterogeneity. For example, they can occur at the site of groundwater emergence in the floodplain inner margin. A.I. Perel'man made distinction between physical-chemical, mechanical barriers (e.g., accumulation on alluvial fans or in bottom sediments), and barriers formed by biogenic migration. Barriers develop due to boundaries between contrasting pH or redox conditions, biological uptake, H_2S accumulation, sorption capacity of clay, humus, metal hydroxides or organic matter, etc. (Perel'man and Kasimov, 1999). Influence of biological uptake is manifested in development of highly productive plant communities able to intercept chemical elements from lateral water flows (Baker et al., 2006). To quantify deposition of elements in phytomass fractions numerous indices were proposed (Bazilevich and Titlyanova, 2008). It is crucially important to choose relevant plant indicators with regard to their bioaccumulation potential (Bargagli, 1998)

To evaluate the ability of chemical elements to accumulate in this or that landscape component (soils, vegetation, water, sediments) a series of coefficients were elaborated. They afford an opportunity to compare concentrations in landscape component (soil, vegetation, water) in relation to that in the Earth crust ("clarke of concentration") or in the other component. For example, coefficient of biological uptake (CBU) is calculated as a content in a plant divided by that in a soil.

In this paper we focus on barrier functions of various spatial elements at the floodplain and in the riparian landscape in the process of river runoff formation. The objectives of the research were established as follows:

- to analyze landscape structure of the floodplains in connection with catena organization of catchments;
- to determine natural and anthropogenic factors that affect lateral matter migration in the background conditions as well as in the agro-landscapes;
- to reveal main flowpaths that ensure geochemical connection between arable lands and floodplains;
- to assess the role that floodplain landscapes play in the ecological network within agro-landscapes.

MATERIALS AND METHODS

The research was performed in the middle taiga of East-European plain (the southern Arkhangelsk region of Russia). The study area (the Zayachya River basin, 154 km^2) is located within the Ustyanskoye plateau (Figure 1) composed of Permian sedimentary rocks (the Sukhona formation). Physical environment of the landscape was shaped by morainic and limnoglacial accumulation in Riss period of Pleistocene. The Zayachya River belongs to the Severnaya Dvina River basin. The upper reaches of the Zayachya River are located within natural taiga forests while middle reaches are within the agro-landscapes. Flow directions of the first- and second-order streams are affected strongly by the system of lineaments stretching northeastward and northwestward (Khoroshev, 2003). Development of the Zayachya River terraces dates back to the late Würm. Alluvial deposits on the terraces commonly overlay Permian marlstones and morainic loams. Terraces and floodplains of the small rivers were formed in Holocene. All the rivers have snow and rain feeding while groundwater contributes to runoff in a much lesser degree. In the upper reaches of the small streams, the oligotrophic and mesotrophic bogs provide runoff in summer. Floods occur in spring while low water is characteristic for summer and winter. Geochemically contrast catenas in the study area were chosen as the main research objects.

To describe landscape structure of the elementary catchments we used the results of landscape mapping in scale 1:10 000 as well as catena studies which comprised all the elementary landscapes from the autonomous to the superaqual ones. Detailed study of the floodplains interior pattern was conducted at key plots to reflect various landscape neighbourhoods. Simultaneously, we sampled plants, humus horizon of soils commonly at the 5-10 cm depth (142 plots), water (20 plots) and bottom sediments (13 plots). Samples of herb aboveground phytomass were collected from the plots 50 $\times$ 50 cm followed by separated determination of mass of gramineous plants, sedges, legumes and forbs. Totally 230 plant samples from 102 plots were collected in which ash content and stock of mineral elements were determined. Concentrations of trace elements in plants and soils were determined by semi-quantitative spectral analysis. To analyze spatial heterogeneity we calculated coefficients of variations. To compare intensity of biological uptake in various landscape units we used clarkes of concentration (calculated in relation to the Earth crust) and coefficients of biological uptake (below referred to as CBU). Sum of clarkes of concentration in plants is below referred to as "biogeochemical activity of a species."

Determination of pH, exchangeable cations (ammonium chloride extraction) and soil humus was performed by conventional methods after sieving through 1 mm mesh net. Total nitrogen content was determined by Kjeldal method, mobile forms of phosphorus and potassium by Kirsanov method.

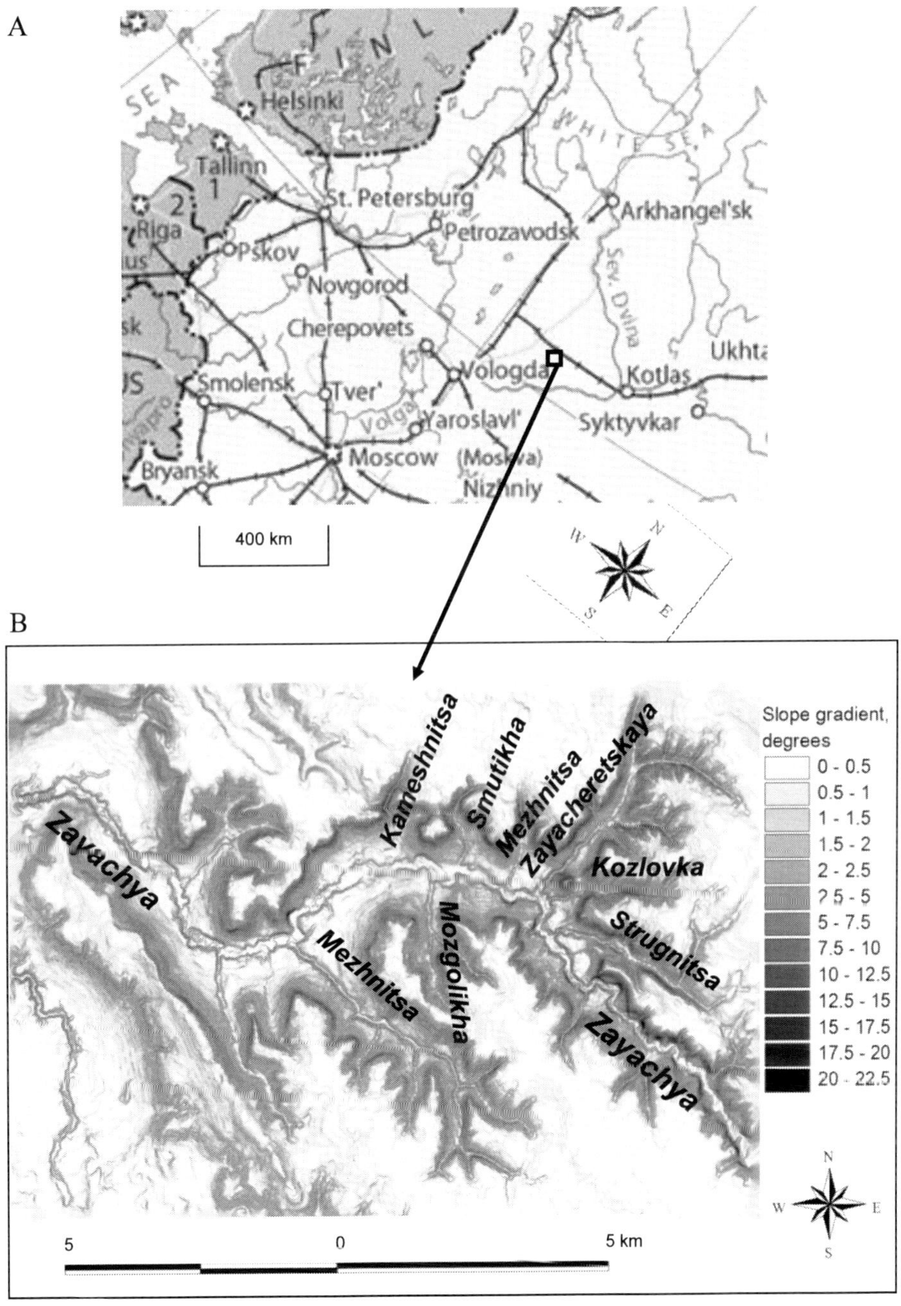

Figure 1. Location of study area (A) and slope gradients from DEM (B).

Water samples for hydrochemical studies were collected during summer low water in 20 plots on the Zayachya River, its tributaries, and in bogs. Samples were immediately filtered through a fine-mesh paper filter into polypropylene flasks. After that the part of filtrate (100 ml) was preserved by addition of 1 ml of chloroform for the analysis of nutrients. The contents of Na, K, Mg, Ca, and SO_4 ions were determined by capillary electrophoresis. The contents of chlorides and value of total alkalinity (overwhelming part of which are hydrocarbonates) were measured using volumetric titration: mercurimetry and acidimetry. The fluorine concentration was determined by ionometry. Dissolved mineral phosphorus (P_{min}) and silicon were analyzed by colorimetric methods using ammonium molybdate with ascorbic acid and ammonium molybdate with Mohr's salt, respectively. The relative error of measurements was less than ± 3%. To assess hydrochemical heterogeneity and man-induced transformation of stream-water chemistry coefficients of lateral differentiation (L) were calculated as a concentration in river water divided by concentration in the background mesotrophic bog water.

RESULTS

Landscape Structure in Floodplains

Natural catenas of the study area include the following sequence of elementary landscapes: (1) autonomous H-class morainic units with *Piceetum myrtilletosum* forests on Haplic Podzols, (2) transeluvial morainic units with close to surface marlstones with *Piceetum oxalidosum* forests, (3) superaqual units on floodplains with *Alneta incanae magnoherbosa forests*.

Most part of plateau is involved in agriculture and below is referred to as agro-landscapes. Catenas with heterogeneous substrate dominate. Autonomous units were formed on morainic or limnoglacial sediments, trans-eluvial ones were formed on Permian marlstones. The lower-lying trans-accumulative units are represented by deluvial trains shaped by mechanical matter migration from plowed slopes. River terraces belong also to autonomous type of units some of them being structural with marlstone base. Anthropogenic activity affects matter migration by means of fertilizers input resulting in involvement of nutrients in water flows and ion discharge. Spatial redistribution of elements along catenas is encouraged by man-induced migration of particulate matter resulting in accumulation in the low-lying units and decrease of sediment discharge.

The Zayachya River basin in the upper reaches belongs to structural plateau with cover of morainic and limnoglacial sediments. Permian marlstones are exposed in the lowest sections of valley slopes as well as in streambeds. The exposures of solid rocks make the floodplains narrower and form riffles which alternate with pools and narrow sandy beaches as well as in alcalinization of water (pH up to 7.2-7.6) (Figure 2 A).

Landscape structure in the valley bottom is relatively simple and consists of low-level and high-level floodplains as well as oxbow depressions at the terrace toe slopes. Narrow low-level floodplain (Figure 2 A, No 1) is developed locally along the streambed. It has a

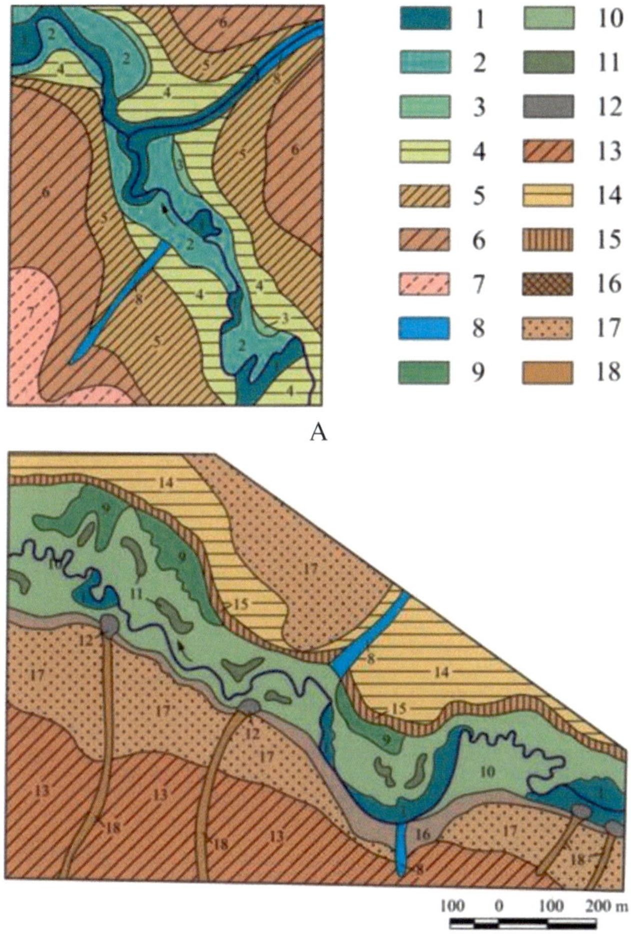

Figure 2. Landscape map of the key plots in the Zayachya River basin in the upper reaches (A) and middle reaches (B). For legend see Table 1.

Table 1. Landscape structure of the Zayachya River basin (legend for Figure 2)

Landscape units	Forests		Meadows
	mixed forests	small-leaved forests	
Floodplains			
Low level floodplain		1	
High-level floodplain	2	9	10
Depressions in floodplain inner margins and oxbows	3		11
Deluvial trains superimposed on floodplains			12
Terraces and interfluve areas			
Terraces	4		
Slopes of valleys and terraces	5, 15, 16		
Slopes of interfluve plains	6, 7		
Alluvial fans			
Gullies		8	

height of 1 m above the river and is composed of sand and gravel with numerous thin humus-rich lenses. *Salix viminalis* and *Alnus incana* dominate in tree layer, *Ribes rubrum* and *R. nigrum* are in understory, *Filipendula ulmaria* and *Aconitum septentrionale* dominate the herb layer (Figure 3 A). Nitrogen-fixing bacteria on alder roots favour enrichment of Umbri-Endogleyic Fluvisols with nitrogen.

High-level floodplain (Figure 2 A, No 2) has a height of 2 m above the river and locally is limited by steep slope adjacent to the river. When steep slope is destroyed by the flood the stream is often dammed by fallen trees. The floodplain is covered by alder, spruce and birch forests. Composition of herb layer is complex since it involves tall forbs (*Filipendula ulmaria, Aconitum septentrionale*) as well as species inherent for boreal forests (*Oxalis acetosella, Trientalis europaea, Majanthemum bifolium*) or broad-leaved forests (*Viola mirabilis, Ajuga reptans, Aegopodium podagraria*). The latter ecological group of species is favoured by close to surface carbonate bedrocks. High humidity enhances detritus accumulation and development of Eutri-Histic Gleysols and Eutric Histosols.

In the inner margins of the floodplains (Figure 2 A, No 3) under the canopy of birch-and-spruce forests abundance of *Filipendula ulmaria* increases and numerous hygrophytes occur (e.g., *Viola palustris*). In bogged spruce forests net productivity (P) accounts for 5 t/ha, phytomass (F) accounts for 170 t/ha, and mortmass accounts for 1700 t/ha (Bazilevich, 1993). This testifies that biological turnover in forest landscape is slowed down (P/F 0.03) and dead matter accumulates actively.

Slow decomposition of organic matter results in growing thickness of peat horizons up to 30-50 cm, neutral gleization (pH 6.8), and anaerobiosis in the lowest horizons of Eutric Histosols. Desulphurization is indicated by H_2S release from soils. Numerous pits from fallen trees accumulate meltwater, rainwater as well as emerging groundwater.

High-level floodplain is adjacent to river terraces (Figure 2 A, No 4) covered by spruce forests with small-leaved trees. Leaching in soils is better pronounced resulting in the development of Histic Albeluvisols. On the valley slopes fall faces with marlstone exposures alternate with forested patches (Figure 2 A, No 5). Within the interfluve plains (Figure 2 A, No 6) bedrocks are covered by quaternary sediments consisting of limnoglacial sands underlain by morainic loams. Due to low permeability of loams the plains are poorly drained and covered by spruce and pine forests with well developed dwarf shrubs and mosses. Oligotrophic and mesotrophic bogs occur of flat areas (Figure 2 A, No 7). Self-organization of middle-taiga landscapes is suppressed by: (1) low productivity due to high humidity, (2) open-ended biological cycle and low rate of organic matter decomposition, (3) nutrients leaching from acid soils (Avessalomova, 2012).

Surface runoff to the Zayachya River valley is ensured by the network of streams from first-order catchments. They can be detected visually by dominance of hygrophilous forbs (e.g., *Filipendula ulmaria*). Groundwater emerges at the slope basis as detected by accumulation of iron-rich jellies.

A

B

Figure 3. The Zayachya River floodplains: meadows with dominance of tall forbs at the low-level floodplain in the Zayachya River upper reaches (A) and wide floodplain with oxbows in the Zayachya River middle reaches (B).

Water streams unite elementary catenas of small forested catchments in the cascade landscape-geochemical systems. The Zayachya River floodplains act as the terminal plots of mass movement. Three processes fulfill barrier functions: (1) accumulation of suspended load at the low level of floodplain, (2) conservation of organic matter at the high level, (3) biological accumulation of elements in forest phytomass.

In the middle reaches the Zayachya River crosses well-drained dissected plains and has a lot of tributaries (the Strugnitsa River, the Kozlovka River, the Mezhnitsa River, the Smitukha River, the Kameshnitsa River, the Mozgolikha River, etc.). The valley at this section is U-shaped with wide floodplains (Figure 3 B), meanders, structural and accumulative terraces (Figure 2 B). Most interfluve areas are cultivated.

Similarly to the upper reaches, the low-level floodplain is covered by forests dominated by *Alnus incana* and *Salix viminalis* (Figure 2 B, No 1). In Umbri-Gleyic Fluvisols oxic conditions are replaced by unoxic ones at the depth of 35-38 cm indicated by iron spots. In phytocoenoses *Filipendula ulmaria* and *Aconitum septentrionale* are accompanied by *Delphinium elatum* which occurs at the westernmost boundary of the areal and is representative of Siberian taiga forbs (Emelyanova et al., 1999). *Alneta incanae magnoherbosa communities occur at the high-level floodplain in the inner margins or encircle oxbow depressions* (Figure 2 B, No 9).

The bulk of the territory in the high-level floodplain is composed of alluvial sandy loams and covered by meadows (Figure 2 B, No 10). Species composition is extremely diverse depending on ecotope humidity. The most species-rich communities (up to 32-37 species, cover up to 90%, height >70 cm) occur in convex patches. The dominating species are *Polygonum bistorta*, *Dactylis glomerata*, *Festuca pratensis*, and *Vicia cracca*.

Most species are eutrophic. Their high nutrients demand is satisfied by periodic nutrient input during flood events. On the other hand, species with various demands for water are present. Though mesophytes dominate, mesohygrophytes (*Filipendula ulmaria*) and hygrophytes (*Galium palustre*) occur also as well as less water-demanding species (*Geranium pratense*, G. sylvaticum, and *Dactylis glomerata*). Co-occurrence of species with various demands for water shows evidence that seasonal and spatial humidity variations are high. The traces of seasonal gleization are detected by iron-rich spots in layered loamy soils. The meadows are valuable for haying but only until the moment of haying cessation which causes decrease of species richness and expansion of *Filipendula ulmaria*.

At the flat portions of floodplains, the spring flood duration is longer. This favours the highest water supply, permanent gleization in the lower soil horizons, and development of wet meadows with domination of *Filipendula ulmaria*, *Deschampsia cespitosa*, and *Scirpus sylvaticus*.

Oxbow depressions are typical for the high-level floodplains (Figure 2 B, No 11) due to the river meandering. Meltwater is sustained there for a long time. In summertime groundwater level is close to surface resulting in incomplete organic matter decay, acid reaction and gleization in Eutri-Histic Gleysols and Eutric Histosols. Wet meadows are dominated by *Carex acuta*, *Deschampsia cespitosa*, and *Calamagrostis canescens*. These species populations generate hummocky terrain, decreased cover (55%) and low species richness. The communities include *Caltha palustris* and *Equisetum palustre*.

More complex landscape pattern is inherent for the sections of the floodplains with deluvial trains superimposed (Figure 2 B, No 12). They are generated by agriculturally induced erosion on valley slopes (Figure 2 B, No 13) and terraces (Figure 2 B, No 14). Natural birch-and-spruce forest stands occur locally on the slope of the 1st terrace (Figure 2 B, No 15) and steep slopes of the 2d high structural terrace (Figure 2 B, No 16). Sheet erosion leads to development of deluvial trains superimposed over the terraces and the floodplains (Figure 2 B, No 17). Rill erosion results in matter accumulation of alluvial fans. High nutrient and water supply is favourable for willow-and-alder stands with tall forbs (*Filipendula ulmaria*, *Heracleum sibiricum*). These communities fulfill important ecological function due to ability to intercept nutrient fluxes on their pathway from fields to floodplains.

The wide valley of the Kokshenga River (Figure 4 A) near the Zayachya River mouth is located within the lowlands developed in sandstones of the Nizhneustyinskaya formation of the Permian period. During Pleistocene deglaciation the lowlands was used by meltwater streams and periglacial lakes. This resulted in formation of a series of terraces widely spread sands, and active lateral erosion. In Holocene free meandering of the Kokshenga River caused development of heterogeneous floodplain landscapes (Figure 5).

Sandy beaches along the streambed (Figure 5, No 1) change their shape quite frequently due to the initial stage of development. Pioneer assemblages consist mainly of *Petasites spurious* and *Calamagrostis epigeios*. The latter is more abundant on levees among the communities of *Salix viminalis* (Figure 5, No 2). Sandy beaches and levees make spatial pattern of the low-level sandy floodplain (1.5 m above the river) more complex (Figure 5, No 3). In comparison with the levees flood have longer duration which leads to high humidity, gleization, and dominance of *Alnus incana* communities and meadows with tall forbs.

A

B

Figure 4. The Kokshenga River valley (A) with oxbow lakes on floodplains (B).

Figure 5. Landscape map of the key plot in the Kokshenga River valley. For legend see Table 2.

Table 2. Landscape structure of the Kokshenga River valley (legend for Figure 5)

Landscape units	Forests		Meadows	Combination of forests, meadows and fens
	mixed forests	small-leaved forests		
Low-level floodplain				
Sandy beaches				
Levees		2		
Segment floodplain		3		
High-level floodplain				
Central floodplain			4	
Sandy ridges		5		
Depressions between sandy ridges				6
Oxbow depressions with lakes				7
Levees		8		
River terraces				
Alluvial sandy terraces	9			

The high-level floodplain has a height of 2.5-3 m above the river. Meadow communities are widely spread (Figure 5, No 4) with cover up to 90%. Species richness is high with large proportion of forbs and legumes. Humus accumulation is well developed. The humus horizon is underlain by well-sorted sands. Meadow floodplains are neighbouring to undulated floodplains with alternating ridges (Figure 5, No 5) and linear depressions (Figure 5, No 6). Sandy ridges with willow communities are oriented along the river. In the depressions high humidity and siltation result in development of wet sedge meadows on Histosols, gleization, and slowed down biological cycle.

High-level floodplains are distinguished by the highest humidity. Oxbow lakes oriented along the river meanders are common for central portions of the floodplain (Figure 5, No 7). The lakes are surrounded by strips of hygrophilous species communities (*Scirpus lacustris*, *S. sylvaticus*, *Typha latifolia*, and *Alisma plantagoaquatica*) which indicate rich nutrients supply (Figure 4 B). Eutrophication of oxbows is easily detected by coloured water and abundance of *Lemnaceae* species. Lentic regime is supported by groundwater level close to surface. High content of organic compounds in water and low decay rate in anaerobic gleyic environment causes sulfate reduction and H_2S release. At the certain distance from the lake in marginal parts of oxbow depressions willow communities with *Filipendula ulmaria* occur.

Multi-dominant and multi-layered small-leaved forests (Figure 5, No 8) on high-level floodplains are of particular interest. Tree layer is composed of aspen (*Populus tremula*) and birch (*Betula pubescens*) as well as of *Alnus incana* and *Padus aviculare*. Undergrowth is represented by *Ribes sp.*, *Rubus idaeus*, *Rosa acicularis*, and *Lonicera xylosteum*. Eutrophic mesohygrophytes (*Filipendula ulmaria*) and mesophytes (*Aegopodium podagraria*) indicate high water and nutrients supply. This kind of floodplain units is in deep contrast with sandy terraces dominated by *Pinetum cladinosum communities* (Figure 5, No 9).

Anthropogenic Sources of Matter Input for Floodplain Landscapes

Floodplains and riparian zones are commonly considered to be extremely sensitive and ecologically significant elements of a landscape. This is explained by accumulation of matter from the entire basin and high diversity of habitats resulting in high biological diversity. Anthropogenic sources of matter input are associated mainly with arable lands in the deeply dissected middle sector of the Zayachya River basin. Plowing on steep slopes causes decrease of humus horizon thickness. In the background forests conditions upper sandy layer is commonly enriched with N, K, H, Al, Sr, Mn, Bi, Mo, and Zr. Deforestation and long agricultural history (since the 15th century) resulted in erosion and topsoil loss, including the sandy layer and, locally, loamy moraine. Therefore, on steep valley slopes matter of underlying marlstone is being involved in water and mechanical migration. This holds true for a group of chemical elements that increase concentration in the plowed soils at slopes and can be transported to the lower sections of a catena: Ca, Li, Ba, Cr, V, Co, Ni, Cu, Zn, Be, Ga, Sc, Y, Yb, and B.

In Figure 6 we illustrate various patterns of accumulation along catena, e.g., Ni – on slopes and floodplains; P – on deluvial trains superimposed over terrace; N – on deluvial trains superimposed over terrace; Mn – on floodplains. Enrichment of plowed soils with mobile forms of P and K as well as Ba, Cr, V, Ni, Cu, Zn, and B is explained by high fertilizers input.

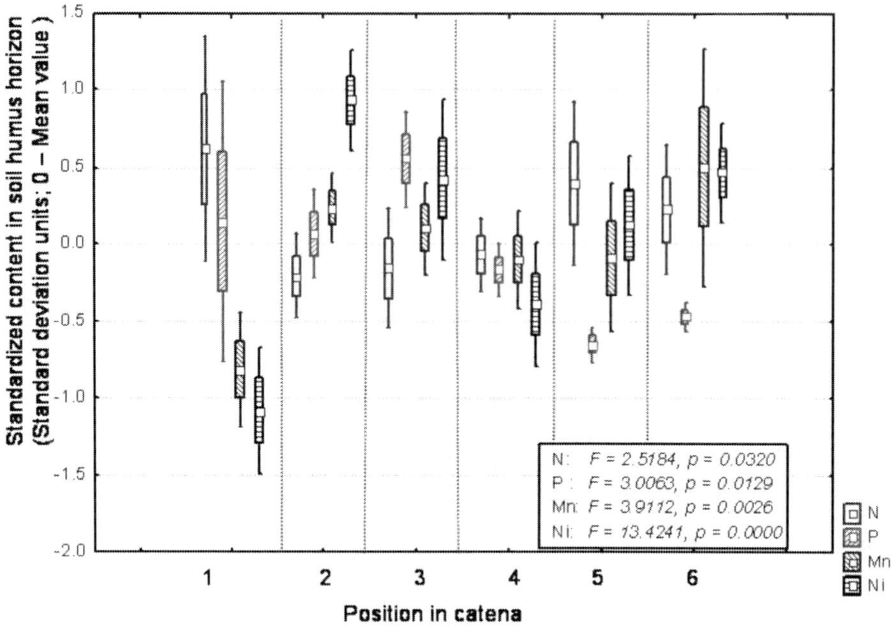

Figure 6. Spatial distribution of chemical elements concentration in soil humus horizon along catena. Position in catena: (1) interfluve (autonomous); (2) valley slope composed of Permian marlstone (trans-eluvial); (3) deluvial train superimposed over terrace (trans-accumulative); (4) terrace (autonomous); (5) deluvial train superimposed over floodplain (trans-accumulative); (6) floodplain (superaqual). N, Mn, Ni represents total content; P is in mobile forms. *F* is Fisher criteria; *p* is statistical significance (*p*-level).

Spatial Distribution of Chemical Elements along Catena in Agro-Landscapes

To reveal zones of concentration and deconcentration of chemical elements in a catena we compared occurrence of content maxima and minima in 35 catenas which have floodplains as the lowest sections. The results show evidence that the group of elements (P, Ba, Cr, Co, Pb, Sn, V, Ga, Y) have the highest concentrations on the deluvial trains superimposed over river terraces and thus cannot reach floodplains (Figure 6). Accumulation on the train quite close to the plowed slope testifies migration in the form of the solid matter as a result of soil erosion. In the narrowest sections of the side valleys (Mezhnitsa Zayacheretskaya, Kozlovka, Smutikha, Mozgolikha – see Figure 1 B) deluvial trains cover the entire terrace. Their distant sectors neighbour the floodplain.

For this reason we will focus attention on the other group of elements that are able to cross river terraces and accumulate on the deluvial trains superimposed over the floodplains. Such trains as a rule have alkaline soil reaction. In those sites we detected accumulation of N (Figure 6), Ca, Zn, Ag, Be, W, and Nb in soil humus horizons. The question of crucial importance is to what extent elements that reached floodplains can either be involved in biological cycle or be immobilized. To answer this question we compared trace elements contents in soil and in leaves of *Filipendula ulmaria* in two types of the deluvial trains superimposed over the floodplain (Table 3). The first one ("A-type" below) adjusts to the steep slope of the terrace which is fully covered by deluvial train. The second one ("B-type" below) adjusts to the slope of terrace that is covered by deluvial train only partially, the distant section of the terrace being isolated from matter input from the plowed slope.

In the humus horizon of the A-type train soil (pH 7.5) Li, V, and B have high concentrations both in soils and in *Filipendula ulmaria* leaves (CBU 6.0, 0.9, and 6.7, respectively – see Table 3) which testify migration from cultivated fields both in immobile and dissolved forms. Zn has relatively low CBU value (3.3) on train but high concentration in the floodplain soils testifies input by colloidal translocation. B, Li, V, Mo, and Ag have high CBU values on the other A-type train where alkaline groundwater emerge and deliver dissolved substance washed out from fields (Table 3). Hence, the above-mentioned trace elements are involved in water migration in cultivated fields and can easily reach floodplains. Biogeochemical barrier provided by *Filipendula ulmaria* communities is significant but cannot exclude transfer of this group of pollutants to the floodplains. However, we did not detected elevated contents in soils of floodplains. This enables us to come to the conclusion that floodplain soils do not act as barriers on the pathway from cultivated field towards water streams. Hence, strips of *Filipendula ulmaria* dominated meadows in the marginal sections of the floodplains provided biogeochemical barrier of crucial importance that is able to protect water streams from pollutants input. Note that this holds true for trace elements that are mobile in alkaline conditions only, i.e., in the impact area of matter input from steep slopes composed of marlstone.

In the humus horizon of soil at the B-type deluvial train superimposed over the floodplain (pH 4.5-5.2) another group of elements is involved in biological cycle. Acid reaction favours high mobility of Zn, Cu, Ni, and Mn (Table 3). This CBU values are 4-10 times as high as that on alkaline soils. Zn, Cu, Ni, and Mn have high contents in floodplain soils as well (Figure 6). This result shows that biogeochemical barrier is not effective for the trace elements with high mobility in acid conditions. In our opinion, two pathways of input to the floodplains are possible: lateral transfer from adjusting terrace slopes and transfer along ephemeral streams followed by alluvial deposition during floods. In contrast, the lower mobility of Li, V, and B in acid soils as compared to alkaline ones (see above) is indicated by low CBU values (Table 3). For Sn, Zr, W, and Ag we detected accumulation in immobile forms without significant biological uptake by *Filipendula ulmaria* communities.

Table 3. Coefficients of biological uptake (CBU) for *Filipendula ulmaria* leaves depending on soils reaction

Element	Soil	
	alkaline	alkaline with emergence of groundwater
Li	6.0	7.5
V	0.9	0.2
B	6.7	10.0
Mo	3.3	2.5
Ag	2.5	3.8
Sr	2.0	1.3
Ba	1.6	1.0
Mn	0.5	0.8
Zn	3.3	3.8
Cu	2.0	0.7
Ni	0.5	0.4

The matter transfer by ephemeral streams (during snowmelt, in particular) along gullies contributes much to soil pollution on floodplains. We hypothesized that trace elements transfer along gullies was regulated by relief and degree on hydrological connection between steep marlstone slopes and floodplains. The comparison of concentrations in various sections of gully bottoms showed that N, Mg, Ni, and Zn can easily reach alluvial fans at neighbouring floodplains. In contrast, transportation of P, Ca, Sr, and Co depend on longitudinal profile of a gully. They can reach alluvial fan either if the distance to the slope is less than 250 m or if the stream is strong enough.

Geochemical conditions on the Zayachya River floodplain indicate intensity of lateral flows in the entire basin. High percentage of arable lands and deforestation within a basin encourage involvement of Ca and Mg in water migration and deposition in the floodplain soils. Concentration of exchangeable Ca and Mg in soils increase sharply if percentage of arable lands in a small river basin accounts for 50-80% and more while percentage of forest cover falls down to 7-20% which is concordance with content in water (Figure 7, No 5-10). In contrast, exchangeable K does not accumulate in floodplain soils. However, its content can be high in bogs at floodplain inner margins, on alluvial fans, and in talwegs of short gullies cut into terraces. This testifies the ability of K to migrate in dissolved forms along catena. Thus, at least two hierarchical levels of spatial organization impose controls over distribution of trace elements in a landscape. Ca and Mg are controlled by the basin-level processes while K is governed by the catena-level ones.

High input of bases to the floodplains significantly affects the pH condition. More acid soils (pH 6.1-6.8) are typical for the floodplain of the Zayachya River which has high forest cover percentage in the upper reaches. In contrast, floodplains in the side valleys are located in close proximity to arable fields and deluvial trains and, hence, have more neutral and weakly alkaline soils (pH 7.0-7.5).

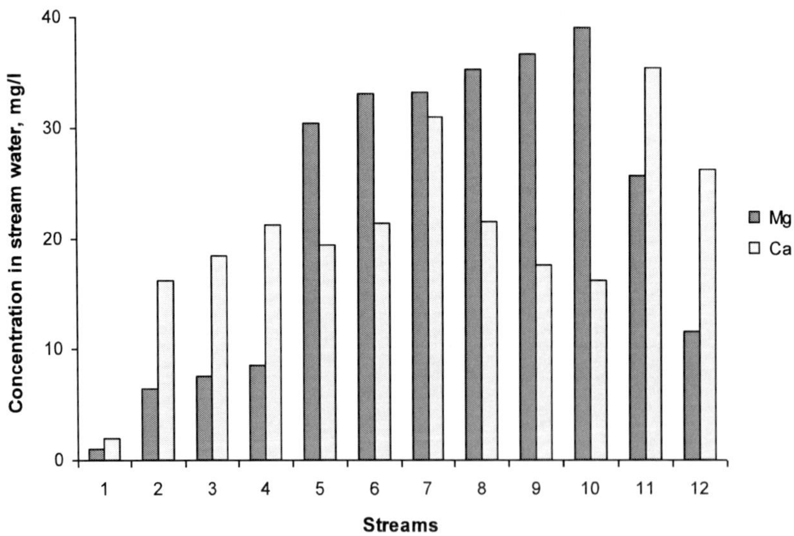

Figure 7. Spatial variation of Mg/Ca ratio in the river waters of the Zayachya River basin streams (for location see also Figure 1 B). The upper Zayachya River basin: (1) Malysh, (2) Zayachya in the upper reaches, (3) Maly, (4) Bolshoy; the middle Zayachya River basin: (5) Kozlovka, (6) Mezhnitsa Zayacheretskaya, (7) Strugnitsa, (8) Smutikha, (9) Mezhnitsa, (10) Mozgolikha, (11) Zayachya in the middle reaches; (12) Kokshenga.

Deposition during flood events ensures enrichment of superaqual soils by N, Mg, Mn, Ba, Ni, Cu, Zn, and W (Figure 6) in comparison with the higher sectors of catenas. It is worth noting that Cu, Ni, Zn, and Mn are delivered to the floodplain soils not with floods only, but with transfer in mobile forms from the terraces across deluvial trains with acid soils. At the same time, we found no evidence of elevated contents in the floodplain soils for trace elements absorbed actively by *Filipendula ulmaria* on deluvial trains with alkaline reaction (B, Li, V, Cr). In our opinion, large fraction of these elements is intercepted at the biogeochemical barriers at the marginal sectors of floodplains adjacent to terraces.

Bottom sediments were analyzed as indicators of the final stage of migration. We compared sediments in basins with different percentage of arable lands. Within agro-landscapes concentration of P turned out to be 2-5 times higher than that in the forested basins. Bottom sediments in the Mozgolikha River basin (percentage of arable lands 55%) have high concentrations of trace elements (Cr, Ni, Cu, Sn, Mo, B), while in the Strugnitsa River basin (percentage of arable lands 2%, percentage of forests 71%) bottom sediments are poor in Mn, Cr, V, Ni, Co, and Zn, i.e., those elements involved in migration due to plowing.

Biogeochemistry of Floodplain Landscapes

The analysis of floodplains landscape structure showed evidence that natural riparian forests still occur in the Zayachya River upper reaches and in the valleys of its tributaries. By this way the natural structure of middle taiga is sustained. Within agro-landscapes of the middle and the lower reaches meadows dominate strongly over fragmented riparian forests. Therefore, we choose aboveground phytomass and nutrients stock in herb cover to characterize biogeochemical contrasts in the floodplains. These parameters are strongly interrelated (correlation coefficient 0.84) and highly variable in space (variation coefficient 0.55 and 0.46, respectively). Edaphic heterogeneity in floodplains affects structural and functional properties of phytocoenoses and encourages species with various phylogenetic specialization.

The share of herb cover in the total phytomass of the middle taiga landscapes increases towards the lower elements of catena. Within the interfluve forests in the upper section of the Zayachya River basin herbal phytomass rarely exceeds 0.4-0.5 t/ha. However, in birch-and-spruce forests of the high-level floodplain it varies from 0.53 to 1.17 t/ha. This tendency is in concordance with the data for the East-European middle taiga (Bazilevich, 1993; Bazilevich and Titlyanova, 2008). More significant increase of phytomass was observed in the low-level floodplain forests with domination of *Alnus incana* due to abundance of tall forbs. The typical species for these communities is *Filipendula ulmaria* which has high biogeochemical activity manifested in biological accumulation of trace elements (B, Ag, Zn, Cu, Mo) delivered to the floodplains from catena by groundwater. This enables us to emphasize specific role of herb cover in involving elements to biological cycle, though the greatest barrier function is provided by trees.

In the upper reaches of the Kozlovka and the Strugnitsa Rivers Gleyi-Histic Fluvisols and Umbri-Gleyic Fluvisols are not saturated by mobile phosphorus (10-40 ppm) and potassium (80-170 ppm). Total nitrogen content (1400-1500 ppm) is higher than in soils of the meadow floodplains. Nitrogen is accumulated at the biogeochemical barrier in organic horizons which is in compliance with the important role of *Alnus incana* in biological uptake.

In the floodplains of the Zayachya River middle reaches biogeochemical parameters are densely related to water supply. Herbal phytomass and supply of metals in ash increase gradually in the following sequence of communities: bogged sedge-dominated meadows – mesophytic multi-dominant meadows – meadows with domination of tall forbs and gramineous plants (Figure 3 B). Neighbourhood with arable lands affects variation of biogeochemical parameters as well.

Aboveground phytomass in species-rich mesophytic meadows achieves 4 to 5 t/ha. The largest share is inherent for forbs (2-3 t/ha) and gramineous plants (0.8-2.7 t/ha). Though phytomass of legumes is lower (0.2-0.5 t/ha) their presence provides higher forage value. Supply of metals in ash accounts for 0.27-0.36 t/ha in the Zayachya River floodplain, spatial variability being dependent on ratios between dominant species and inner mosaics.

Trace elements content in herbal phytomass is related to biogeochemical activity of species belonging to various taxa. Forbs and legumes accumulate B more actively (150-200 ppm, coefficient of concentration (CC – here and below) 12-17) as compared to gramineous plants (30-60 ppm, CC 2.5-5). Most species concentrate Cu and Zn (100-300 ppm, CC 3-6), Mn (300-800 ppm, CC 0.3-0.8) which provide important biological functions (photosynthesis, etc.). Their mobility and bio-availability is explained by favourable acid reaction in Umbric Fluvisols (pH 5.0-5.6). Published data show that in the phytomass of boreal floodplains the increased share of gramineous plants ensures the contents of macroelements as follows (kg/ha): Si 133, N 116, Ca 70, K 64, P 11.5 (Bazilevich and Titlyanova, 2008). High Si content is associated with well-pronounced concentration capacity typical for gramineous plants.

In contrast to mesophytic meadows, bogged sedge meadows in oxbow depressions produce less phytomass (2.4-2.7 t/ha, rarely up to 4.5 t/ha) and have simpler composition. Sedges are the main constituents with low ash content (3.8-4.9%). Simultaneous decrease of both phytomass and ash content results in sinergetic effect, that is decrease of phytobarrier retention capacity with stock of mineral elements in herbs 0.12-0.21 t/ha. Permanent gley conditions in acid Umbri-Gleyic Fluvisols enforces mobility of Mn which is involved into biological cycle rapidly. Its content in ash of sedges and hygrophilous forbs accounts for 1500 ppm (CC 1.5-2.0).

Biogeochemical properties of the floodplain meadows have been strongly affected by recent cessation of haying (Figure 8). Time series testifies that structural and functional parameters have undergone significant changes in 1997-2011 due to succession. Species richness in mesophytic meadows decreased from 32 to 10 species. We detected extinction of some gramineous plants (*Dactylis glomerata*, *Festuca pratensis*, *Phleum pratensis*), lower abundance of *Polygonum bistorta* and other forbs. These changes are explained by expansion of *Filipendula ulmaria* and *Deschampsia cespitosa* with associated increase of projective cover and herbs height. Succession caused increase of herbal phytomass from 4.2 to 6.8 t/ha followed by decrease of forage value and extinction of forage species.

In bogged depressions increase of phytomass from 2.3 to 5.1 t/ha is related to expansion of *Carex acuta* and appearance of *Deschampsia cespitosa* while share of waterlogging-resistant forbs and legumes is low. Phytomass on the deluvial trains superimposed on the inner sections of the floodplains changed only slightly due to permanent dominance of *Filipendula ulmaria*. This trend reflects growing similarity of biogeochemical parameters of abandoned haying meadows and tall forbs dominated meadows prevailing on high-level floodplains and deluvial trains.

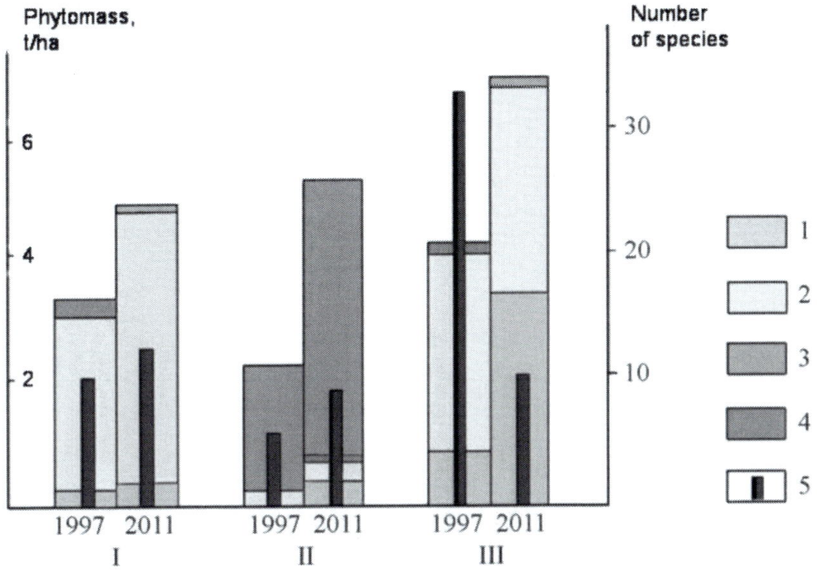

Figure 8. Changes in structural and functional properties of floodplain after cessation of haying: (1) gramineous plants, (2) forbs, (3) sedges, (4) legumes, (5) number of species. I – alluvial fan superimposed over floodplain, II – oxbow depression on the high-level floodplain, III – high-level floodplain.

Changes in biological productivity of wet tall forbs meadows depend on both species composition and various neighbourhoods to arable lands. In agro-landscapes neighbourhood has a crucial importance since sheet and linear erosion provides lateral input of nutrients to the lower sections of catenas.

Meadows adjusting to arable fields (floodplain of the Zayachya River between the mouths of the Smutikha and the Kameshnitsa rivers, floodplain of the Kozlovka River) have the largest phytomass (7.4-8.9 t/ha) and mineral elements stock (0.45-0.58 t/ha). Tall forbs constitute the bulk of the phytomass and provide high retention capacity of phytobarrier. The maximal possible phytomass and mineral elements stock (8.9 and 0.58 t/ha, respectively) was detected on floodplains of small rivers surrounded by cultivated fields (the Kameshnitsa River).

If cultivated deluvial trains adjust floodplains with high water and nutrients supply (mobile P 114 ppm, mobile K 175 ppm), the meadows with tall *Umbellifearae* species (*Heracleum sibirica* and *Anthriscus sylvestris*) are developing. High values of phytomass and mineral elements stock (7.4 and 0.45 t/ha, respectively) indicate high retention capacity of the phytobarrier (Table 4). Despite the seasonal fluctuations, involvement of chemicals in biological cycle and the consequent accumulation in organic soil horizons act as powerful constraints for output from the landscape with river discharge. This holds true for potassium and phosphorus, in particular, because of the active accumulation by *Umbellifearae* species.

High biological productivity was observed for meadows in gully bottoms and alluvial fans where *Filipendula ulmaria* dominates totally. This species with high biogeochemical activity indicates transit of moisture. The gully bottoms act as the main pathways for nutrients washed out from fields which results in the highest phytomass and stock of mineral elements (8.1 and 0.74 t/ha, respectively). In the lowest sections of catenas such meadows adjust tall forbs meadows on floodplains and provide barrier function of ecological network.

Table 4. Phytomass and mineral elements stock in ash for herb layer of floodplain meadows and forests of the Zayachya River basin

Plant associations	Aboveground herb phytomass, t/ha	Stock of elements in ash, t/ha
Floodplain forests in the Zayachya River upper reaches		
Birch-and-spruce forests with forbs (the Zayachya River)	0.53-1.17	0.07-0.19
Alder forests with tall forbs (tributaries)	2.00-3.76	0.18-0.19
Floodplain meadows in the Zayachya River middle reaches		
Meadows with co-dominance of legumes, forbs and gramineous plants (the Zayachya River)	4.18-5.24	0.27-0.36
Tall forbs meadows with *Filipendula ulmaria* (the Zayachya River)	3.52-7.44	0.15-0.39
Tall forbs meadows (tributaries)	6.04-8.92	0.42-0.58
Tall forbs meadows on deluvial trains superimposed on the floodplains of the Zayachya River and its tributaries	3.64-7.44	0.30-0.45
Wet sedge meadows in oxbows	2.36-4.52	0.12-0.21
Statistics		
Average content	4.00	0.28
Standard deviation	2.18	0.13
Coefficient of variation	0.55	0.46
Coefficient of correlation	0.84	

Stream-Water Chemistry

In cascade landscape-geochemical systems of river basins, the runoff acts as a system-forming factor. One-way direction of water movement ensures functional integrity of a catena. The amount of liquid and particulate matter in the river mouth characterizes the output from a basin and provides integrated information concerning migration (Antipov and Fedorov, 2000; Korytny, 2001; Aseyeva et al., 2004; Glazovskaya, 2007). Floodplain landscape units contribute to regulation of matter output by means of barrier zones due to the development of mechanical and physical-chemical barriers as well as phytocoenoses with high demand in nutrients and water. The barriers' functioning restricts release of chemicals to water streams. Geochemical traps play important role for elements with high migration ability.

Conditions of ion migration differ in various parts of the Zayachya River basin and increase spatial variability of geochemical parameters (Table 5). In the upper reaches of the Zayachya River and its tributaries small streams contribute to nutrients supply in the warm period. They receive runoff from the oligotrophic and mesotrophic bogs located within the interfluve areas. Bog waters are acid (pH 3.7-3.8), poor in dissolved salts and belong to Cl-SO_4-Ca type. These waters are poor in biologically active macroelements (Ca, Mg) and are hydrocarbonate-free. Prevalence of chlorides and sulphates in bog waters is explained by atmospheric input. Dissolved F, Si, and mineral P have low content with the exception of small lakes located in depressions.

Table 5. Mineralization (*M*), ion composition and content of dissolved F, Si, and mineral P (P_{min}) in surface and ground waters

Water bodies (samples No)	*M*, mg/l	Concentrations in solution, mg/l									
		Na	K	Mg	Ca	Cl	SO_4	HCO_3	F	Si	P_{min}
The Zayachya River upper reaches											
Streams of elementary catchments and the Zayachya River											
Bolshoy (No 22)	136	1.13	0.46	8.46	21.2	3.28	9.47	91.5	0.061	3.54	0.033
Maly (No 23)	122	1.25	0.76	7.46	18.4	1.97	11.0	81.6	0.050	2.81	0.010
Malysh (No 24)	14.8	0.84	0.92	0.99	1.97	6.57	3.47	0	0.008	5.19	0.012
Zayachya (No 25)	109	1.01	0.36	6.47	16.1	3.28	3.79	77.8	0.055	2.15	0.009
Hollows and lakes											
Hollows within mesotrophic bogs (No 27)	15.1	0.47	5.26	0.67	0.69	7.88	0.12	0	0.009	0.98	0.004
Lake in suffusion depression (No 28)	51.6	0.57	0.57	3.65	7.55	3.94	0.23	35.1	0.040	0.79	0.041
Hollows within the Zayachya River floodplain (No 26)	283	2.03	0.41	17.9	41.0	1.97	0.94	219	0.168	3.12	0.010
The Zayachya River middle and lower reaches											
The Zayachya River and its tributaries											
Strignitsa (No 33)	314	3.31	1.33	33.2	31.0	2.30	14.4	229	0.208	5.26	0.014
Kozlovka (No 34)	247	1.74	0.83	30.4	19.4	2.30	4.59	188	0.097	3.59	0.011
Mezhnitsa Zayacheretskaya (No 36)	280	2.27	1.83	33.1	21.3	5.25	4.27	212	0.187	5.24	0.015
Smutikha (No 37)	264	1.97	1.45	35.3	21.5	12.5	7.98	183	0.194	4.22	0.012
Mozgolikha (No 39)	285	2.11	1.23	39.0	16.2	6.57	4.79	215	0.146	5.09	0.013
Mezhnitsa (No 40)	291	2.74	1.76	36.7	17.6	6.89	5.87	220	0.214	5.63	0.015
Zayachya (No 38)	296	1.90	1.29	25.7	35.4	4.92	3.97	223	0.199	4.18	0.012
Mean	*282*	*2.29*	*1.39*	*33.3*	*23.2*	*5.82*	*6.54*	*210*	*0.178*	*4.74*	*0.013*
Groundwater											
Spring at the Zayachya River valley slope (No 29)	250	1.50	1.73	31.8	14.1	4.27	4.18	192	0.217	5.27	0.015
Spring at the Kozlovka River valley slope (No 35)	244	1.63	0.98	32.9	14.3	1.97	2.52	190	0.118	4.50	0.017
Well (No 30)	261	4.63	0.74	31.7	16.3	5.58	8.98	193	0.110	3.75	0.012
Mean	*252*	*2.59*	*1.15*	*32.2*	*14.9*	*3.94*	*5.23*	*192*	*0.148*	*4.51*	*0.015*
The Kokshenga River valley											
Koshenga (No 31)	186	2.87	0.89	11.5	26.2	3.94	15.2	125	0.131	3.20	0.011
Oxbow lake at the Kokshenga River floodplain (No 32)	2937	169	1.16	82.8	621	198	1766	99.1	0.508	0.18	0.062

While small streams cross forested middle-taiga plains the water is gradually being transformed due to leaching from canopy and soils. The changes involve (1) increase of mineralization up to 109-136 mg/l and growing content of most ions, (2) increase of sulphate content due to leaching from conifers needles and soil organic horizons, (3) appearance of hydrocarbonate ions and corresponding change in water type to HCO_3-SO_4-Ca and further to HCO_3-Ca type.

In general, in the Zayachya River upper reaches the stream-water chemistry is influenced mostly by biological cycle under high humidity conditions in interfluve areas, slowed down decay of organic compounds, and output of biogenic elements by means of acid and acid gleyic leaching. The highest mobility is inherent for Cl and S and high mobility is for for Ca and Mg. The elements that are flushed from the upper sections of a catena are intercepted by geochemical traps developing in the marginal parts of the high-level floodplains. In these locations we detected maximal values of mineralization (more than 200 mg/l), and the highest contents of Na, Ca, Mg, and F. The number of such depressions is not high; their low retention capacity does not prevent elements export with streamflow (Avessalomova et al., 2013).

Agro-landscapes in the Zayachya River middle reaches affect stream-water chemistry in a different way as compared to the middle taiga landscapes of the upper reaches. First, geologic factor influences much stronger since bedrock are closer to the surface in the core area of the plateau. Second, anthropogenic loads increase greatly and generate specific sources of matter input such as fertilizers. Mechanical migration becomes much more intensive due to plowing. Whilst the mineralization increases, contents of elements grow due to various reasons.

Correlation coefficients (0.77-0.99, $p < 0.01$) indicate dense linkages between, on the one hand, mineralization and content of HCO_3 (alkalinity), and, on the other hand, content of Ca, Mg, and F. Content of Ca, Mg, and HCO_3 in agro-landscapes is 4, 2 and 3 times higher, respectively, than in the upper reaches. In parallel, hardness of water increases from 1.2 to 3.9 mg-eq/l, with carbonate constituent being dominant (3.6 mg-eq/l).

These values are similar to that of groundwater emerging in the inner margin of floodplains (pH 7.4). Hardness and alkalinity of waters affect strongly conditions of water migration in Umbri-Gleyic Fluvisols under meadows with dominance of tall forbs. Humus horizons preserve neutral or weakly alkaline reaction (pH 6.5-7.6). Migration of elements occurs in conditions of neutral leaching. At the boundary between floodplains and terraces the emergence of groundwater is commonly indicated by narrow strips with dominance of *Filipendula ulmaria*. The strip has phytomass up to 5.0 t/ha and ensures interception of elements coming from the upper sections of catena (Table 6). By this, lateral phytobarrier can retain up to 0.3 t/ha of mineral substances (Avessalomova, 2014).

Gleyi-Histic Fluvisols occurring in bogged oxbow depressions have pH 5.9. Release of organic acids during sedges remnants decay favours soil acidity. Contrast of geochemical conditions in the floodplains of the Zayachya River middle reaches create prerequisites for the development of geochemical barriers in contact zones with close neighborhood of oxic and reduced or acidic and alkaline conditions.

Table 6. Phytobarrier at the contact of floodplains and agro-landscapes

Agro-landscape units	Contact zone	Floodplain
Deluvial train		
on the structural-accumulative terrace	on floodplain	
Trifolium sp. and *Phleum* sp. communities	Tall forbs meadows with *Heracleum* sp.	Meadows with *Calamagrostis* sp. and *Filipendula ulmaris*
Phytomass, t/ha		
3.08	7.44	3.37
Stock of elements in ash, t/ha		
0.16	0.45	0.29

Regional specifics of stream-water chemistry within the structural plateau is manifested in Ca/Mg ratio. During summer low water small watercourses have base alimentation resulting in Mg prevalence over Ca (33.3 and 23.2 mg/l, respectively) and HCO_3-Ca-Mg water type (Figure 7). This is in concordance with stream-water chemistry in the springs at the toe slopes (32.2 and 14.9 mg/l, respectively) affected by the Sukhona formation rocks. The bedrock influence is well-pronounced during dry summer periods and much less – in rainy time. Marlstone substrate causes prevalence of exchangeable Mg over Ca in soils as well (Gavrilova and Gorbunova, 2002). Outside the plateau, towards the Zayachya River lower reaches, concentrations of these base ions becomes more similar. The Kokshenga River water belongs to HCO_3-Ca type.

Fluorine can be treated as the other indicator of the Permian carbonate rocks influence on stream-water chemistry. Its accumulation is supposed to take place on alkaline barrier during the rocks deposition. According to F content increase affected by growing influence of bedrocks the stream waters of the Zayachya River basin the following order is formed (Figure 9): bogs and streams in the upper reaches (0.01-0.06 mg/l) – hollows at the floodplain (0.17 mg/l) – the Zayachya River and its tributaries in the middle reaches (0.10-0.21 mg/l) – springs and wells (0.11-0.22 mg/l) – the Zayachya River lower reaches and the Kokshenga River (0.13 mg/l) – oxbow lakes on the Kokshenga River floodplain (0.51 mg/l). This sequence testifies high variability of F concentration (coefficient of variation over 70%) and its increased water migration in the middle reaches. Accumulation of dissolved F at floodplains is typical for the traps in oxbow lakes and hollows in the inner margins. Traps retention capacity increases where the floodplains have old meander belts, on the Kokshenga River wide floodplain (Figure 4 B) in particular (Avessalomova et al., 2012).

Besides natural factors, migration of Cl, Na, K, and P in streams is strongly affected by anthropogenic factors as well. Fertilizers input results in growing concentrations of P and K, wastewaters provide elevated contents of Cl and Na in the rivers within the agro-landscapes. The highest Cl concentration (12.5 mg/l) was detected in the rivers within settlements in the Zayachya River upper reaches. Maximal Na concentration occurs in the wells (4.6 mg/l and more) and in streams with livestock farms in the upper reaches (2.7-3.3 mg/l). In comparison with forested catchments, concentration of K is elevated in the streams flowing among arable fields (0.4 0.9 and 1.3-1.8 mg/l, respectively). As the Zayachya River receives runoff from the first-order tributaries, K concentration increases 3.6 times relative to the upper reaches.

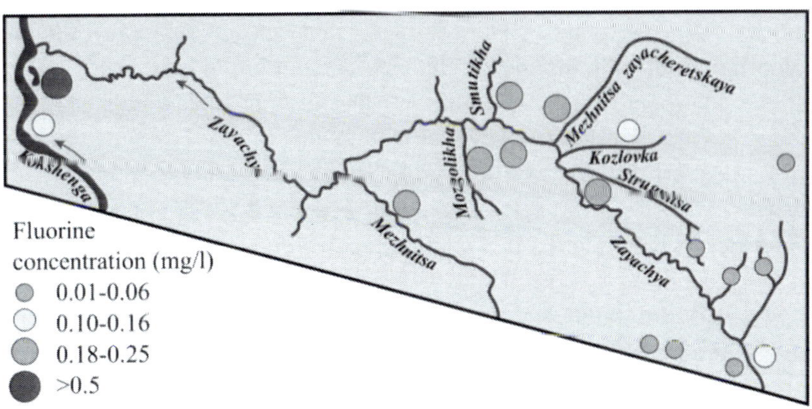

Figure 9. Spatial distribution of fluorine in the Zayachya River basin waters.

At the same time one cannot exclude natural controls over ions contents in the agro-landscapes stream waters. Their contribution is explained by possibility of emergence of the groundwater with high ions content, Cl in particular, from the water-bearing strata of the Sukhona formation. Permian silicate sandstone can contribute to elevated K concentrations. In these strata K concentration exceeds clarke, and its leaching is evidenced by K-rich minerals presence in solid residue of aquifer.

Phosphorus migration is not strongly affected by geologic factor. Its total concentration in the Sukhona formation rocks varies greatly but does not exceed 300 ppm which is lower than clarke (CC 0.3). Arable soils are commonly phosphorus-rich (mobile P on average accounts for 267 ppm). Concentration decreases in the lower soil horizons down to zero in marlstone eluvium. This shows evidence that fertilizers serve as the main source of P input in the agro-landscapes with two principal pathways. The first one is erosion-generated and results in mechanical migration, matter accumulation in the lowest sections of the catena and P involvement in sediment discharge. Comparison of chemical composition of deposits collected immediately after flood event showed evidence that P concentration in soils within the agro-landscapes is 2-5 times as high as in that of the Zayachya River upper reaches. Migration in particulate form favours enrichment of floodplain soils with P. Phosphorus supply in soils varies from low to moderate (mobile P 30-140 ppm), but anyway is higher than in the floodplain soils within forested catchments (10-40 ppm).

The second migration pathway is connected to water migration of P in dissolved form. The downhill motion of meltwater across arable fields and groundwater serves as main input into the stream water.

Surface water in the Zayachya River upper and middle reaches differ in mineral P content with coefficient of variation over 90%. To evaluate spatial contrasts we calculated coefficient of lateral differentiation (*L*). Mineral P content in water of hollows in mesotrophic bogs at the interfluve area (0.004 mg/l) was taken as background value (*L* 1). The increase of *L* up to 5 in stream waters of the agro-landscapes and up to 16 in the floodplain geochemical traps indicated elevated output from cultivated fields (Figures 4 B and 10 A). Phosphorus is involved in biological cycle by meadow communities at the sites of groundwater emergence near the inner margins of floodplains, on the deluvial trains, and on the main floodplain surface. In sites where *Filipendula ulmaria* dominates, the biologic uptake of P can account for 20600 ppm (CC 22). High phytomass value in tall forbs meadows encourages P interception from waters and soils due to high retention capacity.

Silica follows the same spatial pattern in agro-landscapes as mineral P though with less variability (coefficient of variation 47%). Low Si concentration was observed in waters of hollows within mesotrophic bogs (0.98 mg/l, *L* 1) and connected small lakes (0.79 mg/l). Within the bogs low decay rate of *Sphagnum* mosses favours Si accumulation in peat which serves as a biogeochemical barrier. Nevertheless, high acidity encourages elevated Si solubility and its involvement in water migration. For small water courses in the Zayachya River upper reaches Si input is provided by biological cycling and leaching from needles and forest floor. In general, Si is considered to be low-active water migrant (Perel'man and Kasimov, 1999). It is accumulated at the floodplains in hollows of their inner margins (3.12 mg/l, *L* 3.2) with birch-and-spruce forests (Figure 10 B) due to input both with groundwater and from biomass. In the agro-landscapes elevated Si concentration was detected in streams with plowed catchments (*L* 4-5). To explain Si migration one should consider its mobility in alkaline environment and possible input from silicate sandstones of the Sukhona formation. In contrast to P, Si concentration in oxbow lakes is minimal (0.18 mg/l, *L* 0.2) which is even less than in oligotrophic mires.

Floodplains and riparian zones as structural elements of ecological network within the agro-landscapes play a crucial role in restriction of matter export with streamflow. At barrier zones on floodplains we distinguish two mechanisms of elements accumulation. The first one causes loss of mobility and nutrients accumulation on geochemical barriers, such as phytobarrier of tall forbs meadows, sorption barrier in soil humus horizons, oxygen barrier at the groundwater emergence sites, etc. The second mechanism involves accumulation in dissolved form at the sites of geochemical traps in oxbow lakes. The most effective traps are located in the lakes at the Kokshenga River floodplain. Under stagnant water regime the waters are characterized by high mineralization (over 2 g/l) and belong to SO_4-Ca type. Besides input with streamflow and groundwater, the sulphate content increase is associated with gypsum-rich red-coloured Permian rocks of the Nizhneustyinskaya formation in the Zayachya River lower reaches and along the Kokshenga River valley. Maximal concentrations of Cl (198 mg/l), Na (169 mg/l), Ca (621 mg/l), mineral P (0.062 mg/l), and other nutrients cause lake eutrophication. *Lemnaceae* species abundance indicates chloride elevated content. Strip of sedges and tall gramineous plants (*Scirpus lacustris*, *S. sylvaticus*, and *Phragmites australis*) developing around the shoreline intercept Si and form circular phytobarrier. Si is the only water migrant which accumulated not in the dissolved form in lake water but at the phytobarrier along shorelines within the riparian zone.

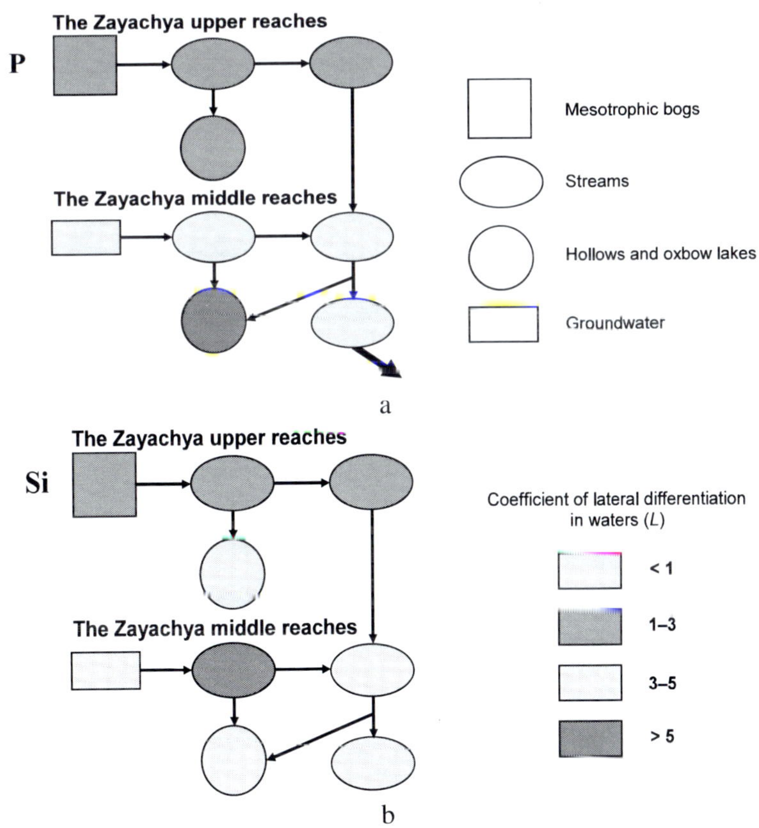

Figure 10. Lateral differentiation of phosphorus and silica in the surface waters.

Discussion

Comparing floodplains in the Zayachya River basin to that close to its mouth shows evidence that landscape structure is getting more complicated towards the Kokshenga River valley. We see the explanation in geologic and geomorphic differences, spatio-temporal organization, and history of these middle-taiga territories, both natural and anthropogenic ones. The above-mentioned factors in combination with climatic ones determine runoff and its effects for morphology of valley landscape units.

Forested floodplains in the narrow Zayachya River valley in the upper reaches have the simplest landscape structure. Within the agro-landscapes interior heterogeneity increases due to high contrasts in water supply and neighboring cultivated areas. Floodplains in the middle reaches with forests and meadows serve as a crucially important constituent of the ecological network since they provide buffer function in the proximity of field margins. The largest contribution of runoff to landscape heterogeneity was detected in the lower reaches of the Zayachya River and in the Kokshenga River valley resulting in the diversity of floodplain landscape units, soil and plant covers.

Thus, analysis of floodplain landscape structure as the lowest section of cascade landscape-geochemical system is urgently needed to reveal their barrier functions in relation to matter migration.

We identified three main migratory flowpaths that affect geochemical structure of floodplain landscapes under agricultural land use in catchments. Each flowpath has specific barriers which can be used or enforced to make ecological network more efficient.

The most intensive flowpath is associated with output of particulate matter or dissolved substance along gullies cut into steep slopes. Longitudinal profile of a gully serves as a migration regulator. Depending on gradient variations the matter can be either partially deposited within the gentle sections of a gully bottom or transferred down to the alluvial fans superimposed over the floodplains. Locally matter accumulation occurs at the intermediate alluvial fans superimposed over terraces.

The second flowpath is developed on plowed slopes during snowmelt which generates hillside flush and sheet erosion resulting in matter migration with final accumulation on deluvial trains superimposed over river terraces. The length of deluvial train acts as a regulator of chemicals transfer. If the marginal sector of the train coincides with the terrace edge (i.e., the entire terrace is covered by train) matter penetrates to the terrace slope and further to the train superimposed over the floodplain. In that case alkaline reaction in the train soils favours high mobility of B, V, Mo, Li, and Ag which can be partially involved in biological cycle and penetrate to the floodplains as well. Acid soils occur on the deluvial trains and alluvial fans if the catchment is located within terraces and not connected to the marlstone slopes. In that case active biological uptake and migration towards floodplains is typical for Cu, Ni, Zn, and Mn. Nitrogen is the most well-expressed pollutant that accumulates in soils of the deluvial trains superimposed over floodplains as well as in soils of the alluvial fans. Tall forbs meadows with high productivity (e.g., communities with domination of *Filipendula ulmaria*, *Aconitum septentrionale*, and *Delphinium elatum*) form biogeochemical barrier which is able to immobilize certain amount of pollutants and prevent them from reaching floodplains and water bodies. This supports the idea that meadow strips along the floodplain inner margins should be treated as crucially important elements of ecological network. Thus pH conditions at the terrace slopes and deluvial trains contribute much to migration regulation and determine risk of pollution of floodplains and water streams.

The third migration flowpath connects plowed slopes and floodplains by means of groundwater. Dissolved substance infiltrates to the alluvial sands on river terraces and emerges on the steep terrace slope at the contact with the watertight morainic loams or marlstones. Soil conditions around the spring sites are commonly alkaline which ensures immobilization of Cu, Ni, Zn, Mn, Sr, Ba and suppresses biological uptake. Migration with groundwater is the most important way of phosphorus transportation from the cultivated fields to the floodplains and water objects. Big share of forest areas within a basin reduces input of mineral forms of phosphorus to the rivers.

Thus, we managed to reveal at least three natural geochemical mechanisms that are able to protect floodplain landscapes and stream water from pollutants washed out from fields. First, plant communities, tall forbs and alder forests in particular, form biogeochemical barriers high retention capacity and deserve careful preservation as the critical elements of ecological network. Second, soils encourage deposition of this or that group of chemical elements dependently on pH and Eh conditions. Alkaline barriers are the most typical. Finally, the deluvial trains form mechanical barriers and reduce amount of pollutants that reaches floodplains. Relatively narrow side valleys have less efficient mechanical barriers since deluvial trains reach the terrace edge. The Zayachya River have much wider flat terraces that provide buffer zone between erosion-affected slope and floodplain.

Our data show that floodplains have crucial importance in the migration structure of the cascade landscape-geochemical systems. Their role is manifested in formation of streamflows. The diversity of floodplain landscapes and contrast migration conditions affect efficiency of barrier zones and by this control stream-water chemistry.

CONCLUSION

Contrast conditions of lateral flows in the small catchments and catenas affect the floodplain landscape morphology which differs much along the cascade landscape-geochemical systems of river basins. Variability in proportions of system-forming processes results in high diversity of floodplain landscape. Within the agro-landscapes, the floodplain diversity is determined by the active gravity-generated slope processes which produce various neighbourhoods of landforms at the sites of matter mechanical accumulation (such as deluvial trains and alluvial fans).

Biological cycle is the main binding factor that determines stream-water chemistry in the forested upper basins. Contribution of geologic factors increases in the deeply dissected middle portion of the basin where carbonate substrate is close to the surface. Anthropogenic impact emphasizes the influence of carbonate rocks. This causes increase of mineralization and content of base ions in stream water. The export of biogenic and biologically active elements (P, K, etc.) by stream water grows significantly due to agricultural activity.

The floodplain landscapes affect migration pathways by means of barrier zones. Barriers are responsible for regulation of geochemical connections between small catchments and the integrating stream. On their pathway toward floodplains the chemical elements can be intercepted by the deluvial trains. Within the floodplain interception functions are inherent for: (1) meadow communities with high productivity and high biogeochemical activity, (2) hydromorphous units providing deposition in soil histic horizons, (3) ecotone units at the contact of contrasting pH and/or Eh conditions of water migration.

The most complicated spatial pattern is developed in the floodplain inner margins where elementary units differ greatly in plant cover (e.g., forests, meadows, and bogs) and in set of ecological and geochemical processes. Biogeochemical, physico-chemical and mechanical barriers neighbour to the geochemical traps located in the oxbow lakes. Traps capacity increases in the lower sections of the cascade landscape-geochemical systems of the river basins. Simultaneous accumulation of a wide range of chemical elements in such floodplain units function as the systems with strong positive feedback loops that enforce eutrophication.

Identification of barrier zones regulating flows of matter is urgently needed for assessment of floodplain significance while projecting ecological network within agro-landscapes.

Acknowledgments

The financial support from Russian Foundation for Basic Research (projects 13-05-00821, 14-05-00170, and 14-05-00624) is acknowledged.

References

Ahern, J., 2004. Greenways in the US: Theory, trends and prospects. In: *Ecological Networks and Greenways*. Concepts, Design, Implementation. Cambridge, Cambridge University Press, 34-55.

Antipov, A. N., Fedorov, V. N., 2000. *Landscape-Hydrological Organization of a Territory*. Novosibirsk, SO RAS Publishing House, 254 pp. (in Russian).

Aseyeva, E. N., Kasimov, N. S., Kroonenberg, S. B., 2004. Basin organization of landscape-geochemical systems. In: *Geography, Society and Environment*. Vol. 2. Contemporary Landscape Processes. Moscow, Gorodets, 489-499 (in Russian).

Avessalomova, I. A., 2012. Catena geochemical organization in the taiga landscapes of East-European plain. In: *Landscape Geochemistry and Soil Geography*. Moscow, APR, 97-117 (in Russian).

Avessalomova, I. A., 2014. Landscape neighborhood as a factor of lateral flows transformation in geosystems. In: *Issues in Geography*. Vol. 138. Horizons of Landscape Science. Moscow, Kodex, 233-250 (in Russian).

Avessalomova, I. A., Dyakonov, K. N., Savenko, A. V., 2012. Geochemical traps on the pathway of water migration of anionogenic elements (on the example of the taiga landscapes of East-European plain). *Vestnik of Moscow University, Ser. 5., Geography*, No 1, 29-35 (in Russian).

Avessalomova, I. A., Savenko, A. V., Khoroshev, A. V., 2013. Landscape-geochemical contrasts of the middle taiga river basins as a factor of the ion discharge formation. *Vestnik of Moscow University, Ser. 5., Geography*, No 4, 3-10 (in Russian).

Baker, M. E., Weller, D. E., Jordan, T. E., 2006. Improved methods for quantifying potential nutrient interception by riparian buffers. *Landscape Ecology*, 21, 1327-1345.

Banaszuk, P., Wysocka, A., Matowicka, B., 2000. Relationships between development of plant communities and habitats in the landscape of river valley. In: *Landscape Ecology. Theory and Applications for Practical Purposes. The Problems of Landscape Ecology*. Vol. 6. Warsaw, 21-30.

Bargagli, R., 1998. Trace Elements in Terrestrial Plants. *An Ecophysiological Approach to Biomonitoring and Biorecovery*. Berlin, Springer-Verlag, 324 pp.

Bazilevich, N. I., 1993. *Biological Productivity of Ecosystems of Northern Eurasia*. Moscow, Nauka, 293 pp. (in Russian).

Bazilevich, N. I., Titlyanova, A. A., 2008. *The Biological Turnover on Five Continents: Nitrogen and Ash Elements in Natural Terrestrial Ecosystems*. Novosibirsk, SO RAS Publishing House, 381 pp. (in Russian).

Chorley, R. Y., Kennedy, B. A., 1971. *Physical Geography: A System Approach*. London, Prentice-Hall Int., 370 pp.

Emelyanova, L. G., Goryainova, I. N., Myalo, E. G., 1999. *The Life of Taiga*. Moscow-Arkhangelsk, 163 pp.

Fortescue, J. A. C., 1980. Environmental Geochemistry: A Holistic Approach. *Ecological Studies*, 35. New York-Heidelberg-Berlin, Springer-Verlag, 347 pp.

Gavrilova, I. P., Gorbunova, I. A., 2002. Iron-illuvial textural-podzolic soils in the area of Arkhangelsk experimental station of Moscow University. In: *Landscape Geochemistry and Soil Geography*. Smolensk, Oikumena, 242-268 (in Russian).

Glazovskaya, M. A., 2007. *Geochemistry of Natural and Technogenic Landscapes*. Moscow, Faculty of Geography of Moscow State University, 350 pp. (in Russian).

Junk, W. J., Bayley, P. B., Sparks, R. E., 1989. The flood pulse concept in river-floodplain systems. *Canadian Spec. Publ. of Fisheries and Aquatic Sci.*, 106, 110-127.

Khoroshev, A. V., 2003. Spatial pattern of the landscape as a function of a block structure of the territory. *Vestnik of Moscow University, Ser. 5., Geography*, No 1, 9-14 (in Russian).

Khromykh, V. S., 2007. Dynamics and functioning of floodplain landscapes. In: *Landscape Analysis for Sustainable Development: Theory and Applications of Landscape Science in Russia*. Moscow, Alex Publishers, 156-164.

Klimo, E., Kulhavy, J., Vavřiček, D., 1996. Changes in the quality of precipitation water passing through Norway spruce forest ecosystem in the agricultural-forest landscape of the Drahanska vysočina Uplands. *Ecology (Bratislava)*, 18, 295-306.

Korytny, L. M., 2001. *The Basin Concept in Nature Management*. Irkutsk, Institute of Geography SO RAS, 163 pp. (in Russian).

Likens, G. E., Bormann, F. H., 1995. *Biogeochemistry of a Forested Ecosystem*. New York, Springer-Verlag, 59 pp.

Malard, F., Tockner, K., Ward, J. V., 2000. Physico-chemical heterogeneity in a glacial landscape. *Landscape Ecology*, 15, 679-695.

Milne, G., 1935. Some suggested units of classification and mapping for East African soils. *Soil Res.*, 4, 183-198.

Opp C., 1991. Actual problems of loadability and load in agro-ecosystems. *Ecology (CSFR)*, 10, 373-388.

Perel'man, A. I., 1977. *Geochemistry of Elements in the Supergene Zone*. Jerusalem, Israel Program for Scientific Translations, 266 pp.

Perel'man, A. I., Kasimov, N. S., 1999. *Geochemistry of Landscape*. Moscow, Astreya, 768 pp. (in Russian).

Pinay, G., Decamps, H., Chauvet, E., Fustec, E., 1990. Functions of ecotones in fluvial systems. In: *The Ecology and Management of Aquatic-Terrestrial Ecotones. Man and the Biosphere Series*, Vol. 4. Paris, UNESCO and Parthenon Publishing Group, 141-169.

Polynov, B. B., 1956. *Selected Works*. Moscow, USSR Academia of Science, 751 pp. (in Russian).

Ratas, U., Puurmann, E., Roosaare, J., Rivis, R., 2003. A landscape-geochemical approach in insular studies as exemplified by islets of the eastern Baltic Sea. *Landscape Ecology*, 18, 173-185.

Röder, M., Syrbe, R.-U., 2000. Relationship between land use change, soil degradation and landscape functions. In: Landscape Ecology. Theory and Applications for Practical Purposes. *The Problems of Landscape Ecology*. Vol. 6. Warsaw, 235-246.

Sugier, P., Czarnecka, B., 1998. Changes of geocomponents in the landscape of Polesie Lubelskie under the influence of antropopression. In: Landscape Ecology Transformation in Europe. Practical and Theoretical Aspects. *The Problems of Landscape Ecology*. Vol. 3. Warsaw, 236-245.

Townsend, C. T., Riley, R. H., 1999. Assessment of river health: Accounting for perturbation pathways in physical and ecological space. *Freshwater Biology*, 41, 393-405.

Turner, M., Gardner, R. H., O'Neill, R. V., 2001. *Landscape Ecology in Theory and Practice: Pattern and Process*. New York, Springer-Verlag, 352 pp.

Vannote, R. L., Minshall, G. W., Cummins, K. W., Sedell, J. R., Cushing, C. E., 1980. The river continuum concept. *Canadian J. of Fisheries and Aquatic Sci.*, 37, 130-137.

Weller, D. E., Baker, M. E., Jordan, T. E., 2007. Progress and challenges in demonstrating riparian buffer effects on nutrient discharges from whole catchments. In: 25 Years of Landscape Ecology. Scientific Principles in Practice. *IALE Publication, Ser.* 4, 1, 483-484.

Wicik, B., 2001. The hydrogeochemical problems in the Campinos national park. In: *Landscape Ecological Methods for Strongly Transformed Areas*. Warsaw, 55-61.

Wickham, J. D., Wade, T. G., Riitters, K. H., O'Neill, R. V., Smith, J. H., Smith, E. R., Jones, K. B., Neale, A. C., 2003. Upstream-to-downstream changes in nutrient export risk. *Landscape Ecology*, 18, 195-208.

In: Riparian Zones
Editor: Oleg S. Pokrovsky
ISBN: 978-1-63484-613-4

Chapter 9

DYNAMICS OF THE IRTYSH RIVER FLOODPLAIN HYDROLOGY AND VEGETATION IN THE PAVLODAR REGION OF THE REPUBLIC OF KAZAKHSTAN

M. A. Beisembayeva, V. A. Zemtsov*, V. A. Kamkin and K. U. Bazarbekov
Tomsk State University, Russia,
S. Toraigyrov Pavlodar State University, Kazakhstan

ABSTRACT

This chapter provides an analysis of environmental water releases from the Upper Irtysh cascade of reservoirs for the period 1964-2014 and the impact of changes in the hydrological regime on the Irtysh floodplain and riparian ecosystems. The study was carried out during 2001-2014 in the valley of the Irtysh River within the administrative boundaries of Pavlodar region (a steppe plain fragment of the river) and its nearest neighborhood. Field studies covered the vegetation of key sites in the floodplain and terraces above it, together forming a river valley. The total length of the surveyed area from north to south is more than 500 km. The length of the cross sections over the right- and left-bank parts of the floodplain is from 6 to 9.3 km. In the profiles, oriented from the river bed in the direction of the original bank, all plant communities in the spatial range were described in detail.

The results of the evaluation of changes in the annual and monthly means of the water flow of transboundary Irtysh River within the Republic of Kazakhstan for the period 1903-2010 due to natural and anthropogenic influences in conditions of water use and intensification of modern climate change are presented. It is shown how the water flow control of the Upper Irtysh cascade of water reservoirs, as well as the uptake of part of Irtysh water flow in China have led to significant changes in the hydrological regime of the Upper Irtysh and anthropogenic modification of the Irtysh floodplain and riparian ecosystems.

The current state of the Irtysh River floodplain under the anthropogenically transformed hydrological regime of the river is shown. The dynamic equilibrium,

* Corresponding author: Email: zemtsov_v@mail.ru.

environmental conditions of formation and reactions of natural geosystems of the Irtysh floodplain on the impact of anthropogenic factors are described.

Keywords: Irtysh River, flow regulation, floodplain, environmental flows, floodplain vegetation, anthropogenic impact

INTRODUCTION

The Irtysh River is the largest in Kazakhstan and one of the largest rivers in the world (length 4248 km) (Figure 1). The state of the environment and the preservation of water resources in the basin of the Irtysh River affects public and economic interests of the three neighboring countries - the Republic of China, the Republic of Kazakhstan and the Russian Federation, which have their own interests in the use of its flow.

The Irtysh valley is a very large and unique area due to its landscape diversity. It stretches south to north as a continuous winding strip of 0.5-1 km to 10-20 km wide, crossing the all steppe and the southern half of the forest area of Western Siberia over 3000 km. Floodplain of Irtysh is characterized by a considerable variety of habitats and plant communities.

In Kazakhstan, large areas of meadows are located in the Irtysh floodplain and riparian zone. The main areas of the floodplain of the Irysh River (about 90%) are concentrated in the Pavlodar region (Figure 2). Currently, more than 90% of the floodplain area is covered with natural vegetation, which is used by households as natural grasslands and pastures having huge resources of forage. Floodplain forests are the most productive in the region and have a number of important biosphere functions.

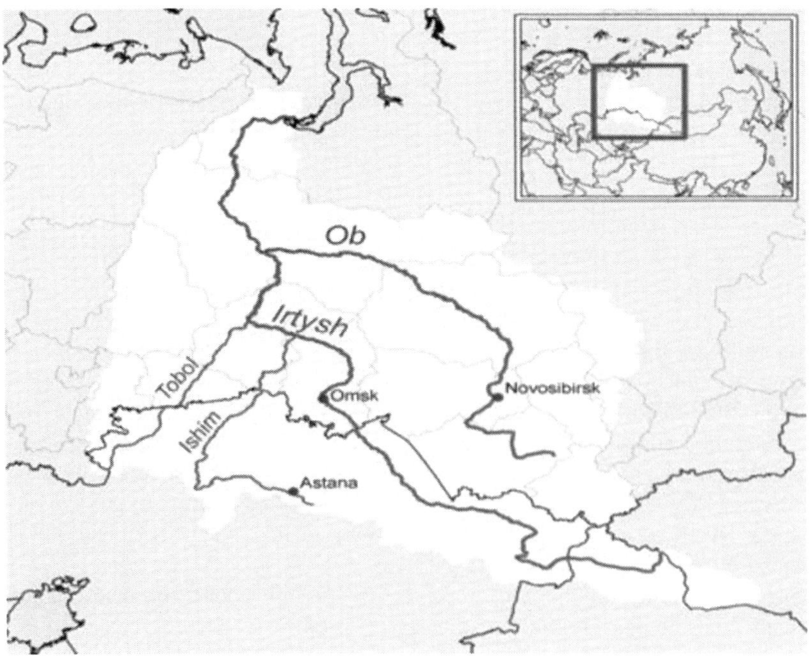

Figure 1. Schematic representation of the Ob and Irtysh River basin.

Figure 2. Irtysh floodplain in Kazakhstan.

Currently, there is an increased interest in the study of the natural regularities of interaction between structural elements of ecosystems and their reaction in response to the change of one or another environmental factor. Of all the natural factors, the plant cover is the most ecologically informative. The dynamic ecosystems in the valleys of major meridianally oriented rivers, where the vegetation is influenced by the imposition of zonal and intrazonal factors, and high speed of fluctuation and succession processes allowing us to study the dynamics of vegetation in a relatively short period of time are of particular scientific interest (Ilyina, 1985). The vegetation of the Irtysh large valley and especially of the floodplain and riparian zone has a huge environmental and economic importance, being the main supplier of green forage, of resource plants, carrying out soil and water protection, recreation, and many other functions.

The problem of rational use and protection of water resources of the Irtysh is complex and multifaceted. One of the most urgent and still unresolved problems is the preservation the floodplain, the total area of which downstream the Shulbinskaya HPP to the border with Russia is over 400,000 ha. The floodplain of the Irtysh River is unique, both in its size and in the richness of its flora and fauna. Its meadows are real wealth, the gold fund, which are especially valuable in a desert, as it allows stabilizing the livestock fodder base. The richest phytocoenosis of these riparian oases is a kind of natural filter, absorbing from the air ash and smoke emissions of industrial enterprises of the region. Simultaneously, the floodplain is a powerful generator of oxygen enriching the air basin of cities and towns in the region. Floodplain over 400 km length is a unique protector, reducing the degree of harmful effects of wind erosion of light soils of the Pavlodar region.

The factor forming the nature of the Irtysh River floodplain is spring water flux es from the Upper Irtysh cascade of reservoirs. Under natural conditions in the past, 89-97% of the Irtysh floodplain area was flooded almost every year. In some years, with a frequency of 1 every 6-8 years only by 60-70% of the floodplain area was flooded and with a frequency of 1 every 12-15 years - up to 10% of the area was flooded. In wet years the entire area of floodplain was inundated for a sufficiently long period. There was natural flushing of numerous channels, oxbows and lakes, as well as waterlogged and saline areas. The yield of the most valuable grasslands reached 5,000–6,000 kg ha^{-1}. Currently floodplain functions through annual spring water flux from the Upper Irtysh cascade. However, the parameters of the actual water releases do not meet the environmental needs of the natural floodplain complex. Further failure to take drastic measures to change the approach to the assessment of necessary water flux and floodplain inundation time may lead to unpredictable consequences.

Due to the orientation to the energy generation priorities and corresponding limitation of the rates and timing of river fluxes from the Upper Irtysh cascade of HPP the ecological balance in the floodplain was disrupted; the degradation of floodplain lands, their salinization, steppe expansion, and waterlogging and paludification in some places started, thus undermined the biological productivity of the floodplain. For example, meadow grass yield fell 4-fold, to 1550 kg ha^{-1}.

Assessment of the agricultural value of the floodplain shows that in the case of absence of floodplain meadows it is very difficult to conduct livestock development in the Pavlodar region. In addition, the flooding of the floodplain is also of great social and demographical significance, because over 75% of the population is concentrated in cities, workers' settlements and villages located in the vicinity of the Irtysh.

To maintain dynamic balance of natural ecosystems and biodiversity conservation data about the patterns of spatial structure of vegetation, ecological conditions of its formation and their responses to the impact of anthropogenic factors are needed. They are required to predict the development of negative trends in the dynamics of ecosystems and timely response to them, the development of measures to optimize the distribution of surface runoff water in different years, as well as rational use and protection of vegetation. The aim of this chapter is to study the effect of the river water fluxes on the state of vegetation cover of the Irtysh River floodplain in its steppe segment.

To achieve this goal the following issues were adressed:

- exploration of long-term options of river water fluxes;
- conducting complex floral and geo-botanical research;
- determination the ecological and cenotic features of dominant plant species;
- identifying factors of anthropogenic transformation of vegetation and to assessing their role in changing phytocenotic diversity;
- forecasting possible dynamics of floodplain ecosystems and developing recommendations for the restoration and protection of vegetation.

STUDY AREA

According to the characteristic features of hydrological regime it is possible to divide a course of the Irtysh River into three parts. The first one represents the Upper Irtysh, the second represents a reach between the cities of Ust-Kamenogorsk and Omsk and the third extends further to the place where the Irtysh flows into the Ob River (Korytniy 1994). The study area encompasses the Irtysh floodplain within the Pavlodar region and its neighborhood where the river flows through the southern part of the West Siberian Plain and has a character of lowland river. The floodplain is well developed and has a length of about 500 km and a width up to 20 km; it spans large area of desert steppe, dry and arid steppe subzones between 51° and 53.8° N. Climate is strong continental, with long and cold more than 5 months winter and 3 months hot summer. The mean temperature of January is about of -17 in the south to -20ºC in the north and July +20 to 22ºC, respectively; annual precipitation is 220 mm in the south to 300 mm in the north.

A main proportion of water inflow the Irtysh River in Kazakhstan receives from the Altai, occupying the northern part of the right bank part of the Irtysh basin and represents a highly dissected mountain massif with deep valleys of many rivers of different size. The rivers here in spring and summer are fed with melt water of snow and glaciers. In contrast to the Altai, where the dense hydrographic network exists, there are no permanent streams in the steppe plains. The Irtysh does not have any considerable tributaries downstream the city of Senay. The main phase of the Irtysh River and its tributaries water regime is a spring flood, which is formed due to melting snow and glaciers, and rainfalls. Low flow seasons take place in the Irtysh River for a long time in late summer - autumn (sometimes interrupted by rain induced floods in autumn) and especially in winter when rivers are fed only by groundwater.

In the Republic of Kazakhstan territory river flow is regulated by the Upper Irtysh cascade of water reservoirs comprising the Bukhtarma (filling and commissioning in 1960-1966); Ust-Kamenogorsk (filling in 1952-1954, reaching full capacity in 1966) and Shulbinsky (start of construction in 1976, running in 1987-1994) reservoirs (Figure 3). The listed reservoirs successively exercise long-term, weekly, seasonal flow regulation, and in accordance with their purpose alter the hydrological regime of Irtysh since their completion and start of work. The K. Satpayev Canal constructed in 1968-1975 transfers a part of the Irtysh River flow for water supply of central Kazakhstan area and Astana, a capital of the Republic of Kazakhstan.

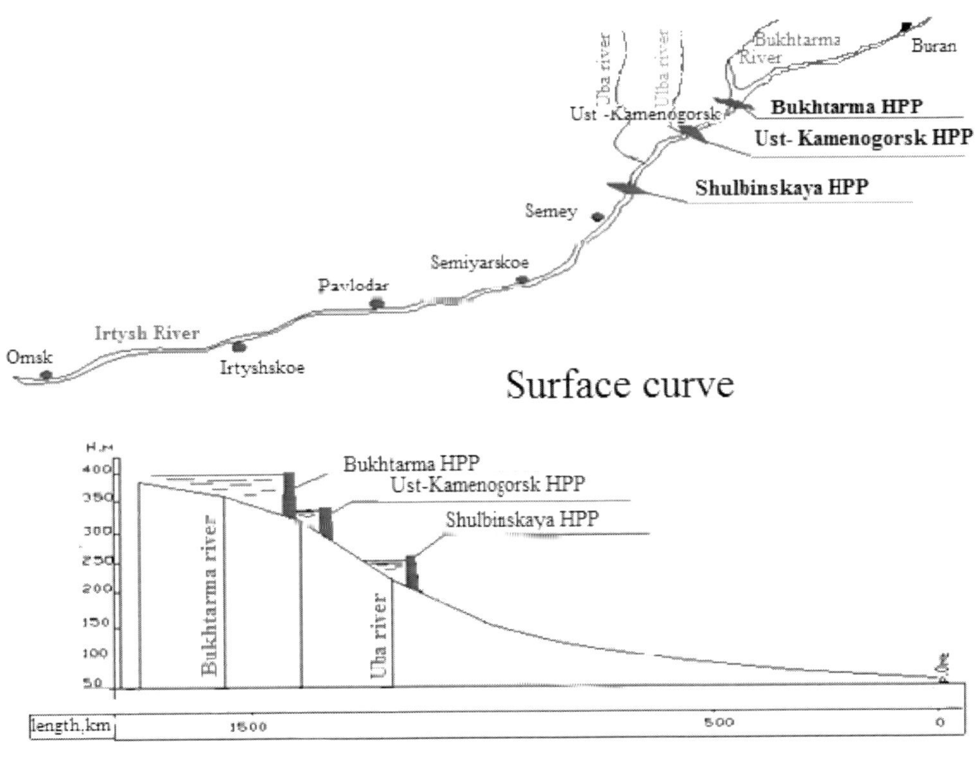

Figure 3. Location of the hydropower plants (HPP) on the Upper Irtysh.

MATERIALS AND METHODS

Our research was carried out during 2001-2014 in the valley of the Irtysh River within the administrative boundaries of Pavlodar region (the steppe plain fragment of the river) and its nearest neighborhoods, the Republic of Kazakhstan. Field studies comprised investigation of the vegetation of riparian zone and floodplain and terraces above floodplain, together forming a valley. The total length of the surveyed area from north to south is about of 500 km. The length of the cross sections of the right- and left-bank part of the valley is between 6 and 9.3 km.

The methodological basis for investigations was the ecosystem or landscape-biogeocenotic approach (Ogar', 1999). Traditional methods of geographical and geo-botanical studies (statistical processing of collected data, geo-botanical descriptions of phytocoenoses, landscape and environmental profiling, mapping, collecting herbarium, etc.) as well as modern methods were used, including determination coordinates of the key sites with a GPS receiver, spatial analysis of satellite images, development of an electronic database, and others.

To study the nature of the impact of the Upper Irtysh cascade of reservoirs on the hydrological regime of the Irtysh River within the Republic of Kazakhstan we used standard hydrological methods of runoff frequency analysis, modeling of intra-annual river flow distribution.

The long-term data of hydrometeorological observations conducted by the National Hydrometeorological Service of the Republic of Kazakhstan ("KazHydroMet") were used in the study including the hydrometric data on runoff at five gauging stations along the Irtysh River to study the impact of the Upper Irtysh cascade of reservoirs on the hydrological regime of the Irtysh River.

Particular attention was paid to the spatial arrangement (structure) of vegetation in its relationship with other components of the landscape (topography, soil, etc.), floristic composition, assessment of phytocenoses state, identification of rare species and communities. In the profiles, oriented from the river bed in the direction of the original bank and crossing the riparian zones, all plant communities in the spatial range were described in detail, their borders were determined and recorded using GPS and the length of the borders was measured. Besides the cross sections, the study area was uniformly covered with a network of geo-botanical sites, covering all the different types of ecosystems. Altogether, about 630 geo-botanical descriptions have been made.

Analysis of the flora is based on personally collected materials (over 1200 herbarium sheets) and published data on the study area. In the analysis of the flora, the focus was not only on its systematic structure, but also the on distribution of species among ecomorphs, biomorphs, florocenotypes and economic groups. For each species its cenotic role and areas of ecological optimum were determined.

In assessing the degree of anthropogenic disturbance of vegetation the conventionally "background" undisturbed communities and their anthropogenic modifications in each type of ecosystem were described.

RESULTS

Hydrological Regime and Its Change Caused by Flow Regulation

The nature of fluctuations in water level in the rivers of the Upper Irtysh basin is determined by sources of their water supply, position and altitude of river watershed areas. The highest water levels are observed during the spring flood (April-May), and much smaller degree in cases of rain floods (September-October). Under natural conditions, before the construction of the Upper Irtysh cascade of reservoirs after the end of river ice break, even more intensive increase in the Irtysh water levels occurred.

The highest daily intensity of the Upper Irtysh and its tributaries water level rise in the area ranges from 5-10 cm to 80-100 cm, depending on the size of the watershed and the main direction of river flow. The duration of lifting does not exceed 10-15 days in average. The maximum water levels of spring flood in the river are usually observed during the second or third decade of April.

In the Irtysh River reach between the Ust-Kamenogorsk HPP and Semiyarskoe village against the background of the main flood wave a number of sharp ups and downs of water level occur, depending on the phase mismatch of the floods in different tributaries and the uneven snowmelt timing in various altitudinal zones of the basin. A pectinate character of water level fluctuations in Pavlodar city is less pronounced due to the regulatory influence of the floodplain and the lack of lateral inflow.

The Irtysh River has different water sources over its course, in the upper flow it is fed by glaciers melt water. River water discharges sharply decrease in downstream direction. The reason for this is the lack of water inflow in the Pavlodar region. Natural regime of the Irtysh River (before the construction of the reservoirs) was characterized by the presence of two floods: the maximum early spring flood originated from the snowmelt in the plains and low hills, and spring-summer flood originated from the melting snow and glaciers in the mountains. The peak of spring flood there was around May 12, the earliest date was April 13 (1938), and latest in June 23 (1956). The flood lasted on average 75 to 80 days, sometimes reaching 100 to 120 days. Very strong, catastrophic floods were quite common.

After the recession of the first flood wave there was high soil moisture and enough heat, thus rapid growth of vegetation occurred. With decrease in soil moisture the vegetation growth has slowed. Then the second flood wave came, which inundated the riparian zone and lowlands, wetlands and filled lakes with water. This produced floodplain soils watering, which favorably affected the state of riparian ecosystems.

During the construction of the Bukhtarma reservoir (1959-1963) Irtysh floodplain inundation by flood waters was discontinued, causing abrupt quantitative and qualitative changes in vegetation and reduction of grasslands productivity by 60-70%. From 1964 until 1988 spring the water fluxes to the river Irtysh from the Bukhtarma reservoir were produced. The total flow volume aimed to support flooding of the floodplain formed from water of the Bukhtarma and Ust-Kamenogorsk reservoirs and runoff of the right tributaries of the Irtysh River - rivers Ulba and Uba. Before creating the first stage of the Shulbinsky reservoir on the Irtysh River there were serious difficulties in providing sufficient water flow. In order to ensure a sufficient flow rate at Semiyarskoe village taking account of the predicted peak rate and timing of the floods on the rivers Uba and Ulba, the daily water discharges of

1400 m^3 s^{-1} from the Bukhtarma reservoir were produced in advance. This led to unsustainable reservoir management during the period of its filling and violation of the regime of Bukhtarma HPP power generation. At the same time, such a scheme of water flux did not allow achieving optimum parameters of a flood wave at the entrance to the Pavlodar region.

After commissioning the Shulbinsky reservoir the water flux allowed the floodplain inundation area of 180 to 350 hectares during spring floods in recent years. In some dry years (1983) there was a lack of floodplain inundation. Actual effect of water flux is largely determined by hydrometeorological conditions in the zone of lateral inflow of the Irtysh formation. Irregular and inefficient floodplain inundation has led to gradual decrease in biomass and recovery potential of the floodplain ecosystems.

Analysis of the water flux in the last decade shows that the area and quality of floodplain inundation is influenced by the total volume of water flux from the Shulbinsky reservoirs, timing and duration of the environmental flows, as well as the maintenance of the daily maximum.

The main hydrological characteristics of water flux over the period 1964-2015 are shown in Figure 4. The 1990 flux of 5.4 km^3 and maximum water discharge of 4200 m^3 day^{-1} with a duration of 20 days were similar to the water releases in 1991, when the area of 361,900 ha was flooded. For this, the optimal duration of flows should be at least 18 days, and the period with the maximum flow rate of at least 5-6 days.

In the analysis of the water volume of flows over the last decade should be noted that the maximum floodplains inundation occurred in 2001: 307,600 ha (92.5% of the total floodplain area). In the subsequent period, the flows decreased the total inundated floodplain area by 58%. In 2012, the Irtysh floodplain virtually was not inundated at all. Nevertheless, since 2013 when 89% of the floodplain area were inundated the spring water releases were close to the natural spring flood rate, and in 2014 81.1% of floodplain was flooded.

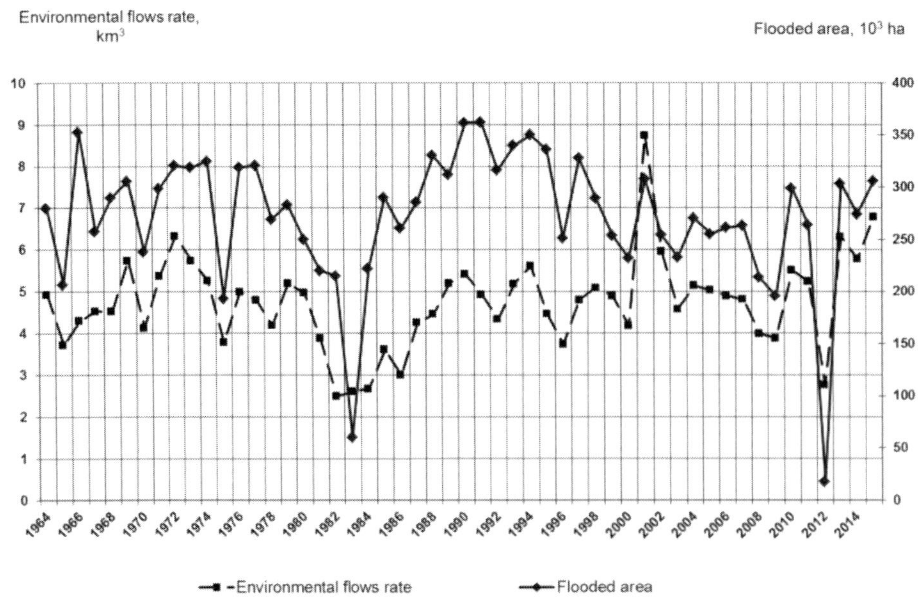

Figure 4. Dynamics of the Irtysh floodplain inundation (1964 to 2015).

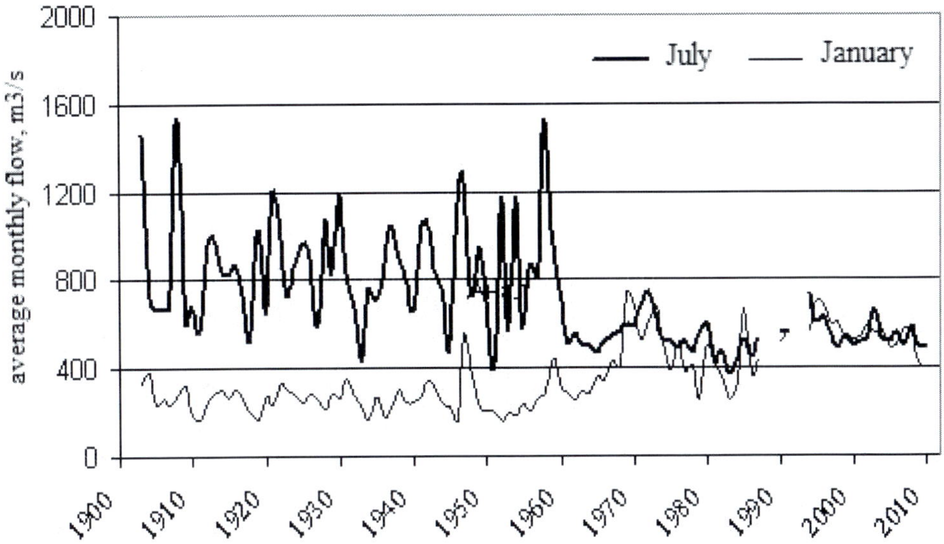

Figure 5. Long-term fluctuations in the average monthly flow in July and January at Ust-Kamenogorsk.

To study the nature of the impact of the Upper Irtysh cascade of reservoirs on the water regime of the Irtysh River floodplain, the seasonal flow rates were considered. Comparing the average monthly discharges "before" and "after" the construction of the Upper Irtysh cascade of water reservoirs allows estimation of the changes in monthly runoff of the Irtysh River distribution at the stations in the city of Ust-Kamenogorsk, Shulba and Semiyarskoe villages, Pavlodar city, Cherlak village. After that we compare them with the conditionally natural regime at Buran village upstream the reservoirs.

The changes in the average monthly water flows appeared immediately after commissioning the respective HPP in the cascade, for example, at Ust-Kamenogorsk, from 1960-1961 and at Semiyarkoe, from 1964-1965 (Figure 5). This might be confirmed by the significance test of the statistical hypothesis of the monthly flows time series homogeneity using the t-test for means and the F-test for variances. Within a year, the redistribution of flow is subject to the following pattern: there is a decrease in monthly discharges during flood, summer and partly autumn low flow periods and increase in winter time. There was an effect of significant reduction of the spring water releases from the reservoirs for the Irtysh floodplain inundation and, respectively, the reduction of inundated areas of floodplain.

In addition, the changes in monthly flow regime were significant. For example, when compared to the average monthly runoff at Ust-Kamenogorsk in the natural and the disturbed (by the HPP cascade operations) periods, the largest increase in the flow rates within the disturbed period occurred in December-February: respectively 105, 95 and 98% of natural monthly discharge (or in absolute terms, 283, 249 and 241 $m^3 s^{-1}$).

According to Figure 6, the distribution of river flow in percent of the total annual runoff shows that the volume of flow between the natural and disturbed periods has changed significantly. It is not possible to trace any considerable changes in the flow distribution at Buran village. In the contrary, there was considerable redistribution of seasonal flow rates: the flow volume in spring-summer decreased significantly from 74% in the period of natural regime up to 50% in the current period of the disturbed regime, i.e., reduction of 24% in runoff is observed. The observed redistribution of runoff during the disturbed flow period is primarily caused by the increase in hydroenergy and water demand of industrial enterprises and settlements located along the Irtysh River, during the low flow period the cascade of reservoirs releases more water, with an aim of a more uniform flow distribution within a year and day.

The reduction of the average monthly discharges for the disturbed period occurred in May to October: 53, 58, 37, 22, 10 and 1%, respectively. The maximum 58% in June corresponds to the reduction in the flow rate of 715 $m^3 s^{-1}$.

In the plains of the Irtysh basin the change in monthly runoff was not so considerable: for example, at Semiyarkoye water flow in May - September decreased by 27, 48, 36, 19, and 12% of the natural rate. The reduction of water discharge by 48% in June is equivalent to the loss of water of about 1049 $m^3 s^{-1}$. The remaining months of the year were characterized by an increase in runoff: the minimum relative increment of flow was observed in October (8%) and the maximum in March (97%).

A wave of water flux propagated along the floodplain during short time. A sharp decline in water discharges from the Shulbinsky reservoir may occur. Because of that, the peripheral parts of the floodplains in the Irtysh and Zhelezinka areas of the Pavlodar region of Kazakhstan loses the necessary amount of water to flood the area, soils do not have time enough to absorb water and accumulate water reserves in the root system. It is clear that over time, negative consequences for the ecological status of riparian landscapes may appear.

The river water quality parameters are dependent on water availability, the seasonal and daily dynamics of processes inside water bodies induced by the activities of the physic-chemical, hydrological and biological factors.

The snow melt water inflow is a reason for low water salinity during the flood period. The river water mineralization increases from 0.5-0.8 $g\ l^{-1}$ in spring and summer up to 1-3 $g\ l^{-1}$ in winter.

The hydrochemical regime of the Irtysh River in headwaters is formed by leaching and dissolution of rocks, surface runoff from the catchment area and pollutants entering the river from outside. After the beginning of the regulation of the river flow has begun the composition of anions and cations in water started to change. Reducing the flow and the water level potential in the river reduces the degree of dilution of pollutants and self-purification, thereby increasing the concentration of water pollutants that accumulate annually in the floodplain during floods, and then accumulate in the soil and living organisms.

During the period of Irtysh flow regulation water salinity has increased by 1.1-1.5 times, the concentration of total iron increased by 17-45 times, phosphates by 2 times, nitrates by 7 times compared to the natural rate. To date, the standard index of Irtysh water pollution within the Pavlodar region is 1.38 that allows to characterize the river as a water source water quality Class III - moderately polluted (Mogilyuk, 2004). Upon cooling, turbine and other industrial heat exchangers produce a huge amount of heated water. Discharges of water with a temperature up to +50 or + 40°C cause thermal pollution of the river which significantly affects the floodplain microclimate (Tsaregorodtseva, 2003).

Thus, the Irtysh flow regulation caused a change in the hydrological regime of the river and lead to a change in conditions of floodplain inundation. The main result of these processes was the steppe formation and salinization of the floodplain.

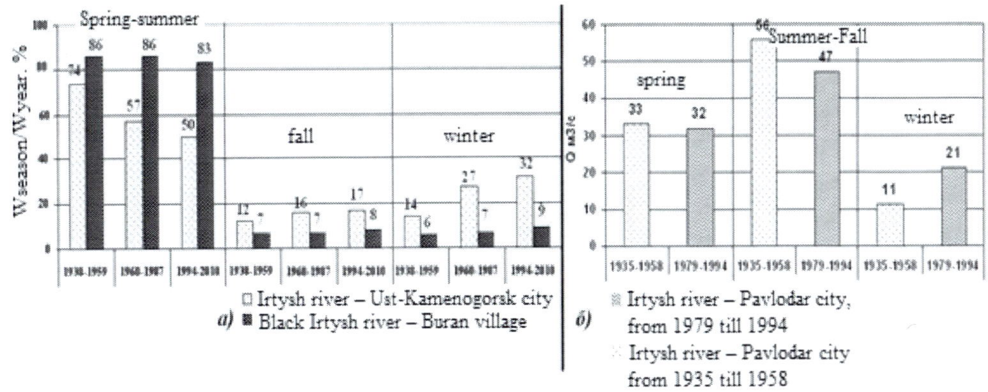

Figure 6. The redistribution of runoff within a year after construction of the Upper Irtysh cascade of reservoirs.

Floodplain and Riparian Vegetation Ecosystems

Floodplain and riparian flora and vegetation are formed under the influence of hydrological regime of the Irtysh River. According to this study the flora of the Irtysh valley within the Pavlodar region is represented by 545 species of vascular plants belonging to 268 genera and 73 families of higher spore, gymnosperms and angiosperms. Most plants belong to 534 species, or 98% are angiosperms, including 413 species of dicots and 121 of monocots. Spore plants are presented by nine species (1.6%), and gymnosperms - only by two species (0.3%).

There are some leading families: *Asteraceae*, 84 species, or 15.5% of the total number of species, *Poaceae*, 57 (10,5%), *Fabaceae*, 34 (6.3%), *Brassicaceae*, 30 (5.5%), *Rosaceae*, 29 (5,3%), *Chenopodiaceae*, 28 (5.1%), *Cyperaceae*, 25 (4.6%), *Ranunculaceae*, 25 (4.6%), *Lamiaceae*, 18 (3.32%) and *Caryophyllaceae*, 17 (3.1%). The remaining 63 families account for 198 species (36% of species total number).

The flora of the investigated area contains the following rare species: *Saussurea robusta*, *Serratula kirghisorum*, *Equisetum sylvaticum*, *Euphorbia microcarpa*, *Hemerocallis lilio-asphodelus*, *Tulipa biebersteiniana*, *Cypripedium guttatum*, *Delphinium elatum*, *Rosa pavlovii*. Of these species, *Rosa pavlovii Chrshan* is listed in the Red Book (Red Book of the Kazakh SSR, 1981). The number of many of these species is extremely small, which makes them particularly vulnerable.

In biomorphological spectrum, the herbaceous plants dominate, belonging to 485 species, of which 358 are perennials, 93 are annuals and 34 are biennials. There are 22 species of shrubs, 16 - trees, 9 - semishrubs, 5 – dwarf semishrubs, and 4 species of dwarf shrubs and lianas.

The ecological structure of flora is dominated by ecomorphmezophilic formation (292 species, or 53.6%), which indicates the general intrazonal character of vegetation. Xerophilic formation includes 157 species (28.8%), hygrophilic include 56 species (10.3%), halophilic include 22 (4%) and hydrophilic include 19 species (3.5%). The flora of the Irtysh valley contains a large number of economically important plant species: forage plants (170), medicinal (147), technical (126), bee plants (111), decorative (87), food plants (72), and others (Kamkin, 2009).

There are some pine forests in the Irtysh valley in the south of the Pavlodar region. According to E. P. Prokopiev (2001), coniferous forests are located here on farewell rocks and fragments of the first above floodplain terrace, sometimes poorly demarcated from the adjacent floodplain.

The literature indicated the only habitat of pine woodlands on hilly sands towering 6-8 m above the surrounding floodplain where grass cover is virtually absent and uneven pine regrowth is observed in the floodplain of the Irtysh River against Chernoye village in Lebyazhye area.

However, the field studies we conducted in August 2006 showed that pine association in the Irtysh floodplain is absent nowadays. Interviewing old-time residents and residents of Chernoye village confirmed our observations; moreover, the locals said that since the 1950s pine tree in the floodplain had not been met. In contrast with Chernoye village, upon the first and second above floodplain terraces there are artificial plantations, which include *Pinus sylvestris* (Figure 7), but it is not abundant and natural regeneration is almost completely absent. Herbal cover consists of seeded grass pastures with estimated area coverage of 95-100%.

Thus, *Pinus sylvestris* in the Irtysh valley within the Pavlodar region is not typical and does not form natural communities. In the floodplain area there are artificially planted pine trees, but they are not numerous, usually oppressed and without natural undergrowth. This is due to the negative reaction of pine on even short-term flooding of floodplain.

The zonal and subclimax steppe and meadow-steppe vegetation of above floodplain terraces is greatly transformed and mostly represented by secondary communities. In the floodplain the following types of intrazonal vegetation are encountered: trees and shrubs (riparian forests), meadows, wetlands, aquatic and steppe vegetation including halophyte options. Pioneer group of plants are also considered separately on the river bars and spits. The main type of vegetation in the floodplain is meadow, which is of the greatest economic interest, since it is the main source of green forage for livestock in the region. Table 1 shows the main dominants of the meadow vegetation type, depending on their position in the longitudinal and transverse sections of the floodplain.

Figure 7. *Pinus sylvestris* in artificial plantations.

Table 1. Distribution and dominants of real meadows within the riparian/floodplain zone of the Irtysh River

Subzone	Riverine and riparian floodplain	Central floodplain	Near-terrace floodplain
Glycophyte real meadows			
Arid steppes	*Agrostisgigantea, Bromopsisinermis, Carex praecox, Galiumboreale, Artemisia procera*	*Agrostisgigantea, Calamagrostisepigeios, Poapratensis, Sanguisorbaofficinalis, Bromopsisinermis, Elytrigiarepens, Filipendulaulmaria, Galiumboreale, Gratiolaofficinalis, Viciacracca, Carexacuta*	*Bromopsisinermis, Equisetum arvense, Sanguisorbaofficinalis, Galiumboreale, Carexacuta*
Dry steppes	*Bromopsisinermis, Calamagrostisepigeios, Elytrigiarepens*	*Alopecuruspratensis, Elytrigiarepens, Sanguisorbaofficinalis, Bromopsisinermis, Glycyrrhisauralensis, Poapratensis, Viciacracca, Galiumboreale, Rumexconfertus*	*Alopecuruspratensis, Elytrigiarepens, Viciacracca, Arctiumtomentosum, Bromopsisinermis, Glycyrrhisauralensis*
Desert steppes	*Bromopsisinermis, Elytrigiarepens, Alopecuruspratensis, Agrostisgigantea*	*Alopecuruspratensis, Elytrigiarepens, Bromopsisinermis, Glycyrrhisauralensis, Galiumboreale, Lathyruspratensis, Poapratensis*	*Elytrigiarepens, Agrostisgigantea, Bromopsisinermis*
Halophyte real meadows			
Arid steppe	-	*Glycyrrhisauralensis, Puccinelliagigantea, Artemisia pontica, Seneciojacobaea*	*Artemisia pontica, Bromopsisinermis, Puccinelliagigantea, Glycyrrhisauralensis*
Dry steppe	-	*Artemisia procera, Elytrigiarepens, Glycyrrhisauralensis, Puccinelliadistans, Alopecuruspratensis*	*Elytrigiarepens, Glycyrrhisauralensis, Puccinelliadistans, Bromopsisinermis, Calamagrostisepigeios, Sanguisorbaofficinalis, Alopecuruspratensis, Sonchusarvensis, Poapratensis, Stellariagraminea*
Desert steppes	*Carexacuta, Elytrigiarepens, Glycyrrhisauralensis, Poapratensis*	*Alopecuruspratensis, Bromopsisinermis, Elytrigiarepens, Leymusramosus, Glycyrrhisauralensis, Galatellapunctata, Sanguisorbaofficinalis, Filipendulaulmaria, Artemisia pontica, Galiumverum, Rumexconfertus*	*Alopecuruspratensis, Elytrigiarepens, Glycyrrhisauralensis, Agrostisgigantea, Puccinelliadistans*

Regularities in a spatial structure of vegetation in the transverse profile of the Irtysh valley are shown in Figures 8-10.

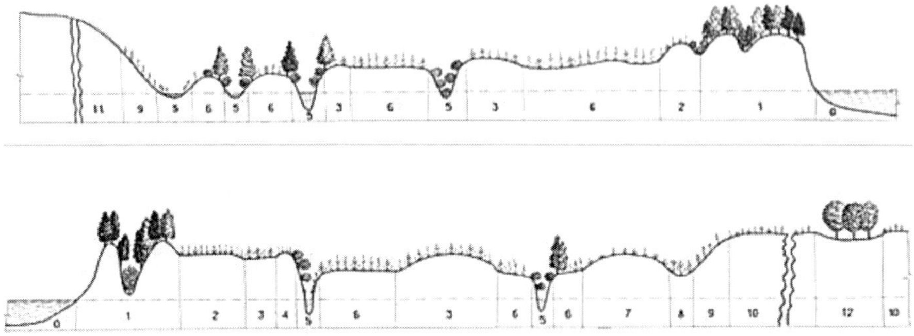

0. Open water without surface vegetation. 1. Willow – poplar riparian forest. 2. Motley grass – cereals meadow. 3. Cereals - large motley grass meadow. 4. Steppefied brome grass meadow. 5. Ecological series of a floodplain side-channel. 6. Large motley grass – sedge - cereal marshy meadow. 7. Medium dense cereal - motley grass meadow. 8. Large motley grass nitrophilous meadow. 9. Sagebrush - wild rye group on a terrace slope. 10. Agrocenoses in complex with sagebrush - motley grass - cereals steppes. 11. Agrocenoses without indigenous vegetation. 12. birch forests.

Figure 8. View across the Irtysh valley within the subzone of arid steppes at the Kaymanachikha - Beregovoe villages cross-section (53°35 'N).

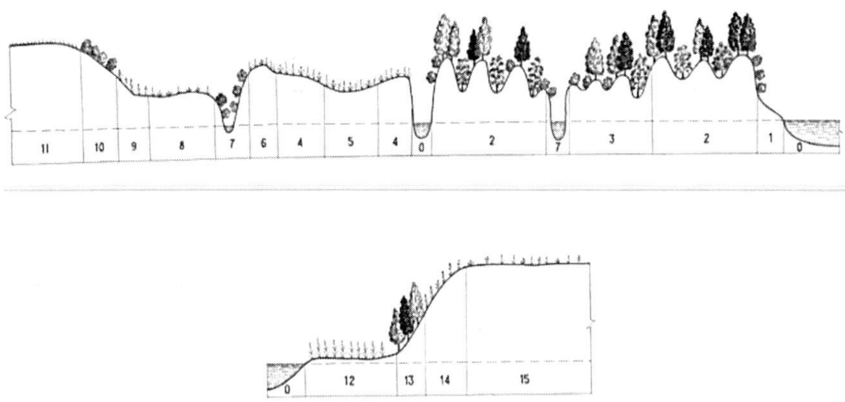

0. Open water without surface vegetation. 1. Pioneering groups on young sandy alluvium. 2. Willow - poplar forest. 3. Shrubs and meadow vegetation complex. 4. Motley grass - cereals real meadow. 5. Motley grass - sedge - cereals marshy meadow. 6. Motley grass - cereals steppefied meadow. 7. Ecological series of a floodplain side-channel. 8. Poor dense halophytic cereals - motley grass meadow. 9. Average dense slightly saline sagebrush - cereals - motley grass steppefied meadow. 10. Dwarf shrubs. 11. agrocenoses without indigenous vegetation. 12. reed fen. 13. poplar forest. 14. sagebrush - wild rye group on a terrace slope. 15. Motley grass – sagebrush – cereal steppe

Figure 9. View across the Irtysh valley within the subzone of dry steppes at the Rebrovka - Shawki villages cross-section (52°25' N).

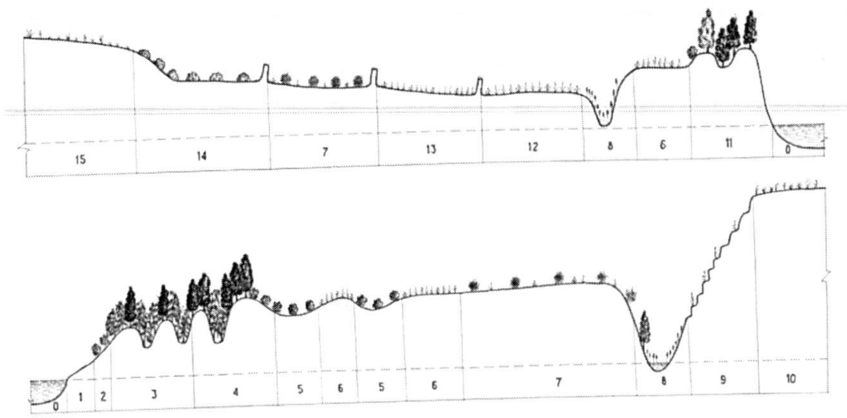

0. Open water without surface vegetation. 1. Pioneering groups on young sandy alluvium. 2. Shrub willows thicket. 3. Willows. 4. willow-poplar forest. 5. Hawthorn – honeysuckle - dog rose shrubs. 6. steppefied halophytic motley grass – sagebrush – cereals meadow. 7. *Halimodendron halodendron* with motley grass, cereals and sagebrush. 8. Ecological series of a floodplain side-channel. 9. Sagebrush - wild rye group. 10. Sagebrush - feather grass - fescue steppe. 11. Poplar forest. 12. Motley grass – cereals – real halophytic meadow. 13. Motley grass – cereals steppefied halophytic meadow. 14. Fescue - *Artemisia austriaca* halophytic meadow steppe with *Nitraria sibirica*. 15. Fescue - *Artemisia austriaca* – weeds strongly knocked steppe.

Figure 10. View across the Irtysh valley within the subzone of desert steppes at the Abay - Podpusk villages cross-section (51°19 'N).

The spatial structure of floodplain vegetation bears the imprint of the dynamic processes taking place in the studied floodplain segments. High dynamics of habitats, which causes considerable dynamic of vegetation, is one of the specific features of the river floodplains. On the one hand, plant communities undergo significant fluctuations under the influence of seasonal and interannual changes in ecological conditions of different habitats. On the other hand, as a result of the erosion-accumulation activity of river channel flow and flood waters, an altitude of a floodplain surface increases that causes irreversible changes in ecological conditions and lead to successional changes in riparian phytocenoses. In addition, one of the powerful factors causing change in phytocenoses is anthropogenic impacts (Prokopiev, 1975).

In a context of the natural evolution of the floodplain and delta landscapes there are three successional and dynamical series of vegetation development: on the riverine natural levees, on plains between channels and in channel depressions (Rachkovskaya, 2003).

As a result of the river channel shift to the opposite bank, the riverine side bar with pioneer sparse vegetation becomes a low ridge, or a site with alternation of fine ridges and hollows, more elevated due to alluvium accumulation segment of a young riverine floodplain inundated by floods each year for a long time. Under these conditions soil-forming processes begin and primitive floodplain wet meadow silty-sandy soils are formed.

S. M. Razumovsky (1981) believed that such young riverine floodplain areas, which are annually flooded and covered with a new silt layer, are generally not available for successional changes. According to this author, they can be made available for pioneer associations only in the case of changes in the abiotic environment and, moreover, the changes not dependent on the vegetation itself.

However, our observations show that even shoots of poplar and particularly willow, appeared on the latest riverine sandy spit, are not destroyed by flood waters and are not killed by the silt layer overlap but plentifully expand gradually reaching 100% of the area. Here, the grass cover is either absent or represented by the same species as the pioneer plant communities.

With a further rise of the ridges surface there is attenuation of the alluvial floodplain regime, improvement of drainage and more intensive soil formation.

Russian willow and particularly white willow fall out of the growing stock, which is even more thinning. Abundance of black and white poplar trees increases and tree stand becomes willow - poplar or poplar. *Carexpraecox*, *Bromopsisinermis*, *Poaangustifolia* with admixture of motley grass appearing in the grass layer.

When poplar forests are destroyed, motley grass - sedge - rhizomatous grasses (brome grass, *Poa angustifolia*, couch grass) well-drained meadows appear, which are replaced due to intensive grazing by bluegrass and finely grass meadows with *Plantago media*, *Polygonum aviculare* and *Trifolium repens*.

At this stage, a riverine floodplain due to the riverbank retreat is transformed into central floodplain, where the processes of succession are going on. Continuing accumulation of alluvium leads to flattening of ridges and leveling of ridge-hollow sections. These regions exempt from the influence of regular floods and steppefied meadow cenoses begin to form. In parallel with decrease of flooding frequency and duration as well as silt layer thickness, monodominant communities are replaced by polydominant (Rabotnov, 1985).

The initial stage of succession of the second type in the Irtysh valley is plantgrowing of floodplain water bodies. Oxbow lakes are open as long as they remain quite deep (up to 3 m) (Prokopiev, 1975). Plant growing of floodplain lakes with large submersed plants begins when a depth reaches 3.5 to 3 m (Figure 11).

Figure 11. Ecological series of plant growing lake.

The next band is a coastal area of temporary flooding. This area is the greater, the larger the amplitude of the fluctuations of water level and the smaller the slope of riverbank. Within the area of flooding conditions are gradually changing from the lowest point, almost all the time under the water, to the highest, flooded rarely and briefly. This is reflected in the vegetation cover formed by a mixture of aquatic plants (tolerant to temporary drying), amphibians and terrestrial plants (tolerant to temporary inundation).

Raising the height of the floodplain gradually leads to steppefying of meadows communities. Xeromesophytic species of meadow grasses and rhizomatous grasses with wide ecological amplitude come first. Couch grass (*Elytrigiarepens*), bluegrass (*Poa pratensis*), licorice (*Glycyrrhisa uralensis*), *Sanguisorba officinalis*, *Filipendula ulmaria*, *Veronica longifolia*, *Galium boreale* and others dominate in communities.

Increasing the height of the terrace in the southern areas of the region is accompanied by salinization associated with the capillary rise and subsequent evaporation of salts rich groundwater from the soil surface. The usual dominant in these conditions is *Glycyrrhisa uralensis*.

The subsequent increase in the height of the terrace and the lowering of the water table up to limits of its possible impact on soil and vegetation leads to the transformation of intrazonal vegetation into solonetz steppe first, and then into a typical steppe.

Finally, the grazing steppe - meadow and steppe cenoses degrade into sagebrush - fescue (*Artemisia austriaca*, *A. frigida*, *A. nitrosa*) and small herbal modifications.

In the study of the vegetation dynamics on floodplain it is not possible to ignore the fluctuating changes in vegetation. Seasonal and interannual variability is most common for meadow and partly steppe communities. Alternation of fluctuating states can change communities drastically (Mirkin, 1981).

When the vegetation fluctuates, different mutual complements of species builders of communities are observed, that are especially strong in habitats with unsteady humidification, abruptly changing from year to year. Under these conditions, plants of dramatically different ecological status grow. Some of them thrive and grow in wet years with high precipitation depth and high groundwater table, the others proliferate during dry years.

Recommendations for Floodplain Ecosystems Maintenance

As important highly productive natural areas, riparian ecosystems are under intense anthropogenic pressure. For maintenance of biodiversity and productivity of floodplain ecosystems the following measures are recommended:

1. Before the release of water flux it is necessary to ensure the filling river channel with the volume of water as large as 0.5 to 1.0 km^3, which allows increasing the Shulbinsky reservoir operational storage capacity to be used. In dry years, filling the river bed within the floodplain extend is only possible by means of shifting the start time of maximum water releases from the Ust-Kamenogorsk HPP (taking account of the lag time) to 8 - 10 days before the start of environmental flows.
2. The total duration of the spring releases must be at least 18 days in the case of maximum discharges at more than 4000 m^3 s^{-1}, and 20 days in case of the power

generation with maximum water discharges 3500 m^3 s^{-1}. The total volume of water releases should not be less than 4.7 km^3.

3. The intensity of the flood water level rise should be at the maximum limit (the maximum allowed water discharge increment is 400 m^3 day^{-1}) and the duration is 7 to 8 days (if the initial water discharge was 1100 m^3 s^{-1}), correspondingly, the water level decline should be 8 to 9 days (the water discharge at the end of flood is 700 m^3 s^{-1}).
4. The duration of the maximum water discharge pass must be at least 6 days and the flow rate should be 3500 m^3 s^{-1} in dry years and 3800-4000 m^3 s^{-1} in years with water flow rate close to the long-term mean.
5. The total duration of the water flux wave should be less than 30 days in wet years. Because the longer duration causes overflooding the downstream floodplain areas, it is more effective to provide a two-peak water flux hydrograph with the greatest possible gap between the peaks.
6. The uncertainty on the start time of the filling channel is about ±6 days, which determines the missed opportunity of the use of 400 - 500 million m^3 of flood water flow if the beginning of natural river flood period delays, however, this risk allows at least 4 times in 5 years yielding additional flooded area increment of 10 - 25 10^3 ha.
7. Phytomelioration in floodplain areas with disturbed vegetation and soil cover (fertilizing of soil after the flood and winter sowing of local forage plants), the fight against weeds should be conducted.
8. Regulation of recreational pressure and control of environmental and fire safety in places of mass recreation are necessary.
9. The annual monitoring of the status of all components of the floodplain landscape is needed.
10. It is also necessary to keep control over synanthropization of the vegetation in the Irtysh valley in order to prevent uncontrolled invasion of new species that can displace the aboriginal flora.

CONCLUSION

Downstream the Upper Irtysh cascade of reservoirs the mean annual water runoff of the Irtysh River in the study area for the period 1960-2010 has changed slightly (4-6%), but significant changes (redistribution) occurred in the monthly and seasonal runoff. Within a year the flow redistribution is subject to the following pattern: in high water season, summer and part of the autumn low flow season it decreased whereas in winter time it increased. This resulted in significant reduction in the volume of spring water releases for Irtysh floodplain inundation and reduction of flooded areas of the floodplain zone.

The main difficulty in the river flux management is to establish the optimal start date for environmental flows, which must be coordinated with the flood timing of lateral inflow. In this respect, it is necessary to ensure water flow rates in the Irtysh between 1200 and 1500 m^3 s^{-1} before the start of water flux with the aim to fill the river channel. Analysis of water release from the HPP shows the following critical factors determining the efficiency of flooding floodplain: the flows volume and maximum discharge, as well as the duration of water flux itself and of the period with the maximum discharges.

The flora of the study area is represented by 545 species, 268 genera and 73 families of higher plants, which characterizes it as of sufficiently high variety. The vegetation cover is composed of 7 types. Phytocenotic diversity is due to spreading the valley through three climatic subzones of steppe zone and complex influence of intrazonal abiotic factors.

Subclimax meadow-steppe and steppe vegetation of the Irtysh valley is located on terraces above the floodplain and strongly transformed by human activities. When moving to the south, the cenotic role of sagebrush increases and the role of motley grass decreases.

Floodplain forests and shrubs confined mainly to the riverine part of floodplain and do not have significant subzonal differences. Riparian forests are located in central and near-terrace part of the floodplain due to the peculiarities of the relief and the presence of floodplain side-channels. Halophyte shrub communities within desert steppes subzone are formed upon near-terrace and ancient central floodplain.

The greatest subzonal differences occur in the composition and structure of floodplain meadows. Within the subzone of arid steppes, the true and wet glikophyte meadows prevail. In dry steppe, the glikophyte and steppefied meadows, and in desert steppe, the steppefied halophyte meadows are common. The process of halophytization of floodplain meadows enhances with the distance from the riverbed toward the original bank.

Marsh vegetation is most prevalent in the near-terrace floodplain and riparian zone. Aquatic and marsh vegetation types does not have any subzonal differences and their characteristics are controlled by hydrological factors.

Some recommendations are proposed to protect the floodplain ecosystems from desertification taking account the hydrological regime regulation and the water flow rates and timing.

ACKNOWLEDGMENTS

The financial support from a BIO GEO CLIM Mega-grant from the Ministry of Education and Science of the Russian Federation and Tomsk State University (No 14.B25.31.0001) and from the D. I. Mendeleev program of Tomsk State University (No 8.1.88.2015) is acknowledged.

REFERENCES

Ilyina, I. S. (1985). *The vegetation cover of the West Siberian plain*, 1985. Nauka, Novosibirsk, 281 p.

Kamkin, V. A. (2009). Regularities of space structure of vegetation of the Irtysh River valley (in the Pavlodar region). PhD thesis. Almaty., 140 p.

Korytniy, L. M. (1994). Problem-oriented water-resource zoning of Siberia. *Geography and Natural Resources (Geographia and Prirodnie Resursi)*, *1*, 32-41.

Mogilyuk, S. V. (2004). *Geo-ecological aspects of water management in transboundary river basins*. PhD thesis. Tomsk., 150 p.

Mirkin, B. M. Theory and practice of phytocenology, 1981. News in Life Science and Technology. *Biology*. Moscow, Znanie., 40 p.

Ogar', N. P. (1999). The vegetation of river valleys of arid and semi-arid regions of continental Asia. Dissertation of Doctor of Biological Sciences. Almaty., 273 p.

Prokopiev, E. P. (2001). Sintacsonomical review of forest vegetation of the Irtysh River floodplain. *Krylovia. V.*, *3*, 1, 13-23.

Prokopiev, E. P. (1975). Dynamical trends in the vegetation cover of the Middle Irtysh floodplain. Questions of morphology and dynamics of vegetation. Scientific notes of the Kuybyshev pedagogical Institute. *Kuibyshev. No.*, *163*, V. 5. 31-42.

Rachkovskaya, E. I., Volkova, E. A. and Khramtsova, V. N. (Eds.), (2003). *Botanical geography of Kazakhstan and Central Asia within the desert region*. St. Petersburg, 424 p.

Razumovsky, S. M. (1981). Regulations of the biocenoses dynamics. Moscow, Nauka., 231 p.

Rabotnov, T. A. (1985). *Ecology of meadow grasses*. Moscow, Moscow State University., 310 p.

Tsaregorodtseva, A. G. (2004). *Ecological sustainability of floodplain landscapes of Pavlodar region under conditions of the Irtysh River flow regulation*. PhD thesis. Almaty., 150 p.

In: Riparian Zones
Editor: Oleg S. Pokrovsky
ISBN: 978-1-63484-613-4

Chapter 10

BIOGEOCHEMISTRY OF ORGANIC CARBON, MAJOR AND TRACE ELEMENTS IN THE FLOODED AND RIPARIAN ZONE OF THE OB RIVER

S. N. Vorobyev, V. V. Drozdov, A. V. Sorotchinskiy, S. N. Kirpotin, L. G. Kolesnichenko, L. S. Shirokova, R. M. Manasypov and O. S. Pokrovsky*

BIO-GEO-CLIM Laboratory, Tomsk State University, Tomsk, Russia
GET UMR 5563 CNRS, University of Toulouse, Toulouse, France
Instititute of Ecological Problems of the North,
Russian Academy of Science, Arkhangelsk, Russia

ABSTRACT

The flood zone of the Ob River, the largest (in watershed area) river of the Arctic Ocean basin, tens of km wide, and, after the Amazon's Varzea, is the world's second largest flooding territory. To better understand the biogeochemistry of the Ob River and adjacent surface waters, we studied, in May and July 2014, the dissolved and colloidal organic carbon and trace metals in small rivers, lakes, and flooded water bodies connected and disconnected with the mainstream, as well as the Ob River itself. All major and trace elements were distributed among two major categories depending on their pattern of dependence on the dissolved organic carbon (DOC) concentration. Dissolved Inorganic Carbon (DIC), Na, Mg, Ca, sulfate, Sr, Mo, Sb and U exhibited a general decrease in concentration with increase the [DOC] reaching the lowest values in upland lakes, which are rich in DOC. These elements marked the influence of underground feeding in July during summer baseflow, which is most visible in flood lakes in the Ob riparian zone and the Ob River itself. In May, the flood lakes were statistically similar to the Ob River. The elevated concentration of DOC (up to 60 mg/L) in the upland lakes was not correlated with groundwater-related elements, suggesting a

* E-mail: soil@green.tsu.ru.

lack of significant groundwater feeding in these lakes. By contrast, insoluble, usually low mobile elements (Al, Fe, other trivalent hydrolysates, Ti, Zr, Hf) and some metals (Cr, Zn, Ni, Pb) demonstrated a steady increase in concentration with increasing DOC, with the lowest values observed in the Ob River and the highest values observed in small tributaries and organic-rich upland lakes in July. It follows that these elements are limited by their main carriers – organic and organo-ferric colloids, rather than by the availability of the source, peat and mineral soil or plant litter. While for the majority of non-colloidal, groundwater-fed elements with high mobility (DIC, Na, Mg, Ca, K, Sr...), the small tributaries can be used as representatives of the Ob main stream, this is not the case for low mobility "insoluble" elements such as Fe, Al, trivalent and tetravalent hydrolysates and metal micronutrients (Cu, Zn, Mn). The low soluble elements and divalent metals exhibit much lower concentration in the river mainstream compared to that in the flood lakes, upland lakes and small rivers. This difference is significantly more pronounced in the baseflow in July compared to a spring flood in May.

A principal component analysis of all elements matrix in all studied rivers and lakes revealed 3 main factors persisting during different seasons. The first factor was linked to organic and organo-mineral colloids and comprises DOC, Al, Ti, V, Cr, Fe, Co, Ni, Ga, As, Zr, Ba, REEs and Hf. The 2nd factor was controlled by DIC and major cations and also included Rb and Sr, assuming important link of these highly mobile elements to atmospheric precipitates and/or carbonate mineral dissolution. Finally, the 3rd factor comprising pH, SO_4, Mo, Sb and U, was probably linked to groundwater leaching of soluble elements and could also reflect the decrease of adsorption of axyanions (Mo, Sb, U carbonate complexes) onto clays with the decrease of pH. The factorial structure was highly stable during different contrasting seasons among small rivers, upland lakes and flood lakes. Presumably, autochthonous processes such as the photo-oxidation and bio-oxidation of organo-ferric colloids and phytoplankton uptake decrease the concentration of these elements in the river mainstream. The riparian and the flood zone of the Ob river are very important regulators of dissolved and colloidal flux of organic carbon and metals from the adjacent watershed (small rivers of the river valley, upland lakes) to the main river, via numerous flood lakes and groundwater reservoirs.

Keywords: Siberia, riparian zone, flood, baseflow, lakes, rivers, underground waters, organic carbon, metals, colloids

INTRODUCTION

Despite the importance of the Ob River, the largest, in terms of watershed area, river discharging to the Arctic Ocean, systematic seasonally resolved studies of this river dissolved load (carbon, metals) are limited compared to those of the Lena, Yenisei and Mackenzie Rivers. A possible reason for this understudy could be weak coverage of the Ob watershed by the permafrost (only 30% of the basin, compared to 75%, 90%, and 100% of the Mackenzie, Yenisei, and Lena watersheds, respectively, Environment Canada, 2015; McClelland et al., 2004) which is in the center of the attention of most "arctic" researchers. Yet, the on-going climate change in the boreal and subarctic region will likely cause not only permafrost degradation and respective river flux change, but also the precipitation regime, and in particular, the relative contribution of the spring flood versus summer and winter base flow (Peterson et al., 2002; McClelland et al., 2006; Shiklomanov and Lammers, 2009; Rennermalm and Wood, 2010). Here, the Ob River, exhibiting the second largest (after the

Amazon's Varzea, Viers et al., 2009) flooding area, may become the major regulator of dissolved carbon and related element transport from the land to the Arctic Ocean. Secondly, the majority of the Ob's watershed lays within highly productive boreal taiga zone. The productivity of the terrestrial boreal biome is the highest on the riparian zone and river banks, especially in the permafrost-free region (Huston and Wolverton, 2009). Because the degradation of plant litter is one of the major factors regulating overall export of DOC and chemical element from the watershed in the boreal zone (Pokrovsky et al., 2012), and the plant litter leaching is very fast (Aerts and Chapin, 2000), the spring flood period on the Ob river is especially important for the overall functioning of the western Siberia plain. And thirdly, the Ob River watershed is likely to be most vulnerable to on-going climate change and permafrost thaw, as the major part of its permafrost coverage is intermittent and sporadic permafrost (0-2°C) rather than continuous and discontinuous permafrost unlike the other Arctic rivers (McClelland et al., 2004). The former is known to be most unstable under ground temperature rise (Romanosvky et al., 2010).

The further importance of the Ob River and its watershed for the Artcic and subarctic region function stems from the very high dissolved Fe concentration in this river; the mean value for the period 1990-1996 is a factor of 5 higher than that of the Yenisey and the Lena rivers (Alexeeva et al., 2001). As a result, the Ob River provides almost 40% of the total flux of this important, potentially limiting micronutrient from the land to the whole Arctic basin (Alexeeva et al., 2001). To better understand the mechanisms regulating the contemporary fluxes of carbon, trace metals and major nutrients at the Ob watershed and to predict possible future changes, the use of the time series only at the terminal gauging station (i.e., Cooper et al., 2008; Holmes et al., 2012) is not sufficient. Rather, detailed, seasonal studies of the different components of the watershed, such as the small tributaries, flooded water bodies and adjacent lakes, are necessary to reveal the contribution of the various sources, notably surface and ground waters, to the main stream. To achieve this goal, we sampled small first order tributaries as well as upland and flooded zone lakes and the Ob River's middle course in the boreal taiga zone during the spring flood (May) and the summer baseflow (July). We addressed the following specific questions:

1. Can we distinguish the control of the groundwater versus surface runoff on the river hydrochemistry for different major and trace elements during two contrasting seasons?
2. How important is the control of the dissolved organic carbon on the TE concentration in various surface water bodies?
3. Can we approximate the main river hydrochemistry by its flooded zone during the spring flood?
4. Which autochthonous processes are responsible for the TE concentration change and colloid transformation in the flood zone, lakes and rivers?

We anticipate that understanding the main features of the dissolved load transformation between the feeding water bodies and the main rivers may help to predict the magnitude and direction of future changes of the Ob River's middle course hydrochemistry in response to the change of the water flooding, river discharge regime and water temperature rise.

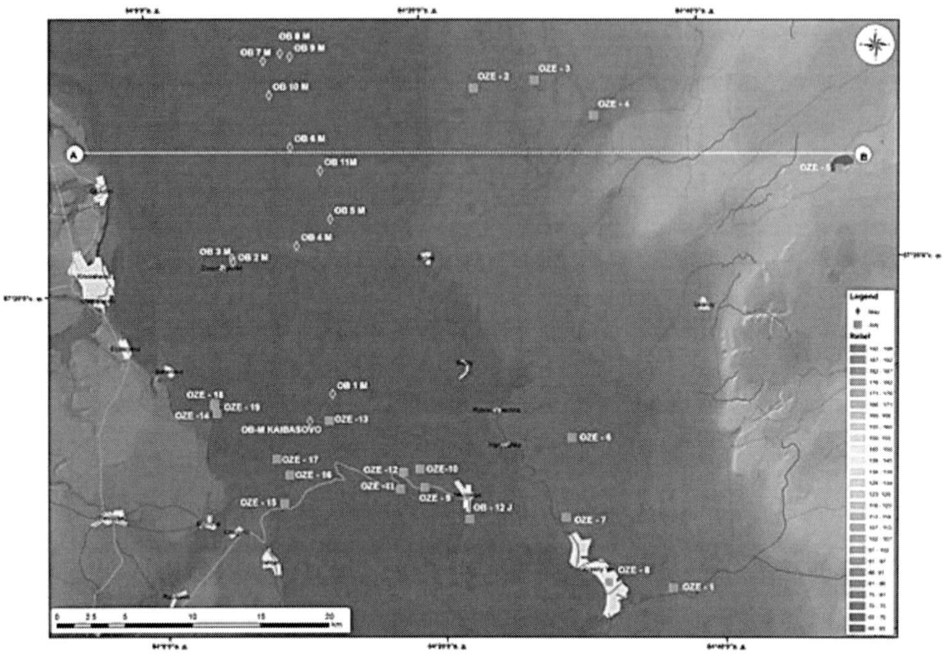

Figure 1A. Map of the region including the Ob River middle course, its flood zone and adjacent lakes and small tributaries.

Study Sites, Sampling and Analyses

The middle course of the Ob River, located within the boreal taiga biome (Figure 1 A), includes (i) isolated and interconnected water bodies that are flooded during the spring period but are persistent during baseflow (called herein "flood lakes"); (ii) the flooded area of the river valley and first terrace that is covered by the river during high flow in May and extends, via a system of interconnected shallow ephemeral lakes and primary and secondary water channels, over 5-10 km from the main water channel (called "flood zone"); (iii) permanent lakes located at the upper terraces of the river that are not flooded during the spring ("upland lakes"); and (iv) small, first order tributaries of the Ob River that have a watershed area of 10 to 100 km² and drain both the terraces and the flood zone ("small rivers"). Here, we operationally define an "upland" as the territory of the 1st to 3rd river terraces that are not subjected to flooding that is at a higher elevation than the alluvial plain, which is considered to be "lowland" (see profile Figure 1 B). To some degree, the normal transect of the Ob River and its flooding zone is similar to that of the Mackenzie Delta lakes, comprising a significant flooding gradient (i.e., Squires et al., 2002; Tank et al., 2009).

These contrasting but representative water objects were sampled during two open water periods in 2014: the spring flood in May and the summer base flow in July. Typical photos of the Ob River riparian and flood zone are given in Figure 2. The isolated lakes were sampled using portable PVC boat whereas the larger, interconnected water bodies including the mainstream of the Ob River were sampled from the motor boat using submersible polycarbonate Aquatic Research® water sampler. Small rivers were sampled from the river bank or the middle of the flow.

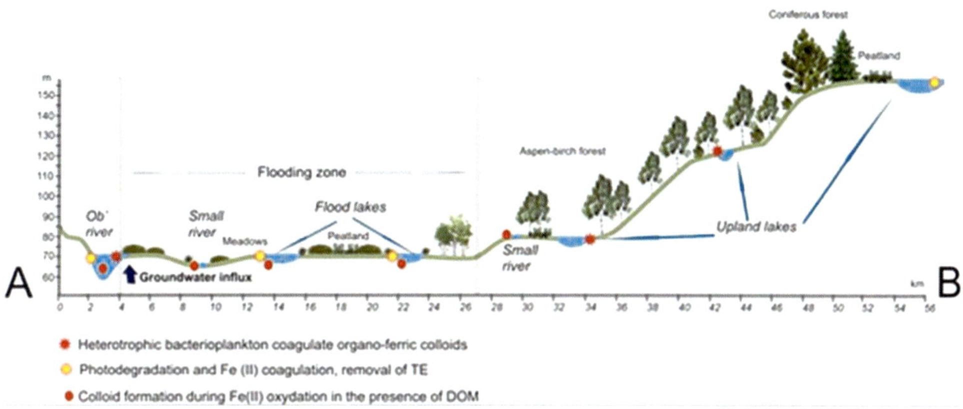

Figure 1B. Geomorphological profile of line A-B (Figure 1 A) including all major studied water objects.

All field procedures were complained to the strict, metal free sampling and handling techniques as described previously (Shirokova et al., 2010). Water samples were collected from the bridge or via a boat near the middle of the stream flow in pre-cleaned 1-L high-density polyethylene (HDPE) containers connected to a non-metallic stick. The value of pH was measured using a combined Schott-Geräte electrode calibrated against NIST buffer solutions (pH = 4.00, 6.86 and 9.18 at 25°C) with an accuracy of ± 0.02 pH units. The collected waters were immediately filtered through disposable 0.45 μm acetate cellulose membranes (25-mm diameter). The first 20-50 mL of the filter-passed permeate (referred to as filtrate) were systematically rejected.

To assess the proportion of colloids, dialysis experiments were performed using 20-50 ml precleaned dialysis bag placed in 2 litre polypropylene containers filled by unfiltered river water (ex-situ dialysis). The duration of the dialysis procedure was between 48 and 72 h, consistent with earlier kinetic studies (Pokrovsky et al., 2005, 2010). EDTA-cleaned, trace-metal pure SpectraPor 7® dialysis membranes made of regenerated cellulose and having 1 kDa poresize were thoroughly washed in 0.02 N bidistilled HNO3, ultrapure water, filled with ultrapure MQ deionized water and put into natural water for dialysis experiments. The efficiency of dialysis procedure was evaluated by comparison of concentration of major anions (i.e., Cl-) or neutral species (H4SiO4°) not associated with colloids in the dialysis bag and in the external solution. These concentrations were always identical within ± 10% suggesting equilibrium distribution of dissolved components. Note that the oxidation of Fe(II) during separation is very unlikely because all Fe in well oxygenated surface waters of studied water objects should present in the Fe(III) form (Lofts et al., 2008).

Filtered and dialysed solutions for major cation and trace element analyses were acidified (pH 2) with ultrapure double-distilled HNO3 and stored in PP or HDPE bottles previously washed with ultrapure 0.1 M HCl and rinsed with MilliQ water. The preparation of bottles for sample storage was performed in a clean room. The MilliQ field blanks were collected and processed in order to evaluate potential sample contamination introduced by our sampling and handling procedure. Organic carbon blanks of filtrate and dialysates never exceeded 0.1 mg/L, which is low for the organic-rich rivers sampled in this study (typically, 10 to 60 mg/L DOC). For all major and most trace elements, concentrations in blanks were below analytical detection limits (≤0.1-1 ng/L for Cd, Ba, Y, Zr, Nb, REE, Hf, Pb, Th, U; 1 ng/L for Ga, Ge, Rb, Sr, Sn, Sb; ~10 ng/L for Ti, V, Cr, Mn, Fe, Co, Ni, Cu, Zn, As).

Figure 2. Ground photo of the Ob River riparian zone (upper) and its flood zone (bottom) in May 2014 (spring flood period).

Aqueous silica concentration was determined colorimetrically (molybdate blue method) with an uncertainty of 2% using a Technicon automated analyzer. Alkalinity was measured by potentiometric titration with HCl to pH = 4.2 using the Gran method, with a detection limit of 10-5 M and an uncertainty of 2%. DOC was analyzed using a Carbon Total Analyzer (Shimadzu TOC 6000) with an uncertainty better than 3%. Major anion concentrations (Cl, SO4, F, NO3) were measured by ion chromatography (HPLC, Dionex 2000i) with an uncertainty of 2%. Trace elements (TE) were measured without preconcentration by ICP-MS using both Ar and He mode to avoid the interferences. The typical uncertainty on element concentration measurement ranged from 5-10% at 1-1000 μg/L to 20-30% at 0.001-0.1 μg/L.

For many samples of surficial fluids and ultrafiltrates having very low concentrations of TE, on the order of 0.001 μg/L comparable with detection limits such as Cd, Hf, Ge, Cs, Ga, Se, Sn, W, Bi, the minimal estimated uncertainty is 30%. During the ICP-MS analyses, the international geostandard SLRS-5 (Riverine Water Reference Material for Trace Metals certified by the National Research Council of Canada) was measured each 20 samples to check the validity and reproducibility of the analyses (Yeghicheyan et al., 2013).

RESULTS

Categories of Major and Trace Elements Exhibiting Dependence on DOC Concentration

The list of sampled water bodies and their main hydrochemical characteristics is given in Table 1. Depending on their affinity to DOC, several groups of major and trace elements were distinguished. The 1st category comprised Dissolved Inorganic Carbon (DIC), Na, K, Rb, Mg, Ca, sulfate, Cs, Sr, Mo, Sb and U. These elements exhibited a general decrease in concentration with an increase in the [DOC] achieving the lowest values in upland lakes, which are rich in DOC (Figure 3 A-H). For the Ob' River itself, the element concentration decreased with an increase in the [DOC]. Within this first category, various parts of the flooded zone sampled in May are particular interesting as they presented a rather contrasting behavior for different elements. First, the major (Ca, Mg, Na and DIC) and trace elements (Sr, Mo, Sb, U) were either independent of the [DOC] or their concentration decreased with the [DOC]. On the other hand, K, Rb, Si and Cu were strongly correlated with the DOC in the flooding zone in May ($R_{Cu, DOC} = 0.93$). The analysis of the element concentration in the flood zone as a function of the distance from the main stream demonstrated an increase of the concentration of a number of elements such as macro (Si, K) and micro-nutrients (Mn), DOC and related metals (Fe, Ni, Co, and Cu) which exhibited high affinity to plant biomass (Figure 4). These elements are likely to be influenced by plant leaching (Si, K, Mn) and are present in the form of uncomplexed ions or organic and organo ferric colloids (Fe, DOC, Cu, Ni, Co). The ground water – fed elements are mostly pronounced in flood lakes during summer baseflow in July (Figure 3 A-E). Though most of them (DIC, Na, Mg, Ca, and Sr) including the major elements, are statistically independent of the DOC at $p < 0.05$, some clearly decrease as the [DOC] increases, marking an abrupt decrease of the groundwater influence (Mo, U, Sb), Figure 3 F, G, and H.

The 2nd category is composed of few elements that are irrelevant to the DOC concentration, depicting no clear pattern in the concentration-DOC trend among various water objects. These are B, Cl, Si (except the flood zone in May) and V. Presumably, elements present as neutral molecules or oxyanions weakly interact with DOM and that may originate from atmospheric deposits (B and Cl), or surface flow (Si and V) rather than groundwater.

The 3rd category of elements comprises those that are controlled by organic and organo-mineral colloids as follows from the results of the size fractionation described in the next section. These are insoluble, usually low mobile elements (Al, Fe, other trivalent hydrolysates, Ti, Zr, and Hf) and some metals (Cr, Zn, Ni, Ba, and Pb). They demonstrated a steady increase in concentration with an increase in the DOC, with the lowest values pbserved in the Ob River and the highest values observed in small tributaries and organic-rich upland lakes in July (Figure 5 A-D, I-L). Typically the lithogenic elements of this category (Al, Fe, Ti, Zr, Hf, Th, and REEs) first increase their concentration by c.a. 2 orders of magnitude from 10 to 30 mg/L of DOC, and then achieve a maximal concentration in most organic-rich rivers and upland lakes in July. At the same time, organic-rich upland lakes presenting c.a. 60 mg/L of DOC are not necessarily enriched in these elements relative to small rivers.

Table 1. List of sampled objects and their main hydrochemical properties

Sample	Description	Date	GPS
	The Ob' River		
OB-M	Ob above Kozurbak river	19.05.2014	N 57°15.615'; E 84°04.206'
OB 1 M	Ob Kaibasovo (island)	19.05.2014	N 57°15'48.3; E 84°12'34.4
OZE - 19	Ob above Kozurbak river	19.07.2014	N 57°15.615'; E 84°04.206'
OB - 12 J	Ob at Nikolskoe	18.07.2014	N 57°10.761'; E 84°21.997'
	Flood zone lakes		
OB 2 M	Entrance to Lake Monatka	19.05.2014	N 57°10'47.6"; E 84°05'12.0"
OB 3 M	Beginning of Lake Monatka	19.05.2014	N 57°21'12.2"; E 84°05'42.9"
OB 4 M	Middle part of Lake Monatka	19.05.2014	N 57°21'32.1"; E 84°10'27.8"
OB 5 M	r. Kulgesnikha	19.05.2014	N 57°22'30.6"; E 84°12'57.8"
OB 6 M	confluence Kulgesnikha - Oskina	19.05.2014	N 57°25'19.3"; E 84°10'19.0"
OB 7 M	confluence Bogonos - Oskina	19.05.2014	N 57°28'41.6"; E 84°08'38.4"
OB 8 M	Small river inlet on the flood zone	19.05.2014	N 57°28'56.5"; E 84°09'53.6"
OB 9 M	Futhest point on the flood zone	19.05.2014	N 57°28'48.2"; E 84°10'34.9"
OB 10 M	Small stream from the flooded zone	19.05.2014	N 57°27'21.3"; E 84°08'55.8"
OB 11M	r. Kuzhenikha, flooded forest	19.05.2014	N 57°24'13.1"; E 84°12'23.5"
	Upland Lakes		
OZE - 1	1st terrace, Lake Schuchie	16.07.2014	N 57°07.680'; E 84°36.358'
OZE - 2	3rd terrace, Lake Maloe Schuchie	16.07.2014	N 57°27.274'; E 84°23.744'
OZE - 5	Lake Bolshoe Schuchie	17.07.2014	N 57°23.398'; E 84°49.879'
OZE - 3	Inlet to Lake Glubokoe (2 km)	16.07.2014	N 57°27.425'; E 84°28.208'
	Small rivers, Ob's tributaries		
OZE - 4	r. Chernaya (5th km), on 3rd terrace	17.07.2014	N 57°25.975'; E 84°32.356'
OZE - 6	r. Adrianova (> 10 km length)	17.07.2014	N 57°13.665'; E 84°29.658'
OZE - 7	r. Talach	17.07.2014	N 57°10.632'; E 84°28.970'
OZE -14	r. Shigarka	19.07.2014	N 57°15.263'; E 84°04.082'
	Flood lakes		
OZE - 9	Protoka Vesna, 31th km Nikolskoe	18.07.2014	N 57°12.059'; E 84°18.786'
OZE-10	Ishtakhta protoka	18.07.2014	N 57°12.714'; E 84°18.548'
OZE -11	Glubokoe lake (1 to 2 km)	18.07.2014	N 57°12.037'; E 84°17.169'
OZE -12	Unconnected lake (Kaibasovo)	18.07.2014	N 57°12.651'; E 84°17.443'
OZE -13	Connected lake, Kaibasovo	18.07.2014	N 57°14.770'; E 84°12.214'
OZE - 15	Beginning of Kozurbak	19.07.2014	N 57°11.696'; E 84°08.703'
OZE - 16	Small inlet from the lake (Kozurbak)	19.07.2014	N 57°12.768'; E 84°09.213'
OZE - 17	Kozurbak, middle course	19.07.2014	N 57°13.399'; E 84°08.446'
OZE - 18	Kozurbak, mouth zone	19.07.2014	N 57°15.627'; E 84°03.819'
	Snow		
SF1	Integral snow sample	10.03.2014	N 56°31'48"; E 84°09'44"
SF3	Integral snow sample	12.03.2014	N 57°06'26.3"; E 83°54'28.4"
SF4	Integral snow sample	15.03.2014	N 57°20'19.4"; E 83°56'32.4"
SF9	Integral snow sample	15.03.2014	N 58°04'26.6"; E 82°49'31.7"

Sample	pH	DIC mg/L	DOC mg/L	Cl- mg/L	SO_4 mg/L	Na µg/L	Mg µg/L	Al µg/L	Si µg/L
OB-M	8.04	20.9	23.9	2.83	9.23				
OB 1 M	8.23	14.4	16.8	1.50	6.95	5079	3260	17.7	2173
OZE - 19	8.20	20.4	4.0	1.88	7.51	4952	4450	4.8	2221
OB - 12 J	8.00	20.3	4.0	1.81	7.36	4803	4298	4.7	2277
OB 2 M	8.09	10.55	13.44	0.88	5.17	3827	2278	16.2	66.56
OB 3 M	8.02	11.01	13.59	0.92	5.27	3977	2363	14.3	79.49
OB 4 M	8.04	11.66	14.4	1.09	5.93	4317	2597	19.8	115.9
OB 5 M	8.07	11.19	16.1	1.04	5.79	4504	2863	27.9	258
OB 6 M	7.98	10.82	16.6	1.00	5.67	4367	2769	32.3	257
OB 7 M	8.00	10.78	16.8	1.00	5.51	4260	2767	32.2	252
OB 8 M	7.91	12.5	26.7	1.49	6.58	4690	3283	33.8	617
OB 9 M	7.68	11.25	29.8	1.43	2.86	4182	3027	32.4	587
OB 10 M	7.81	10.37	17.96	0.79	3.29	3766	2630	17.5	142
OB 11M	7.97	11.07	14.51	1.03	5.81				
OZE - 1	6.15	1.88	56.6	0.17	1.10	2674	1374	497	2515
OZE - 2	5.90	1.05	27.0	0.17	0.61	1315	704	134	2180
OZE - 5	5.80	0.64	12.0	0.12	0.96	1044	383	107	1326
OZE - 3	4.94	0.83	61.5	0.64	0.17	2777	822	623	5545
OZE - 4	6.00	1.87	33.0	0.18	1.17	4123	962.2	534	10820
OZE - 6	7.45	18.0	17.1	0.14	0.39	4697	4217	16	10990
OZE - 7	7.05	20.5	29.0	0.14	0.38	5665	4717	32	10360
OZE -14	8.00	58.3	9.0	6.15	3.74	16340	14280	0.05	5542
OZE - 9	6.90	21.6	12.6	0.33	1.33	3860	5780	0.426	2541
OZE-10	7.70	34.0	14.2	36.8	0.48	32370	8190	<d.l.	4581
OZE -11	6.80	14.1	11.8	1.61	1.03	4327	4008	2.81	1458
OZE -12	6.70	19.7	12.2	0.33	0.38	2909	5362	2.11	2179
OZE -13	7.12	33.0	8.6	60.1	0.09	36370	9426	2.64	2631
OZE - 15	8.17	29.0	11.2	15.6	5.53	16650	7450	2.48	2960
OZE - 16	7.40	28.2	13.1	25.1	6.33	22080	7706	1.94	3654
OZE - 17	8.75	24.2	8.0	12.9	6.50	12620	5860	2.01	2280
OZE - 18	8.40	23.4	6.9	9.20	6.53	9892	5404	3.61	2402
SF1	8.73	0.80	0.86	0.53	0.99	354	78	15.4	3.5
SF3	6.82	2.12	0.69	0.10	0.54	160	35	6.6	13.9
SF4	6.59	0.40	0.79	1.72	0.58	1160	39	10.8	27.8
SF9	5.11	0.40	0.75	0.37	0.74	244	22	33.7	19.4

The divalent metals and micronutrients Mn, Cu and Zn should be considered separately from other elements (Figure 6 A, B, and C). The highest Mn concentrations are encountered in small rivers and upland lakes in July, in which Mn is independent of the DOC. The lowest concentrations of all three metals are observed in the Ob' River in July. Zn exhibited highest the concentrations in DOC-rich upland lakes and rivers in July, whereas Cu demonstrated the highest concentrations in the flood zone in May, at the furthest distances from the main stream.

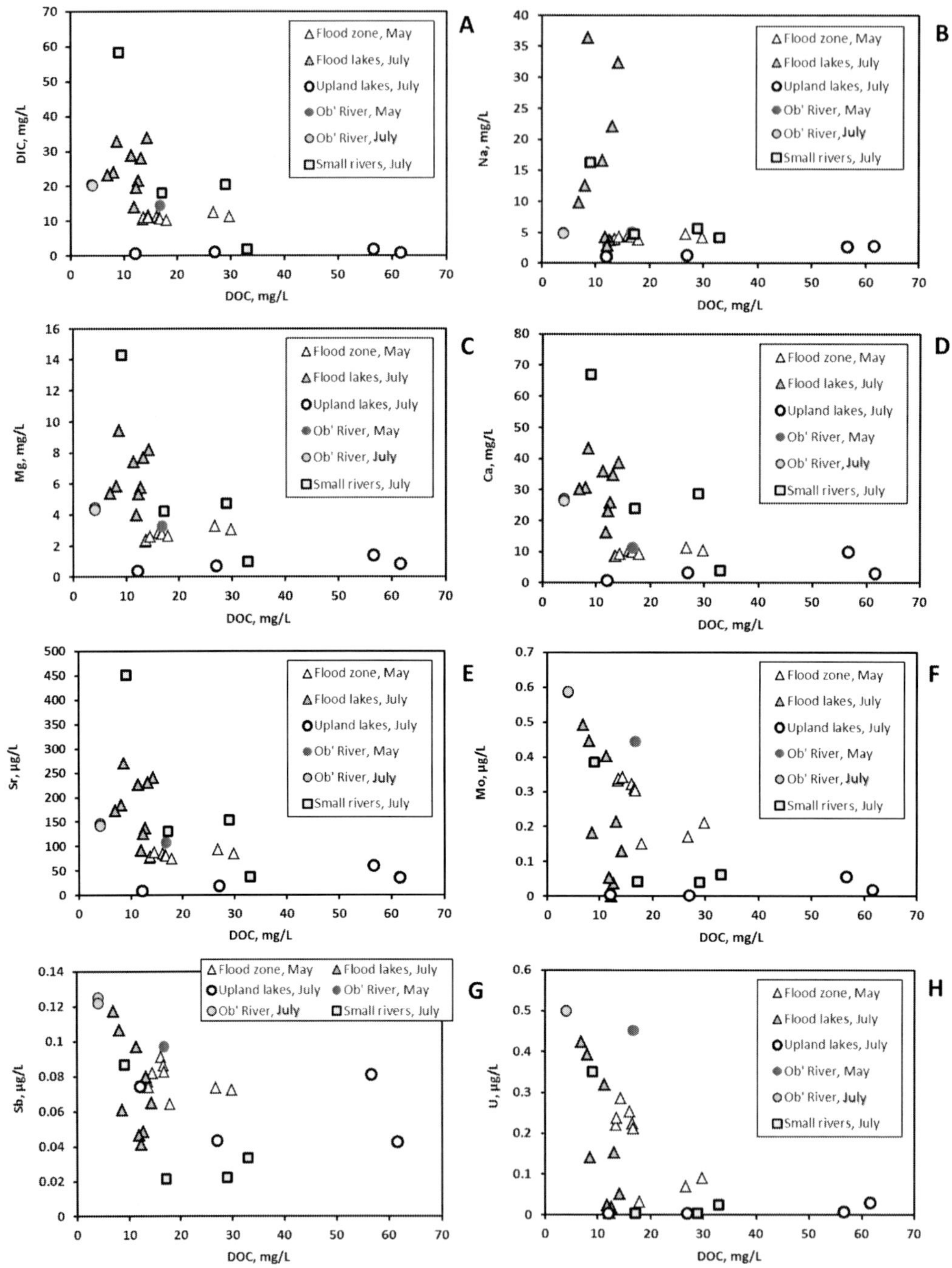

Figure 3. Dissolved (<0.45 μm) concentration of DIC (A), Na (B), Mg (C), and Ca (D) as a function of dissolved organic carbon (DOC) for the Ob'river flood zone, rivers and lakes during baseflow and spring flood.

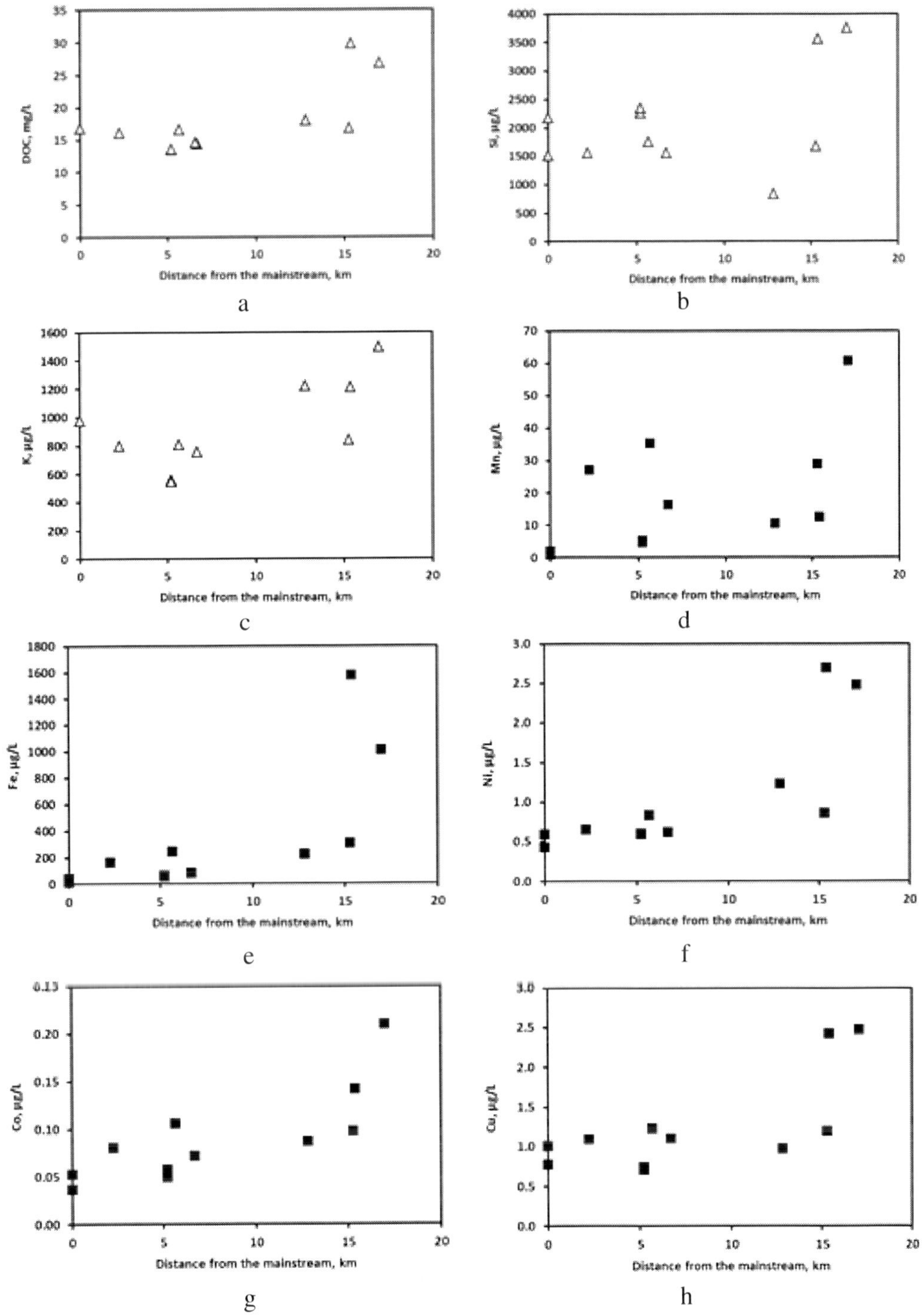

Figure 4. DOC (A), Si (B), K (C), Mn (D), Fe (E), Ni (F), Co (G) and Cu (H) concentration in the flood/riparian zone as a function of the distance from the mainstream.

Given the large geographic coverage of samples from the spring flood zone, we tested, whether the water objects of the flooded zone can be considered to be surrogates for the main river chemical composition. For this, we calculated the ratio of the element mean concentration in the Ob' River and that in the flood zone in May (Figure 7). Only a few elements deviate significantly from unity. These are U and Mo, which are enriched by a factor of 1.5-2.5 in the river relative to the flood zone water objects, and Zn, Fe and Mn, which are depleted by a factor of 3 to 10 in the Ob River. The other ~ 40 major and trace elements present in the flood zone waters fall within ± 40% of their riverine concentrations in May.

Principal Component Analysis of Major and Trace Elements

A principal component analysis of all elements matrix in all studied rivers and lakes objects revealed 3 main factors (24.5, 9.2 and 5.2 eigenvalues, respectively) persisting for different seasons and contrasting water objects.

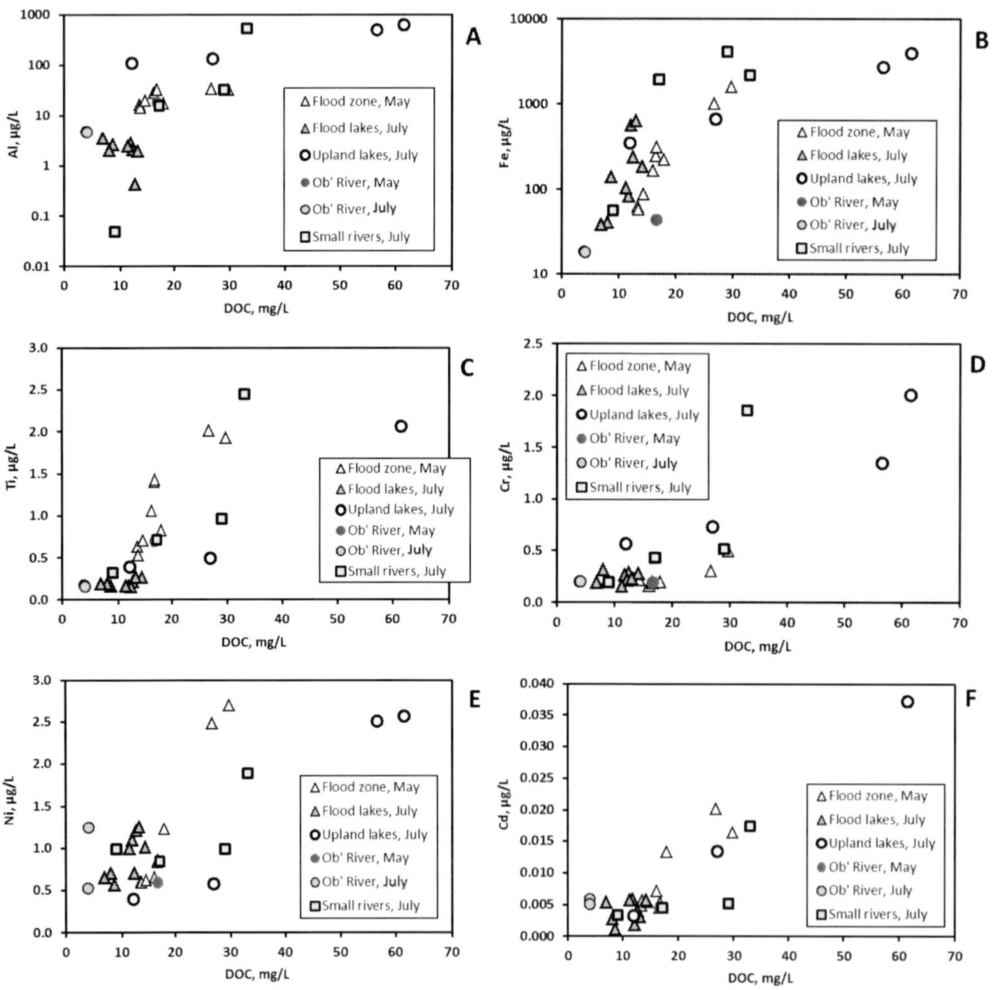

Figure 5. (Continued).

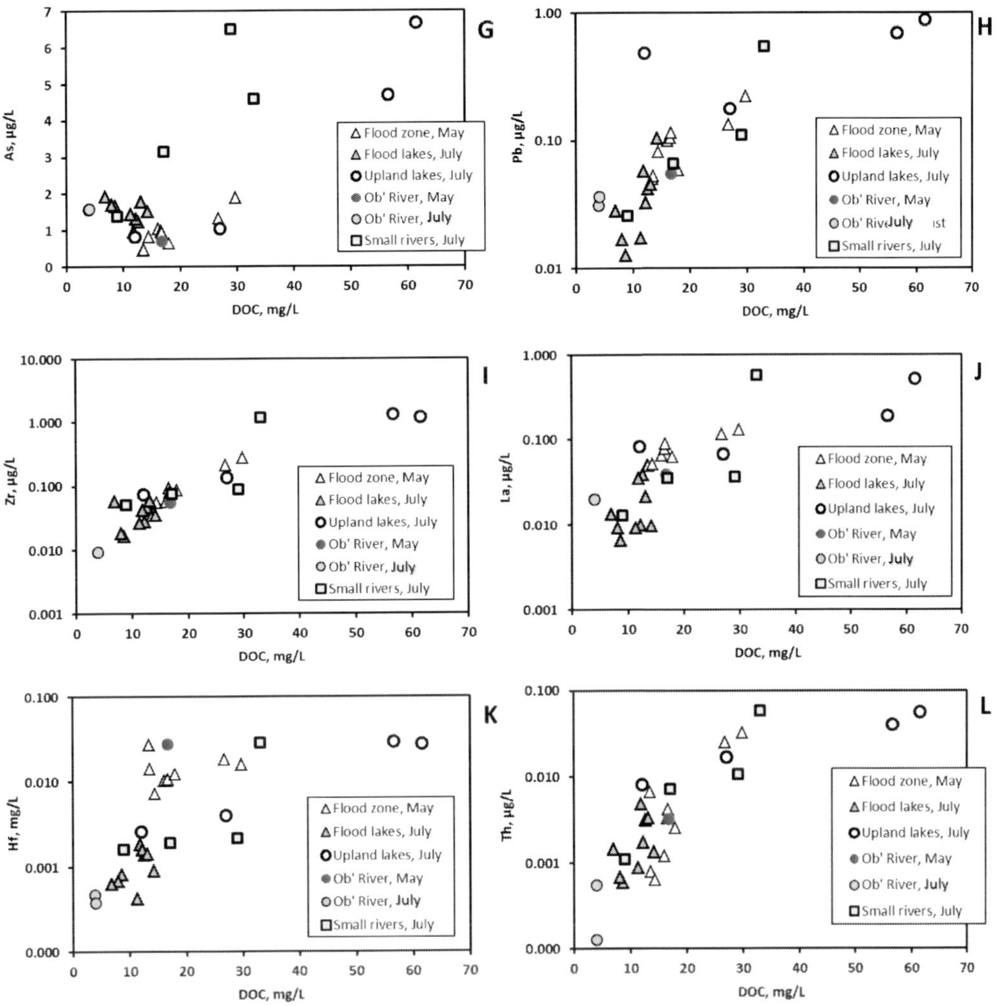

Figure 5. Dissolved (<0.45 μm) concentration of Al (A), Fe (B), Ti (C), Cr (D), Ni (E), Cd (F), As (G) and Pb (H) as a function of dissolved organic carbon (DOC) for the Ob'river flood zone, rivers and lakes during baseflow and spring flood. Dissolved (<0.45 μm) concentration of Zr (I), La (J), Ti (C), Hf (K), and Th (L) as a function of dissolved organic carbon (DOC) for the Ob'river flood zone, rivers and lakes during baseflow and spring flood.

The diagram of PCA results is given in Figure 8, and the formulae of element attribution to various factors is as follows:

F1 = $(DOC_{0.86}\ Al_{0.92}\ Ti_{0.80}\ V_{0.76}\ Cr_{0.95}\ Fe_{0.74}\ Co_{0.81}\ Ni_{0.75}\ Ga_{0.91}\ As_{0.77}\ Zr_{0.94}\ Ba_{0.91}\ La_{0.95}\ Ce_{0.95}\ Pr_{0.96}\ Nd_{0.96} \ldots Yb_{0.97}\ Lu_{0.96}\ Hf_{0.71})$

F2 = $(DIC_{-0.91}\ Cl_{-0.75}\ Na_{-0.84}\ Mg_{-0.95}\ K_{-0.77}\ Ca_{-0.95}\ Rb_{-0.78}\ Sr_{-0.96})$

F3 = $(SO4_{-0.91}\ Mo_{-0.86}\ Sb_{-0.83}\ U_{-0.82})$

The first factor is linked to organic and organo-mineral colloids and comprises DOC, Al, Ti, V, Cr, Fe, Co, Ni, Ga, As, Zr, Ba, REEs and Hf. There are two families of trace elements within this factor: the first group is controlled by organic complexes (Al, Co, Ni, Ba) with dissolved fulvic and humic acids and the second one is represented by organo-ferric colloids of trivalent and tetravalent hydrolysates (Ti, Ga, Zr, REEs, Hf). Arsenic also belongs to the family of organo-ferric colloids as it can be adsorbed at the surface of Fe (III) oxyhydroxide.

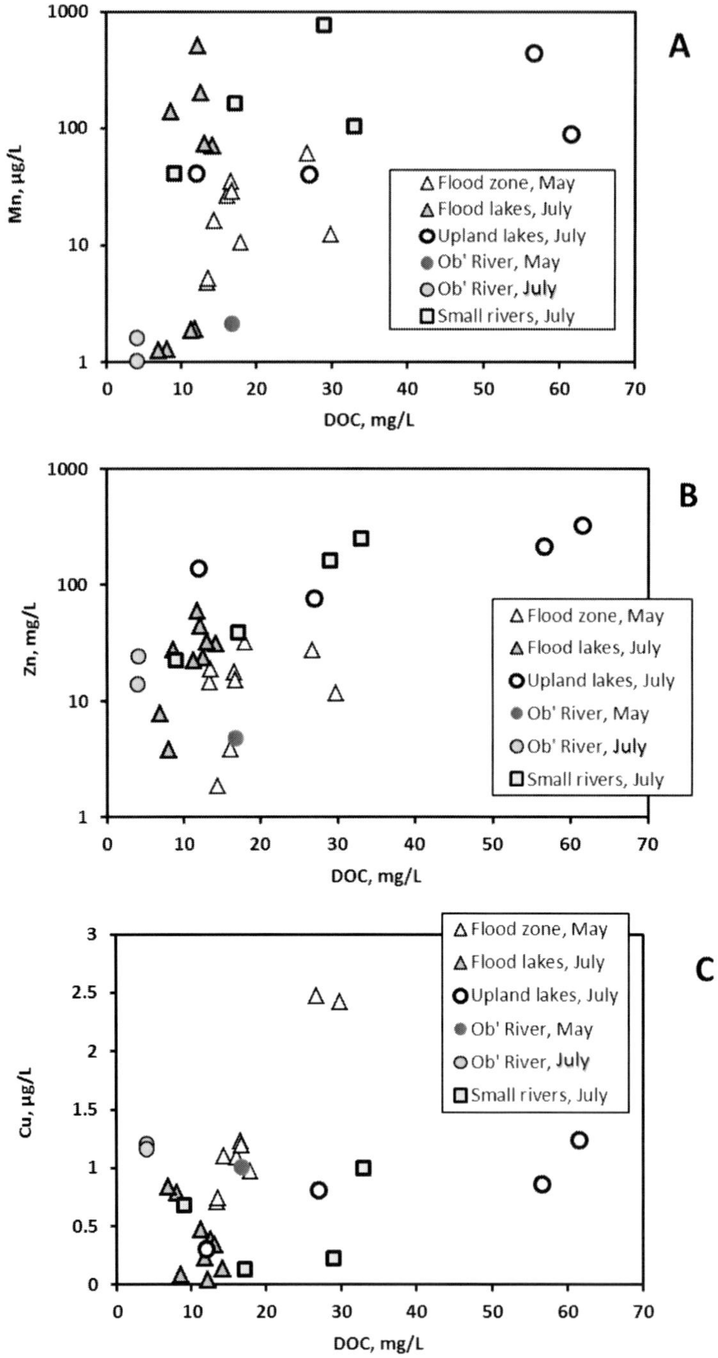

Figure 6. (Continued).

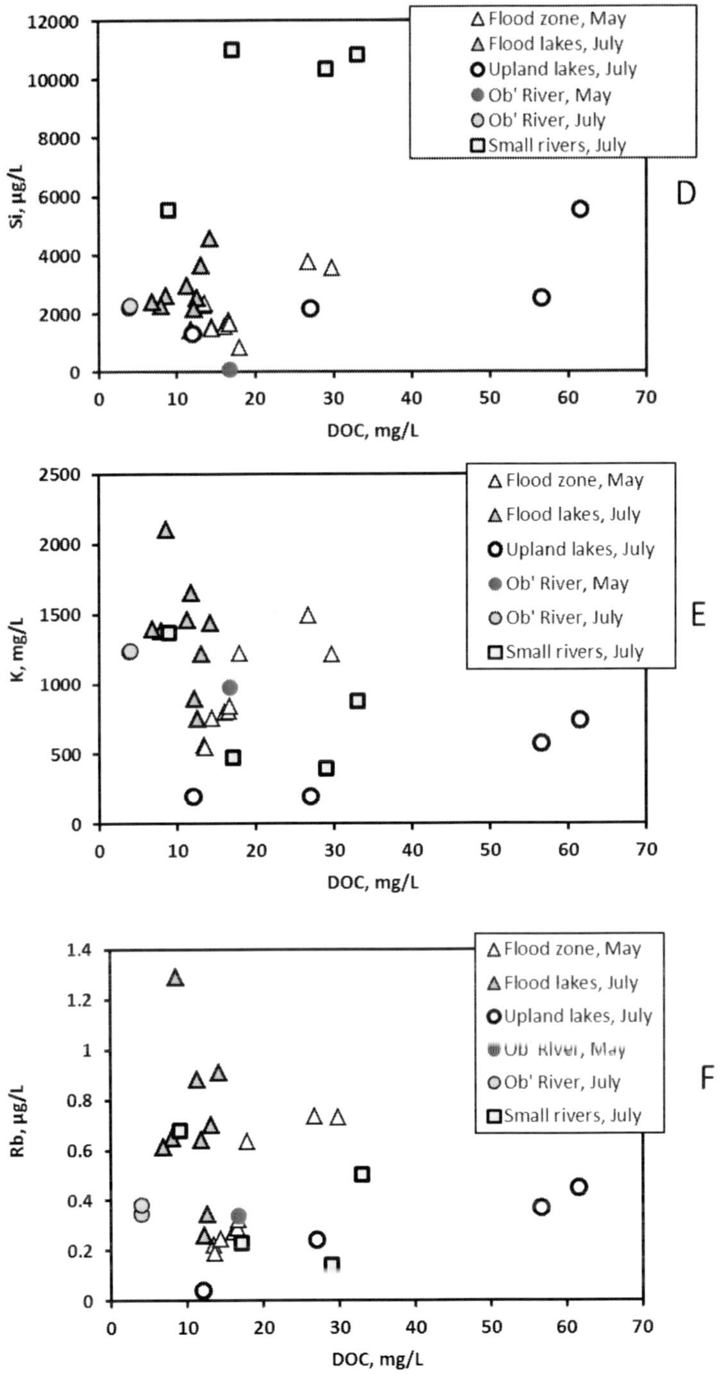

Figure 6. Dissolved (<0.45 µm) concentration of Mn (A), Zn (B), and Cu (C) as a function of dissolved organic carbon (DOC) for the Ob'river flood zone, rivers and lakes during baseflow and spring flood. Si (D), K (E) and Rb (F) concentrations as a function of dissolved organic carbon (DOC) for the Ob'river flood zone, rivers and lakes during baseflow and spring flood.

The 2nd factor was acting negatively on the DIC and major cations and also included Rb and Sr, assuming important link of these highly mobile elements to atmospheric precipitates and/or carbonate mineral dissolution. Finally, the 3rd factor negatively affected pH, SO_4, Mo, Sb and U, and was probably linked to groundwater leaching of soluble elements. It could reflect the decrease of adsorption of oxyanions (Mo, Sb, U carbonate complexes) onto clays with the decrease of pH. This factorial structure was highly stable during different contrasting seasons and among small rivers, upland lakes and flood lakes. It was fully consistent with anti-correlation between DOC (1st factor) and groundwater-linked elements (Mo, Sb, U) and between highly mobile alkali and alkaline-earth elements. Moreover, the PCA structure of all lakes samples (not including rivers and snow) yielded almost identical distribution of elements among 3 factors (not shown). This strongly suggested that upland lakes, flood plain lakes and flooded zone bear the signature of all three major processes controlling dissolved load of surface waters in the Ob River basin, i.e., the organic matter leaching from surface vegetation, atmospheric input and groundwater feeding.

Colloidal Transport of Major and Trace Elements in the Ob River, Its Riparian Zone and Upland Lakes during Summer Baseflow and Spring Flood

The proportion of colloidal fraction was evaluated using in-situ dialysis in small, organic-rich upland lake and in the Ob river mainstream, in May and July (Figure 9 and Figure 10). For the Ob River, proportion of colloidal form was <10-20% for anions and weakly complexed ions (Si, HCO_3, K, Na, Ca, Mg, Rb, Cl, Cs, Li, SO_4, Mo, Sb, V, W). Divalent metals (Cu, Ni, Zn, Co) exhibited moderate proportion of colloids during both seasons. Trivalent and tetravalent hydrolysates - Al, Fe, all REEs, Ti, Zr, Hf, Th- were strongly bound to organo-ferric colloids (>70%). There was an increase of the $LMW_{<1\ kDa}$ fraction in July relative to May for Fe, Cr, Ti and Cd. The presence of the small size suspended fraction of the size range from 1 kDa to 0.45 μm containing lithogenic elements such as Fe, Cr and Ti may explain their high colloidal proportion in May. For the other elements, we observe a slight increase of colloids in July relative to May, probably because of the dominance of LMW forms from plant leachates, which are highly abundant in the flood zone in May.

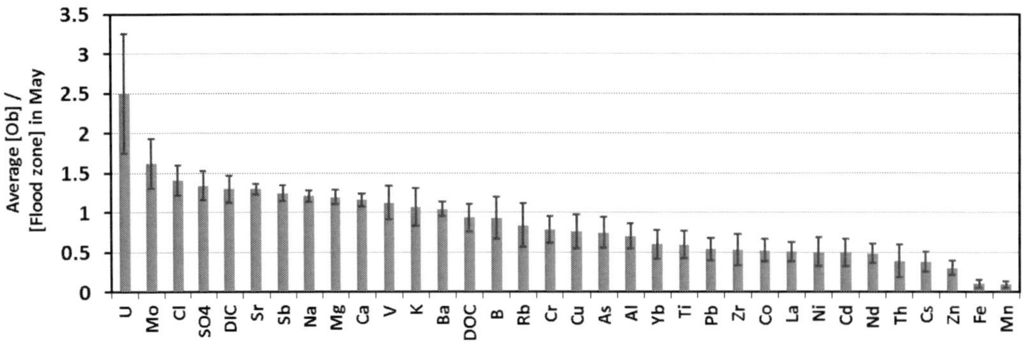

Figure 7. The ratio of the element concentration in the Ob' River to that in the flood zone in May.

Colloidal speciation in the upland lake is similar to that measured in the humic, organic-rich (30 mg/L), low-TDS lakes of western Siberia (i.e., Pokrovsky et al., 2011, 2013). The majority (>60%) of trace elements and also alkaline-earth metals (Ca, Mg, Sr, and Ba) present as colloids and only Cl, Si, Rb, Cs and Sb are present in a small proportion of the colloidal forms (<30%, as illustrated in Figure 10).

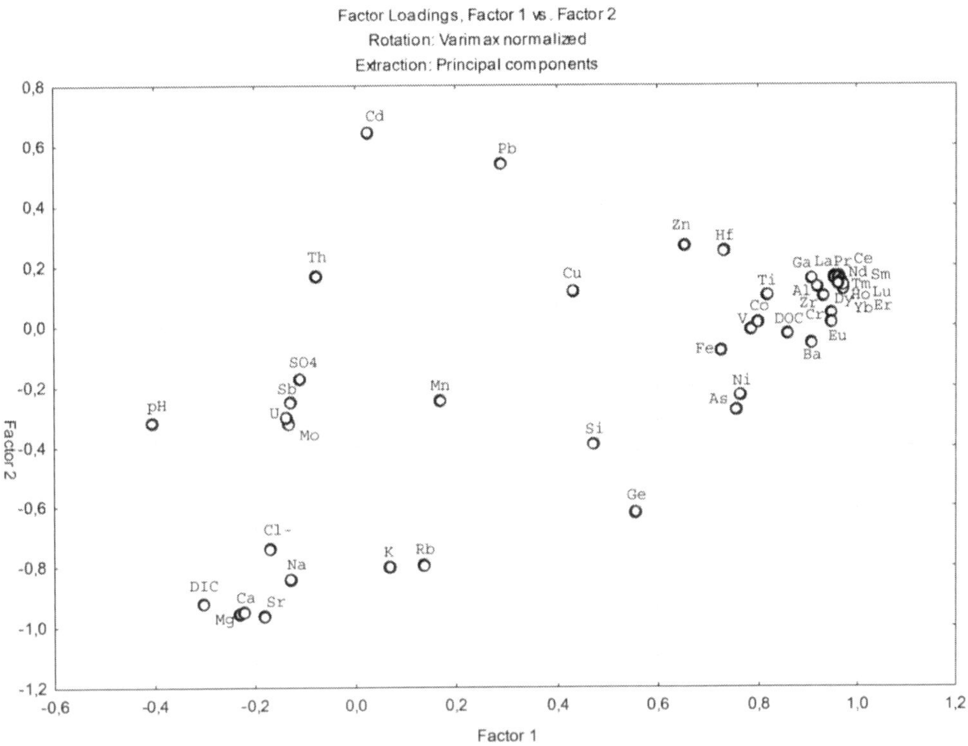

Figure 8. PCA diagram of all studied water objects in the flood and riparian zone of the Ob River.

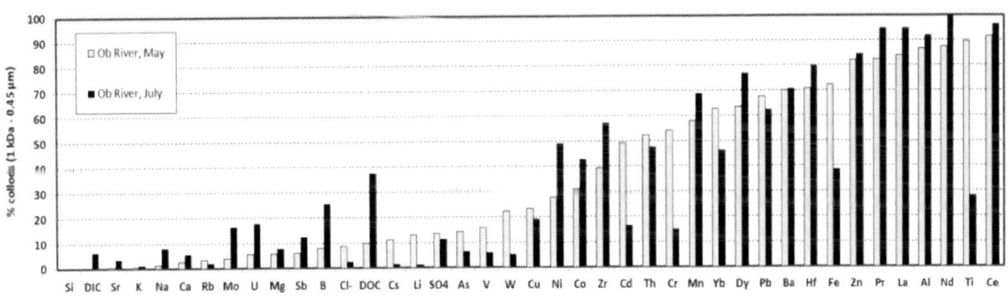

Figure 9. Proportion of colloidal fraction in the Ob River in May and July.

DISCUSSION

Previous Studies of the Ob River Watershed and Element Biogeochemistry in the Flood Zone

The geochemistry of the suspended matter of western Siberia's large river (Irtush, the largest tributary of the Ob River) was studied ten years ago (Gordeev et al., 2004) during the summer baseflow period; however, no data on the dissolved river water fraction were collected. Occasional data of some major element and nutrients (Gordeev et al., 1996) and trace metals, such as Fe, Cu and Zn (Dai and Martin, 1995; Telang et al., 1991; Shiklomanov and Skakalsky, 1994; and Alexeeva et al., 2001), are available for the mouth zone of the Ob River. Moran and Woods (1997) reported the Cd, Cr, Cu, and Ni concentration in the water column of the Ob and Irtush Rivers during summer baseflow. The range of metal concentrations reported in this previous study (0.001-0.015 for Cd, 0.06-0.34 for Cr, 1.8-4.8 for Cu and 0.8-2.8 μg/L for Ni) is in good agreement with our baseflow and spring flood measurements (0.005, 0.2, 1.0-1.2 and 0.5-1.3 μg/L for Cd, Cr, Cu and Ni, respectively). Over the past decade, a significant amount of information on the major and trace element concentrations in the mouth zone of the Ob River was collected in the course of PARTNERS and ARCTIC GRO programs (Holmes et al., 2012). All of the aforementioned studies provided the major features of dissolved and suspended load chemical composition either at the terminal gauging stations or at the main river channel stations, without any insights into the mechanisms controlling the element transport from the source (soil, peat, or litter) to the main stream via primary and secondary river channels and intermediate water bodies (lakes, small tributaries, and flood zones). The present study distinguished between the contribution of the different sources and assessed the first-order autochthonous processes in the water bodies during the two main open-water seasons by taking a "snapshot" of the spatial variability of all of the components of the flood zone and adjacent territories.

Highly soluble, highly mobile elements, such as major anions (DIC, sulfate), cations (Na, K, Mg, Ca), other alkali (Rb, Cs), alkaline earth traces (Sr), trace oxyanions (Mo, Sb) and U, marked the influence of underground feeding in July, which was most visible in flood lakes on the Ob riparian zone and the Ob River itself. In May, the flood lakes were statistically similar to the Ob River. The elevated concentration of the DOC (up to 60 mg/L) in the upland lakes was not correlated with groundwater-related elements, suggesting a lack of significant groundwater feeding in these lakes.

Presumably, the major (Ca, Mg, Na and DIC) and trace elements (Sr, Mo, Sb, and U) that are independent of the DOC or that decrease their concentration with the DOC are strongly influenced by groundwater feeding. This is further accentuated in flood lakes during summer baseflow (July), demonstrating an abrupt decrease of the groundwater influence for Mo, U, and Sb with an increase in the DOC (Figure 3 F, G, H).

In contrast, K, Rb, Si and Cu demonstrate strong increase in concentration with the DOC in the flooding zone with the maximal concentrations achieving within the flooded grass field (sample OB-9M, at the most remote sampling point), Figure 4. Most likely, these elements are controlled by leaching from the grass biomass. For example, Rb and K correlations with the DOC are the highest the flood zone (R^2 = 0.82 and 0.72, respectively). While K is an essential component of grass and other ground vegetation covered by flood water in May, isomorphic Rb follows tightly this macronutrient as it is known from other environments (Stepanova et al., 2014; Kabata-Pendias, 2004).

Insoluble, usually low mobile elements (Al, Fe, other trivalent hydrolysates, Ti, Zr, and Hf) and some metals (Cr, Zn, Ni, Ba, and Pb) demonstrated a systematic increase in concentration with an increase in the DOC in all of the studied water objects. Presumably, they are limited by the availability of their main carriers, organic and organo-ferric colloids, rather than the availability of the source, i.e., peat and mineral soil or plant litter. It is important to note that the majority of these trace elements significantly increase their concentration in the flooding zone, especially in the stagnant waters at a large distance from the river mainstream.

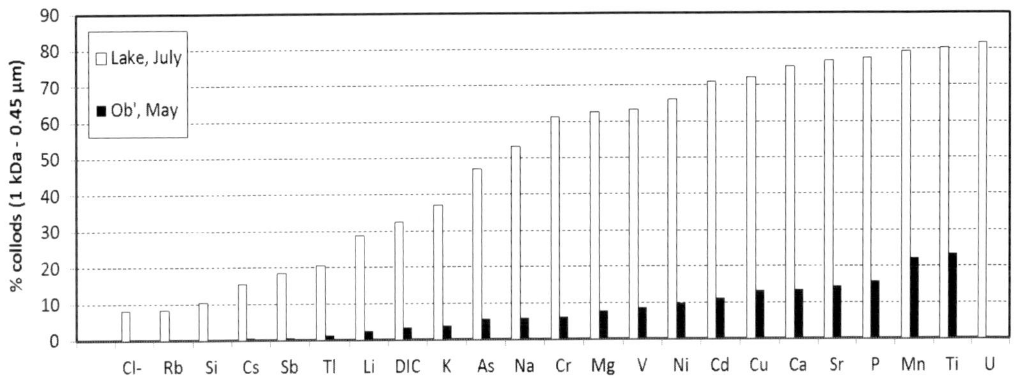

Figure 10. Proportion of colloidal forms (1 kDa-0.45 μm) in upland lakes in July (white columns) in comparison to that of the Ob' River in May.

The role of the dissolved organic matter on rare earth elements' transfer in surface waters is evidenced by the positive correlation between the DOC and the REE concentration. Taking together, in all of the sampled water bodies, the correlation coefficient for DOC–heavy REEs is higher than 0.6, whereas the same coefficient for the DOC–light REE correlation is <0.6. This result confirms the relative affinity of HREE to organic ligands and LREE to iron colloids, which were established from several previous studies in boreal and temperate zones (Andresson et al., 2006; Sholkovitz, 1995; Leybourne and Johannesson, 2008). Consistent with that result, the percentage of colloids (essentially organo-ferric entities) is higher for LREE, and the low molecular weight fraction (essentially organic ligands) is higher for HREE at $p < 0.05$ (see Figure 9).

Mn, Zn and Cu present a concentration-DOC dependence pattern, which is the opposite of the other elements. These potential micronutrients are known to be strongly linked to the dissolved organic carbon in temperate and subarctic rivers and lakes (Rember and Trefry, 2004; Ponter et al., 1992), as also follows from our dialysis results (Figure 6). However, their concentrations do not follow the same general trend as that of the other insoluble traces. The lowest Mn concentration in the Ob River in July may be linked to a strong insulation and the high residence time in this system, thus leading to strongly pronounced photo-oxidation, especially in the course of diurnal photosynthesis (Scott et al., 2002; Nimick et al., 2003; Pokrovsky and Shirokova, 2013). The elevated Mn concentration in certain flood lakes during baseflow (Figure 6 A) may be due to both underground discharge and sediment respiration (upward flux, Audry et al., 2011). Additionally, an element may be mobilized from silt deposits from the RSM (Björkvald et al., 2008), because flood plains are known to play an important role in the storage of fine suspended sediment (Bradley and Cox, 1990), which is especially important for the Ob River (Smith and Alsdorf, 1998). In contrast, in small rivers and organic-rich upland lakes, the influence of photo-oxidation and phytoplankton uptake is minimal and Mn concentration here is maximal. Likewise, for Zn (Figure 6 B), the biouptake in the main river course in July decreases its concentration relative to the May flood by approximately 5-fold. The highest concentrations in the organic-rich upland lakes and small rivers in July ($R^2_{Zn, DOC}$ = 0.71 and 0.91, respectively) illustrate the importance of organic colloids but also the lack of uptake by the phytoplankton or biodegradation in these small humic water bodies. Cu exhibits no visible link to the DOC in small rivers and flood lakes in July but exhibits pronounced control by the DOC in the flood zone in May and in the upland lakes in July ($R^2_{Cu, DOC}$ = 0.93 and 0.79, respectively, Figure 6 C). This result indicates the importance of the organic binding of this metal and its mobilization from surrounding peat deposits in lakes and ground vegetation in the flooding zone. By contrast, the mineral and underground feeding that are detectable for Zn in lowland flood lakes and small rivers during summer baseflow, are virtually unimportant for Cu. The latter presumably originated from the organic (surface litter and grass of the riparian zone) horizon rather than the deep (mineral) horizon.

The comparison of the Ob River's water dissolved load to the isolated and interconnected water bodies from the flooding zone in May demonstrated a high similarity, within ± 30-40% of the dissolved element concentration in the flood zone and the main stream (Figure 7). Significant enrichment of the Ob River in U and Mo may suggest pronounced underground feeding in the main water channel, which is virtually absent for the surface flood water bodies. By contrast, a factor of 10 depletion of the Mn and Fe in the water of the Ob River relative to the surface water bodies suggests intensive in-stream processes, such as Mn^{2+} oxidation and Fe^{3+} colloidal coagulation, due to heterotrophic bacterioplankton respiration activity. These processes are even more pronounced during summer baseflow in July because the concentration of DOC, trivalent (Al, Fe, REEs) hydrolysates and tetravalent hydrolysates (Ti, Zr, Hf, Th), as well as some divalent metals, significantly decreases relative to the spring flood.

Note that the Mn enrichment in the Ob River in May does not exceed a factor of 1.6 relative to July, whereas the enrichment of the flood zone relative to the main stream in May is a factor of 10 for this element. Therefore, the most likely cause of such a difference between the different water reservoirs for this element could be the significant release of Mn from degrading plant litter (flooded grass and litter fall in the riparian zone, photo Figure 11) rather than in-stream processes depleting its concentration in the main river channels relative to the inlet surface waters in May. Similarly, Zn and Si are also enriched in terrestrial vegetation and have superior concentrations by a factor of 2 to 3 in the flooded zone compared to the main stream because of their intensive leaching from the terrestrial vegetation litter and grass.

Allochthonous and Autochtonous Processes Controlling Element Transfer in the Flood Zone and the Main River

It can be hypothesized that the water residence time, which is known to be the main control of organic carbon (Algesten et al., 2003; Hanson et al., 2011; Kohler et al., 2013) and, presumably, related trace elements in the watersheds, is quite short in the flooded zone. This can strongly increase the relative impact of the surface (organic) source, such as fast plant litter leaching in the flood zone versus deep (groundwater) input related to slow mineral dissolution for rivers and permanent lakes. On the other hand, element removal from the water column of the river or lakes via phytoplankton uptake and heterotrophic and photo degradation of colloids, followed by precipitation to the bottom sediments, may also be tightly linked to the water residence time (WRT) in these water bodies. A preliminary evaluation of the hydrological balance of all of the studied water objects allowed them to be classified as follows in terms of the WRT: flood zone in May (days to 1-2 weeks) ≤ small rivers in July < Ob River < flood lakes in July < upland lakes (≤1-2 years). This order cannot be directly translated to the DOC and element concentration in various parts of the ecosystem measured in this study, as other enriching and depleting factors compete for the resulting concentration. The processes enriching the surface water bodies in various solutes are groundwater feeding, plant litter leaching, and peat (soil) leachate. The degree of groundwater feeding (visible only for the 1st group of highly labile elements) follows the order:

flood zone in May < upland lakes < small rivers in July < Ob River in May < Ob River in July < flood lakes in July.

The main processes affecting the surface bodies of the water chemical composition for various components of the lateral profile, from the main stream to the upland lakes, are shown in Figure 1 B.

The factors depleting the element concentration are phytoplankton uptake and photo and bio-degradation of colloids, leading to DOM and related element coagulation and precipitation, as shown in other studied continuums of the boreal watersheds (Agren et al., 2014; Ilina et al., 2014). Note that the vulnerability of the riverine DOC to bacterial mineralization is highly sensitive to the origin of organic matter, and it is the highest for forest carbon at high flow and the lowest for mire carbon (Berggren et al., 2009).

The influence of the processes removing the elements from the water column, such as plankton uptake and photo and biodegradation, is largest in flood lakes in July and the Ob River itself during baseflow and smallest in small rivers, the latter having the lowest water residence time. Note that even during few days of exposure of fresh peat-derived organic matter, it can be photo-oxidized with a rate of 10 mg C L^{-1} day^{-1} (Moody et al., 2013). However, we do not expect that the photo-oxidation and bacterial degradation will be significant in the organic-rich Ob's tributaries (samples OZE-4, 6, 7 and 14, Table ESM-1). In a similar boreal stream network, the loss of the DOC due to photo- and biodegradation from headwaters to the outlet was shown to be less than 1 mg L^{-1} (Tiwari et al., 2014).

Role of Organic and Organo-Ferric Colloidal Transport of Trace Elements in the Ob Watershed

The colloidal speciation of most major cations and all of the trace elements (≤80% in $LMW_{<1}$ kDa forms) in the humic, low TDS upland lake is consistent with the peat feeding of this lake. The majority of low-mineralized water bodies of western Siberia are very rich in colloids, with even major elements (Ca and Mg) often being present in a >50% colloidal form (Pokrovsky et al., 2013; Shirokova et al., 2013). This contrasts with the organic-rich boreal rivers, which rarely have more than 30% of colloidal Ca (i.e., Dahlqvist et al., 2004, 2007; Pokrovsky et al., 2010). Such rivers have a much higher TDS at an otherwise similar DOC and, as a result, lesser stability of the organic and organo-mineral colloids in solution because of the compacting of the electric double layer (see discussion in Pokrovsky et al., 2013).

Figure 11. Leaching of grass in the riparian zone of the Ob river secondary channels (May 2014).

Generally, the proportion of the TE colloidal forms in the Ob River in July is similar to that in May, with a relatively smaller fraction of organic colloids during the spring flood at an otherwise higher DOC concentration. Presumably, the plant leachates are in the $LMW_{<1\ kDa}$ forms during the spring flood, whereas an increase in the organic colloids and other elements in July may be linked to the bio- and photo-degradation of the DOC, leading to the coagulation of the $LMW_{<1\ kDa}$ of organic and organo-mineral compounds. It was reported, however, that the photooxidation of the river's DOM decreases the concentration of the organically bound Cu and Fe complexes (Brooks et al., 2007; Shiller et al., 2006) and that UV irradiation is also capable of significantly decreasing the concentration of Al and Fe bound to complexes, with allochthonous OM leading to metal hydroxide coagulation (increase of the molecular weight) and their subsequent sedimentation (Kopacek et al., 2005; Kopacek et al., 2006).

The existence of a certain limit on the concentration increase of Al, Fe, Ti, Zr, Hf, Th, and REs after a threshold of the DOC concentration (Figure 5 A, B, C, I, J, K, and L) is evidenced in the organic-rich (c.a., 30 mg/L DOC) small rivers sampled in July, without a further increase in the concentration of TE in the upland lakes containing up to 60 mg/L of the DOC. This strongly suggests that in organic-rich (i.e., >20 mg/L) surface waters, the concentration of these elements is limited by their source (mineral soil or groundwater, most visible in small rivers during summer baseflow) rather than the availability of the carriers (soluble soil/peat humic and fulvic compounds). At the same time, the significant increase in the Fe, Ti, Pb, Zr, and Th with the DOC ($10 < DOC \leq 30$ mg/L) in the flood zone in spring along with the increase in the distance from the river suggests their mobilization is from plant litter and grass in the riparian zone and that their transfer is in the form of organo-ferric colloids. This mobilization is, however, almost an order of magnitude smaller than that occurring in small rivers draining podzol mineral soils and peatlands during summer baseflow. In the latter, the concentration increase of the insoluble elements ranges between 2 and 4 orders of magnitude for a ~3-fold increase in the DOC. The most likely mechanism of organo-ferric colloids generation, which is well established in other boreal zones (c.a., Pokrovsky and Schott, 2002), is TE coprecipitation with the Fe oxyhydroxide subjected to oxidation at the surface redox front. This most likely occurs within the riparian or hyporheic zone of the stream, where underground, partially anoxic water meets well oxygenated, organic-rich surface waters. Note that the enrichment of hyporheic sediments with Fe-oxyhydroxide is fairly well known in other pristine boreal rivers (cf. Siergiegiev et al., 2014). Therefore, given the rather high DOC concentration in the surface western Siberian context, including Ob's flood plain/riparian zone, we suggest that the element source (including Fe) is of primary importance for colloid generation in the riparian zone of the small rivers. The limited underground feeding of small, organic rich upland lakes (~60 mg/L of DOC) precludes the further buildup of the trace element concentration relative to the flood lakes and organic-rich rivers, despite the doubling of the DOC from the surrounding peatlands. Various pathways of colloid generation are shown schematically in Figure 12.

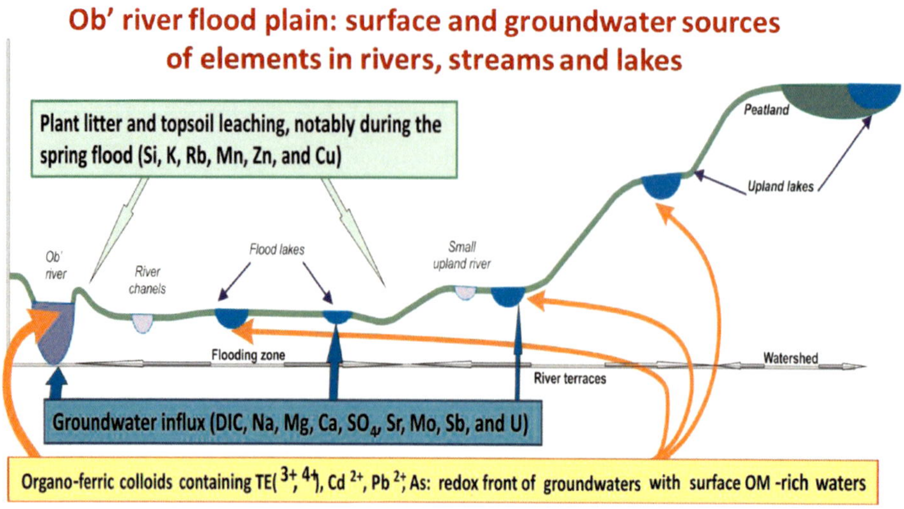

Figure 12. Pathways of colloids generation and trace element influx within the flooding and riparian zone of the Ob River middle course.

An overwhelming role of riparian and flood zone consists in providing large surface area for mixing of Fe(II)-rich underground waters with surface DOM-rich waters, the latter being originated essentially from riparian vegetation leaching and degradation. In this regard, the colloids of the Ob River flood plain may be different from colloids of other boreal rivers, for which the main source of DOM are adjacent bogs and mires. For the Ob River, chemical nature of organic matrix stabilizing amorphous Fe(III) oxy(hydr)oxides in colloidal form may include higher amount of carboxylates and proteins leached from fresh leaves of trees and abundant grass biomass of the floodplain (Figure 12). Besides, this organic matter should by much younger than that originated from the peat mires.

Projections of Element Concentration and Speciation in the Ob River Watershed under Climate Change Scenario

The general picture of river water hydrochemistry in western Siberia under a climate warming scenario is that the relative role of groundwater feeding will increase as long as the thawing depth thickens and the permafrost degrades (Frey et al., 2007). Therefore, in the permafrost-bearing zone, the role of surface flow, peat leaching and plant litter degradation versus groundwater feeding will decrease. However, there is no current projection on the evolution of the permafrost-free boreal taiga zone, such as the majority of the Ob River watershed. Unlike European boreal and subarctic regions, which are currently recovering from a past acidification impact and demonstrating a Zn, Ni, and Cu decrease over the previous few decade(s) (Huser et al., 2011, 2012), western Siberia was only weakly affected by local atmospheric pollution, as also follows from the moss analyses across the WSL (Stepanova et al., 2015). As such, the evolution of the hydrochemical parameters in this region will be mainly linked to the natural factors.

In the middle course of the Ob River studied in this work, the main factors influencing the element input from the surrounding mineral and organic substrate to the Ob River are i) the groundwater influx during the baseflow and ii) plant leaching, which depends on the water surface area coverage of the flooding zone during high flow. There is no reason to assume any significant modification of the groundwater chemical composition and discharge under climate warming; rather, the modification of the winter precipitation regime will influence the water discharge and flood area during the spring period. The exact projection of the precipitation change in western Siberia with sufficient seasonal resolution that is necessary to assess both small watersheds and the middle course of the main river is not available. Therefore, the degree of element delivery to the river is not possible to predict. By contrast, the main factors influencing element uptake from the water column and limiting their transport in the form of organic complexes and colloids can be evaluated. The DOC of the river water and small streams is believed to increase in response to on-going climate change in western Europe and Canada (Oni et al., 2013; Vuorenma et al., 2006). However, the range of this increase is likely to be below 20 mg/L, whereas the most significant increase of the insoluble element concentration in the small rivers of the Ob region occurs at DOC > 20 mg/L. Therefore, the mire-dominated feeding of the majority of the studied water bodies via peat leaching, which provides significant DOC concentration in headwater streams, are unlikely to be modified by the temperature rise.

The processes removing the TE from the water column are expected to be more sensitive to climate change. However, their influence will be mostly manifested during the summer baseflow period. First, there will be an earlier onset of spring phytoplankton growth in temperate lakes in a warmer climate (Peeters et al., 2007). Second, the heterotrophic bacterioplankton degradation of colloids will increase as the water temperature rises, provided that the inorganic nutrients are not limited (i.e., Berggren et al., 2010). Finally, photodestruction should be enhanced at elevated temperatures (Leifer, 1988; Likens, 2010). The increase of both the primary productivity and heterotrophic mineralization of the DOM in the water column will certainly decrease the concentration of the total dissolved (<0.45 μm) DOC and trace elements in upland lakes, the Ob River and flood lakes, similar to what was reported during short-term water heating in other boreal lakes in European Russia (i.e., Shirokova et al., 2013) and western Siberia's thermokarst lakes (Pokrovsky et al., 2013).

CONCLUSION

A hydro-chemical study of the Ob River's middle course, first order small tributaries, persisting and temporary flood lakes and upland lakes as well as the large flood zone in May and July established first order factors controlling the dissolved organic carbon and related metal sources and sinks in this environmentally important ecosystem. Considering two contrasting seasons, spring flood and summer baseflow, helped distinguish the elements controlled by the groundwater influx (DIC, Na, Mg, Ca, SO4, Sr, Mo, Sb, and U) and those controlled by surface winter runoff via plant litter and topsoil leaching, notably during the spring flood (Si, K, Rb, Mn, Zn, and Cu). The main carriers of many insoluble trivalent and tetravalent elements and some divalent metals (Cd, Pb, Cu) are organo-ferric colloids stabilized by dissolved organic matter. The main river stream's dissolved chemical composition can be approximated within ± 30-40%, by that of the flood zone in May. It is hypothesized that the main autochthonous processes controlling the DOC and related TE transformation between different water bodies are phytoplankton uptake and the microbial heterotrophic and photo-degradation of organo-mineral colloids, which in turn strongly depend on the water residence time. It is possible that these autochthonous processes will mostly affect the removal of colloidal TE from the water column of the flood lakes, large and small rivers and upland lakes of the Ob River's middle course watershed, which is under ongoing climate warming.

Acknowledgments

We acknowledge financial support from a BIO-GEO-CLIM Mega-grant from the Ministry of Education and Science of the Russian Federation and Tomsk State University (No 14.B25.31.0001), and the CAR-WET-SIB II framework program.

References

Aerts, R., Chapin, III F. S. The mineral nutrition of wild plants revisited: A reevaluation of processes and patterns. *Adv. Ecol. Res.* 1999;30:1–67. doi:10.1016/S0065-2504(08)60016-1.

Agren, A. M., Buffam, I., Cooper, D. M., Tiwari, T., Evans, C. D., Laudon, H. Can the heterogeneity in stream dissolved organic carbon be explained by contributing landscape elements? *Biogeosciences* 2014;11:1199–213.

Audry, S., Pokrovsky, O. S., Shirokova, L. S., Kirpotin, S. N., Dupré, B. Organic matter mineralization and trace element post-depositional redistribution in Western Siberia thermokarst lake sediments. *Biogeosciences* 2011;8:3341–58. doi:10.5194/bg-8-3341-2011.

Alexeeva, L. B., Strachan, W. M. J., Shluchkova, V. V., Nazarova, A. A., Nikanorov, A. M., Korotova, L. G. et al. Organochlorine pesticide and trace metal monitoring of Russian rivers flowing to the Arctic Ocean: 1990-1996. *Mar. Pollut. Bull.* 2001;43:71–85. doi:10.1016/S0025-326X(00)00166-1.

Algesten, G., Sobek, S., Bergstro, A. K., Agren, A., Tranvik, L. J., Jansson, M. Role of lakes for organic carbon cycling in the boreal zone. *Glob. Chang. Biol.* 2003;10:141–7. doi: 10.1046/j.1529-8817.2003.00721.x.

Andersson, K., Dahlqvist, R., Turner, D., Stolpe, B., Larsson, T., Ingri, J. et al. Colloidal rare earth elements in a boreal river: Changing sources and distributions during the spring flood. *Geochim. Cosmochim. Ac.* 2006;70:3261–74. doi:10.1016/j.gca.2006.04.021.

Berggren, M., Laudon, H., Jansson, M. Hydrological control of organic carbon support for bacterial growth in boreal headwater streams. *Microbial Ecol.* 2009;57:170–8. doi: 10.1007/s00248-008-9423-6.

Berggren, M., Laudon, H., Jonsson, A., Jansson, M. Nutrient constraints on metabolism affect the temperature regulation of aquatic bacterial growth efficiency. *Micorbial Ecol.* 2010; 60:894–902.

Björkvald, L., Buffam, I., Laudon, H., Mörth, C.-M. Hydrogeochemistry of Fe and Mn in small boreal streams: The role of seasonality, landscape type and scale. *Geochim. Cosmochim. Ac*. 2008;72:2789–804.

Bradley, S. B., Cox, J. J. The significance of the floodplain to the cycling of metals in the River Derwent, UK. *Sci. Total Environ*. 1990;97/98:441-454.

Brooks, M. L., McKnight, D. M., Clements, W. H. Photochemical control of copper complexation by dissolved organic matter in Rocky Mountain stream, Colorado. *Limnol. Oceanogr*. 2007;52:766–79.

Cooper, L. W., McClelland, J. W., Holmes, R. M., Raymond, P. A., Gibson, J. J., Guay, C. K. et al. Flow-weighted values of runoff tracers ($\delta^{18}O$, DOC, Ba, alkalinity) from the six largest Arctic rivers. *Geophys. Res. Lett*. 2008;35(18):L18606. doi: 10.1029/2008GL0350 07.

Dahlqvist, R., Benedetti, M. F., Andersson, K., Turner, D., Larsson, T., Stolpe, B. et al. Association of calcium with colloidal particles and speciation of calcium in the Kalix and Amazon rivers. *Geochim. Cosmochim. Ac*. 2004;68:4059–75.

Dahlqvist, R., Andersson, K., Ingri, J., Larsson, T., Stolpe, B., Turner, D. Temporal variations of colloidal carrier phases and associated trace elements in a boreal river. *Geochim. Cosmochim. Ac*. 2007;71:5339–54.

Dai, M., Martin, J. First data on trace metal level and behaviour in two major Arctic river-estuarine systems (Ob and Yenisey) and in the adjacent Kara Sea, Russia. *Earth Planet Sc. Lett*. 1995;131:127–41.

Environment Canada [Internet]. *MacKenzie River Basin*. [updated 2004 May 6; cited 2015 Jan. 30] Available from: http://www.usask.ca/geography/MAGS/MRBasin_e.htm.

Frey, K. E., Smith, L. C. Amplified carbon release from vast West Siberian peatlands by 2100. *Geophys. Res. Lett*. 2005;32(9):L09401. doi:10.1029/2004GL022025.

Frey, K. E., Siegel, D. I., Smith, L. C. Geochemistry of west Siberian streams and their potential response to permafrost degradation. *Water Resour. Res*. 2007;43:W03406. doi: 10.1029/2006WR005149.

Gordeev, V. V., Martin, J.-M., Sidorov, I. S., Sidorova, M. V. A reassessment of the Eurasian river input of water, sediment, major elements, and nutrients to the Arctic Ocean. *Am. J. Sci*. 1996;296:664–91.

Gordeev, V. V., Rachold, V., Vlasova, I. E. Geochemical behaviour of major and trace elements in suspended particulate material of the Irtysh river, the main tributary of the Ob river, Siberia. *Appl. Geochem*. 2004;19:593–610.

Ilina, S. M., Drozdova, O. Y., Lapitsky, S. A., Alekhin, Yu. V., Demin, V. V., Zavgorodnaya, Yu. A. et al. Size fractionation and optical properties of dissolved organic matter in the continuum soil solution-bog-river and terminal lake of a boreal watershed. *Org. Geochem*. 2014;66:14–24.

Hanson, P. C., Hamilton, D. P., Stanley, E. H., Preston, N., Langman, O. C., Kara, E. L. Fate of allochthonous dissolved organic carbon in lakes: a quantitative approach. *PLoS ONE* 2011;6(7):e21884. doi: 10.1371/journal.pone.0021884.

Holmes, R. M., McClelland, J. W., Peterson, B. J., Tank, S. E., Bulygina, E., Eglinton, T. I. et al. Seasonal and annual fluxes of nutrients and organic matter from large rivers to the Arctic Ocean and surrounding seas. *Estuar. Coast*. 2012;35:369–82. doi: 10.1007/s 12237-011-9386-6.

Dickens, A. F., Baldock, J., Kenna, T. C., Eglinton, T. I. A depositional history of particulate organic carbon in a floodplain lake from the lower Ob' River, Siberia. *Geochim. Cosmochim. Ac.* 2011;75:4796–815.

Huston, M. A. H., Wolverton, S. The global distribution of net primary production: resolving the paradox. *Ecol. Monogr.* 2009;79:343–77.

Huh, Y., Panteleyev, G., Babich, D., Zaitsev, A., Edmond, J. M. The fluvial geochemistry of the rivers of Eastern Siberia: II. Tributaries of the Lena, Omoloy, Yana, Indigirka, Kolyma, and Anadyr draining collisional/accretionary zone of the Verkhoyansk and Cherskiy ranges. *Geochim. Cosmochim. Ac.* 1998;62:2053–75.

Huh, Y., Edmond, J. M. The fluvial geochemistry of the rivers of Eastern Siberia: III. Tributaries of the Lena and Anabar draining the basement terrain of the Siberian Craton and the Trans-Baikal Highlands. *Geochim. Cosmochim. Ac.* 1999;63:967–87.

Huser, B. J., Fölster, J., Köhler, S. J. Lead, zinc, and chromium concentrations in acidic headwater streams in Sweden explained by chemical, climatic, and land-use variations. *Biogeosciences* 2012;9:4323–35.

Huser, B. J., Köhler, S. J., Wilander, A., Johansson, K., Fölster, J. Temporal and spatial trends for trace metals in streams and rivers across Sweden (1996–2009). *Biogeosciences* 2011;8:1813–23.

Ingri, J., Widerlund, A. Uptake of alkali and alkaline-earth elements on suspended iron and manganese in the Kalix River, northern Sweden. *Geochim. Cosmochim. Ac.* 1994;58: 5433–42.

Ingri, J., Widerlund, A., Land, M., Gustafsson, Ö., Andersson, P. S., Öhlander, B. Temporal variations in the fractionation of the rare earth elements in a boreal river, the role of colloidal particles. *Chem. Geol.* 2000;166:23–45.

Ingri, J., Widerlund, A., Land, M. Geochemistry of major elements in a pristine boreal river system, Hydrological compartments and flow paths. *Aquat. Geochem.* 2005;11:57–88.

Johannesson, K. H., Tang, J., Daniels, J. M., Bounds, W. J., Burdige, D. J. Rare earth element concentrations and speciation in organic-rich blackwaters of the Great Dismal Swamp, Virginia, US. *Chem. Geol.* 2004;209:271–94.

Kabata-Pendias, A. Soil–plant transfer of trace elements - an environmental issue. *Geoderma* 2004;122:143–9.

Köhler, S. J., Kothawala, D., Futter, M. N., Liungman, O., Tranvik, L. In-lake processes offset increased terrestrial inputs of dissolved organic carbon and color to lakes. *PLoS ONE.* 2013;8(8):e70598. doi:10.1371/journal.pone.0070598.

Kopáček, J., Klementová, S., Norton, S. A. Photochemical production of ionic and particulate aluminium and iron in lakes. *Environ. Sci. Technol.* 2005;39:3656–62.

Kopáček, J., Marešova, M., Norton, S. A., Porcal, P., Vesely, J. Photochemical source of metals for sediments. *Environ. Sci. Technol.* 2006;40:4455–59.

Land, M., Öhlander, B. Seasonal variations in the geochemistry of shallow groundwater hosted in granitic till. *Chem. Geol.* 1997;143:205–16.

Leifer, A. *The kinetics of environmental aquatic photochemistry: theory and practice.* 1st ed. Washington, D.C.: Am. Chem. Soc.; 1988. 304 p.

Likens, G. E., editor. *Biogeochemistry of inland waters.* Academic Press; 2010.

Leybourne, M. I., Johannesson, K. H. Rare earth elements (REE) and yttrium in stream waters, stream sediments, and Fe-Mn oxyhydroxides: Fractionation, speciation, and controls over REE + Y patterns in the surface environment. *Geochim. Cosmochim. Ac.* 2008;72:5962–83.

Lobbes, J. M., Fitznar, H. P., Kattner, G. Biogeochemical characteristics of dissolved and particulate organic matter in Russian rivers entering the Arctic Ocean. *Geochim. Cosmochim. Ac.* 2000;64:2973–83.

Lofts, S., Tipping, E., Hamilton-Taylor, J. The chemical speciation of Fe(III) in freshwaters. *Aquat. Geochem.* 2008;14:337–58.

McClelland, J. W., Holmes, R. M., Peterson, B. J., Stieglitz, M. Increasing river discharge in the Eurasian Arctic: Consideration of dams, permafrost thaw, and fires as potential agents of change, *J. Geophys. Res.* 2004;109:D18102. doi:10.1029/2004JD004583.

McClelland, J. W., Déry, S. J., Peterson, B. J., Holmes, R. M., Wood, E. F. A pan-Arctic evaluation of changes in river discharge during the latter half of the 20th century. *Geophys. Res. Lett.* 2006;33:L06715. doi: 10.1029/2006GL025753.

Moody, C. S., Worrall, F., Evans, C. D., Jones, T. The rate of loss of dissolved organic carbon (DOC) through a catchment. *J. Hydrol.* 2013;492:139–50.

Moran, S. B., Woods, W. L. Cd, Cr, Cu, Ni and Pb in the water column and sediments of the Ob-Irtysh Rivers, Russia. *Mar. Pollut. Bull.* 1997;35:270–9.

Nimick, D. A., Gammons, C. H., Cleasby, Th. E., Madison, J. P., Skaar, D., Brick, C. M. Diel cycles in dissolved metal concentrations in streams: Occurrence and possible causes. *Water Resour. Res.* 2003;39(9):1247. doi: 10.1029/2002WR001571.

Peeters, F., Straile, D., Lorke, A., Livingstone, D. M. Earlier onset of spring phytoplankton growth in lakes of the temperate zone in a warmer climate. *Global Change Biol.* 2007; 13:1898–909.

Peterson, B. J., Holmes, R. M., McClelland, J. W., Vorosmarthy, C. J., Lammers, R. B., Shiklomanov, A. I. et al. Increasing river discharge to the Arctic Ocean. *Science* 2002; 298:2171–3.

Pokrovsky, O., Schott, J. Iron colloids/organic matter associated transport of major and trace elements in small boreal rivers and their estuaries (NW Russia). *Chem. Geol.* 2002;190: 141–79.

Pokrovsky, O. S., Dupré, B., Schott, J. Fe-Al-organic colloids control of trace elements in peat soil solutions. *Aquat. Geochem.* 2005;11:241–78.

Pokrovsky, O. S., Schott, J., Dupré, B. Trace element fractionation and transport in boreal rivers and soil porewaters of permafrost-dominated basic terrain in Central Siberia. *Geochim. Cosmochim. Ac.* 2006;70:3239–60.

Pokrovsky, O. S., Viers, J., Shirokova, L. S., Shevchenko, V. P., Filipov, A. S., Dupré, B. Dissolved, suspended, and colloidal fluxes of organic carbon, major and trace elements in Severnaya Dvina River and its tributary. *Chem. Geol.* 2010;273:136–49.

Pokrovsky, O. S., Shirokova, L. S., Kirpotin, S. N., Audry, S., Viers, J., Dupré, B. Effect of permafrost thawing on the organic carbon and metal speciation in thermokarst lakes of western Siberia. *Biogeosciences* 2011;8:565–83.

Pokrovsky, O. S., Shirokova, L. S., Zabelina, S. A., Vorobieva, T. Y., Moreva, O. Y., Klimov, I. et al. Size fractionation of trace elements in a seasonally stratified boreal lake: control of organic matter and iron colloids. *Aquat. Geochem.* 2012;18:115–39.

Pokrovsky, O. S., Shirokova, L. S., Kirpotin, S. N., Kulizhsky, S. P., Vorobiev, S. N. Impact of western Siberia heat wave 2012 on greenhouse gases and trace metal concentration in thaw lakes of discontinuous permafrost zone. *Biogeosciences* 2013;10:5349–65.

Pontér, C., Ingri, J., Burmann, J., Boström, K. Temporal variations in dissolved and suspended iron and manganese in the Kalix River, northern Sweden. *Chem. Geol.* 1990; 81:121–31.

Pontér, C., Ingri, J., Boström, K. Geochemistry of manganese in the Kalix River, northern Sweden. *Geochim. Cosmochim. Ac.* 1992;56:1485–94.

Rember, R. D., Trefry, J. H. Increased concentrations of dissolved trace metals and organic carbon during snowmelt in rivers of the Alaskan Arctic. *Geochim. Cosmochim. Ac.* 2004; 68:477–89.

Rennermalm, A. K., Wood, E. F. Observed changes in pan-arctic cold-season minimum monthly river discharge. *Clim. Dyn.* 35;923-39.

Shiklomanov, I. A., Shiklomanov, A. I., Lammers, R. B., Peterson, B. J., Vorosmarty, C. J. The dynamics of river water inflow to the Arctic Ocean. In: Lewis, E. L., Jones, E. P., Lemke, P., Prowse, T. D., Wadhams, P., editors. *The fresh water budget of the Arctic ocean*. Dordrecht: Kluwer; 2000. p. 281–96.

Shiklomanov, A. I., Lammers, R. B. Record Russian river discharge in 2007 and the limits of analysis. *Environ. Res. Lett.* 2009;4:045015. doi:10.1088/1748-9326/4/4/045015.

Shiklomanov, I., Skakalsky, B. G. Studying water, sediment and contaminant runoff of Siberian rivers: Modern status and prospects. *Arctic Research of the United States* 1994; 8:295–306.

Shiller, A. Syringe filtration methods for examining dissolved and colloidal trace element distributions in remote field locations. *Environ. Sci. Technol.* 2003;37:3953–7.

Shiller, A. M., Duan, S., van Erp, P., Bianchi, T. S. Photo-oxidation of dissolved oganic matter in river water and its effect on trace element speciation. *Limnol. Oceanogr.* 2006; 51:1716–28.

Shirokova, L. S., Pokrovsky, O. S., Viers, J., Klimov, S. I., Moreva, O. Yu., Zabelina, S. A. et al. Diurnal variations of trace elements and heterotrophic bacterioplankton concentration in a small boreal lake of the White Sea basin. *Ann. Limnol. – Int. J. Lim.* 2010;46:67–75.

Shirokova, L. S., Pokrovsky, O. S., Kirpotin, S. N., Desmukh, C., Pokrovsky, B. G., Audry, S. et al. Biogeochemistry of organic carbon, CO2, CH4, and trace elements in thermokarst water bodies in discontinuous permafrost zones of Western Siberia. *Biogeochemistry* 2013;113:573–93. Sholkovitz, E. R. The aquatic chemistry of rare earth elements in rivers and estuaries. *Aquat. Geochem.* 1995;1:1–34.

Siergieiev, D., Wilderlund, A., Ingri, J., Öhlander, B. Flow regulation effects on the hydrogeochemistry of the hyporheic zone in boreal rivers. *Sci. Total. Environ.* 2014;499: 424-436.

Smith, L. C., Alsdorf, D. E. Control of sediment and organic carbon delivery to the Arctic Ocean revealed with space-borne synthetic aperture radar: Ob' River, Siberia. *Geology* 1998;26(5):395-8.

Stepanova, V. M., Pokrovsky, O. S., Viers, J., Mironycheva-Tokareva, N. P., Kosykh, N. P., Vishnyakova, E. K. Major and trace elements in peat profiles in Western Siberia: impact of the landscape context, latitude and permafrost coverage. *Appl. Geochem.* 2015;53:53–70.

Tank, S. E., Lesack, L. F. W., Hesslein, R. H. Northern Delta lakes as summertime CO_2 absorbers within the Arctic Landscape. *Ecosystems* 2009;12:144–57.

Telang, S. A., Pocklington, R., Naidu, A. S., Romankevich, E. A., Gitelson, I. I., Gladyshev, M. I. Carbon and mineral transport in major North American, Russian Arctic, and

Siberian rivers: The St. Lawrence, the Mackenzie, the Yukon, the Arctic Alaskan rivers, the Arctic basin in the Soviet Union, and the Yenisei. In: Degens, E. T., Kempe, S., Richey, J. E., editors. *Biogeochemistry of major world rivers*. Chichester and New York: John Wiley and Sons; 1991. p. 57–74.

Tiwari, T., Laudon, H., Beven, K., Agren, A. M. Down-stream changes in DOC: Inferring contributions in the face of model uncertainties. *Water Resour. Res.* 2014;50:514–25.

Viers, J., Barroux, G., Pinelli, M., Seyler, P., Oliva, P., Dupré, B. et al. The influence of the Amazonian floodplain ecosystems on the trace element dynamics of the Amazon River mainstem (Brazil). *Sci. Total Environ.* 2005;339:219–32.

Vuorenmaa, J., Forsius, M., Mannio, J. Increasing trends of total organic carbon concentrations in small forest lakes in Finland from 1987 to 2003. *Sci. Total Environ.* 2006;365:47–65.

Yeghicheyan, D., Bossy, C., Bouhnik Le Coz, M., Douchet, Ch., Granier, G., Heimburger, A. et al. A Compilation of Silicon, Rare Earth Element and Twenty-One other Trace Element Concentrations in the Natural River Water Reference Material SLRS-5 (NRC-CNRC). *Geostand. Geoanal. Res.* 2013;37:449–67. doi: 10.1111/j.1751-908X.2013.002 32.x.

In: Riparian Zones
Editor: Oleg S. Pokrovsky
ISBN: 978-1-63484-613-4

Chapter 11

ALLUVIAL SOILS OF THE OB RIVER FLOODPLAIN AND THEIR SIGNIFICANCE IN THE FORMATION OF GEOCHEMICAL FLOW FROM WESTERN SIBERIA

S. N. Vorobyev**, S. N. Kirpotin, T. E. Vorobyeva, L. G. Kolesnichenko and L. A. Izerskaya
National Research Tomsk State University, Russia
University of Toulouse, France

ABSTRACT

The soil cover of the Ob floodplain and its tributaries play an important role in the conceptual model of the hydrochemical flow. Little research was conducted on the most significant function of the soil cover of the Ob floodplain as a source of methane, organic carbon, huge filter, sedimentator and geochemical barrier regulating the input of substances from the catchment of Western Siberia to the World Ocean. Given its global functions, in this chapter we describe the Ob floodplain as the best site for developing approaches to studying global processes in the floodplains of rivers in the Northern Hemisphere.

Keywords: review, Ob River, basin, floodplain, soils, Siberia

INTRODUCTION

The Ob is the main river in Western Siberia. It originates in the Altai Mountains, flows from south to north, and enters the Kara Sea. The river Ob Irtysh is the seventh largest river in the world and third largest in Asia. In terms of drainage area it is the 5th largest river in the world. Its length is 3,676 km. The drainage area is 2,990,000 km^2. The main part of the river basin (about 85%) is in the West Siberian Plain. Its average long-term run-off is 405.0 km^3

* Email: soil@green.tsu.ru.

(Rybalskyi et al., 2007). The hydrological system of the Ob is usually divided into three main parts. The Southern Ob is the area of the river system from the riverheads to the Tom river outlet. The Middle Ob is the area from the Tom river outlet to the Irtysh river outlet. The Lower Ob is the area from the Irtysh river outlet to the Gulf of Ob. The main part of the hydrochemical flow of the Ob is formed on paludified watershed areas during snow thawing (Alenkin and Brajnikova, 1964). Chemical elements in the snow accumulate mostly from the aerosols. The input of snow flow into the river flow can be more than 70% for some elements. The aerial flow component is 37% for macro-components (Ca, Mg, Na, K, Fe, Mn) and for organic carbon it is 16% (Savichev and Ivanov, 2010).

During other periods the chemical water load on the river is 2-3 times lower than during the spring snow melt flood (Temerev, 2006; Agafonov, 2010). During summer and autumn the flow has minimum concentrations of biogenic elements (i.e., nitrate nitrogen, ammoniacal nitrogen, and total phosphorus).

The most significant factor of river flux formation is the soil cover of the watershed. In soils there are the best conditions created for both dissolving mineral and organic components and adsorption of heavy metals and trace elements (Temerev, 2006; Shvartsev, 1998). The greatest influence of soils on the hydrochemical flow is evident in the plains with a small-angle surface slope and paludified watersheds. There is a close connection between the river basin hydrology, erosion, sediment transport, vegetative ground cover and land use. Forest clearing, agriculture, mining operations, and all other activity on the watershed affect the composition of the river water (Black, 1997).

Figure 1. The Ob and Irtysh basin watersheds together with the study site.

Today the basin approach to studying runoffs is the main one, and the role of drainage soils in the runoff formation is doubtless. At the same time, floodplain and alluvial soils have equal significance in the formation of the geochemical flux. The importance of alluvial soils is thought to be obvious, but there is no research proving this well-known fact.

According to rough estimates given in the system SAS. Planet (2015), the floodplain area of the Ob from the confluence of the Biya and the Katun rivers to the Gulf of Ob is more than 60 thousand km^2 (Figure 1). This is an enormous flooded area, which both accumulates substances weighed and dissolved in water from the whole catchment and modifies them significantly. The floodplain of the Ob in Western Siberia is a gigantic geochemical barrier, which regulates the inflow of the substances from the catchment into the World Ocean. Soils have great importance in this barrier. However, there was little effort to study the floodplain as the geochemical barrier and **alluvial** soils as the main component of this barrier. This chapter aims to fill this gap by providing a thorough pedological and geochemical analysis of soil cover of the floodplain and the processes of river flux formation occurring within this territory.

Soil Cover of the Basin and the Formation of Geochemical Runoff of the Ob

The soil cover of the Ob basin consists mostly of paludified soils and swamp, whose properties and composition determine the input of substances from the catchment area into the floodplain (Schipper and Loos, 2003; Dyukarev, 2005). The River Ob and the peat bogs of Western Siberia are closely interconnected and should be viewed as a single hydrological and landscape geochemical system in global climatic processes and forecast scenarios. The basin of the Upper Ob with mount-forest soils (Puzanov et al., 2007) differs from the basin of the Middle and Lower Ob, but its area is not large in comparison with the whole basin.

The peculiarity of the Ob basin is its location on the territory with freeze through rocky subsoils. Seasonal freeze through subsoils are widespread in the Southern Ob, Middle Ob and partially Lower Ob. In the middle part of the Lower Ob there are regularly freeze through subsoils, and in its northernmost and lowest part, the Ob basin has permanently frozen subsoils. The specific character of the rivers located in the area of permafrost subsoils and soils is that snowmelt water flows over frost-bound subsoil. According to Frey et al. (2007), mineralised groundwater in the north of Western Siberia is the source of soluble materials in the areas without permanent freezing. On the areas of drainage with permafrost headwaters are not connected to the high-TDS groundwater, and the prechannel flow is formed on peats, which lack mineral substances.

However, the influence of soils on the chemical composition of snowmelt water, which flows into the river, even on drainages with permafrost is significant. When snow melts on the warm areas of drainage, the soils there start to contribute to the flux. We can view the unfrozen parts of soils as point or diffuse sources of substances input into the river water. Black (1997) has shown models of substances input from these sources.

It should be noted that one significant difference of the Ob from the rivers of forest and steppe European zones is that the peak of river suspended matter (RSM) content in the Ob coincides with the peak of snowmelt flood (Lopatin, 1952). This allows us to suppose that there is a greater influence of drainage soils on the Ob floodplain and its flow than of the drainage of rivers in the forest and steppe zones in Europe.

According to rough estimates by Temerev (2006), the input of water extract from the soils into the snowmelt flow during floods can be great. For example, in 2005 the upstream input of soils into the chemical flow of the Ob tributary, the Barnaulka, in the outlet was 71% for copper, 56% for lead, 43% for zinc, and 18% for cadmium.

During a snowmelt flood, the input of soils into the chemical flow of the Upper Ob near Barnaul was 63% for iron, 5% for copper, and 25% for zinc. In September, in the same area the input of soils was 62% for iron, 20% for manganese, 15% of copper, 24% of lead, and 8% of zinc.

Smaller modification of flow and accumulation of aqueous run-off in the swamps of the Vasyugan are typical of bogged drainages. In September in the Middle Ob near Kolpashevo, the input of soils into the chemical river flow was 39% for iron, 17% for manganese, 18% for copper, 25% for lead, and 3% for zinc. In September, in the Lower Ob near the village of Belogorie, the input of soils into the chemical river flux was 20% for iron, 12% for manganese, 4% for copper, 4% for lead, and 1% for zinc (Figure 2).

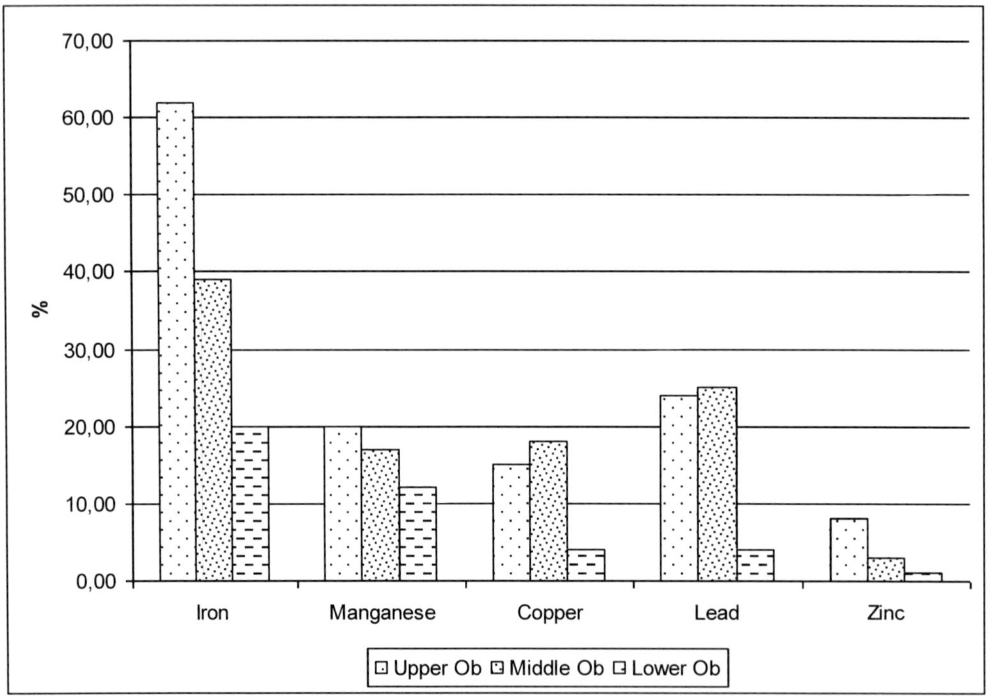

Figure 2. The input of soils into the chemical flow of the Upper and Lower Ob (September), estimated by Temerev (2006).

The amount of input of basin soils into the flow is rather significant. Groundwater brings much less than 50% of trace elements, nutrients and organic matters into the Ob flow (Savichev, 2007). Thorough research of soils input into the chemical flow of the Ob has further prospects. In the Ob basin, there are around 150,000 rivers, which are diverse Ob tributaries (Atavin et al., 2000). There are more than 400,000 lakes in the Ob basin

(Rybalskyi et al., 2007). Their catchments are natural model reservoirs of varying size with natural landscapes, which are useful when studying factors affecting surface water composition. For example, if the composition of dissolved organic matter of the water body is high in p-hydroxybenzenes (aromatic compounds), this means that there is a larger input from moss and swamps. If the composition of dissolved organic matter of the river is high in lignin, this means that the flow is formed with forest vegetation and forest soils (Amon et al., 2012). Unfortunately, such studies in Western Siberia are limited and there is very little information on the lakes (Semenova et al., 2012).

INFORMATION ON THE SOIL COVER OF THE OB FLOODPLAIN

The main data on the soil cover, physical, chemical and agrochemical properties of soils in the Ob floodplain were obtained from 1960 to 1980. The soil cover of the Ob floodplain was not studied equally. Major research was carried out during inventories and soil ratings for agricultural needs. That is why there was detailed research of those parts of the floodplain, which were prime for reclamation. The Middle Ob is essential for agriculture and is the most carefully examined area (Nepryakhin, 1963; Smetanin, 1963; Dobrovolsky et al., 1973; Ioganzen, 1968; Pashneva, 1981; Slavnina et al., 1981; Abramova and Pashneva, 1996; Afanas`eva et al., 1984; Dobrovol'skiy et al., 1984; Slavnina et al., 1986; Shepelev and Shepeleva, 1995; Abramova et al., 1996; Shepelev, 1999; Avetov et al., 2008). There is a much smaller number of papers on the research of soils in the floodplain of the Upper Ob (Gantimurov et al., 1974; Dobrovolsky et al., 1974; Nechayev and Scrubs, 1993; Burlakova and Kazanceva, 1999; Kazantseva, 2007; Petrov, 1979; Gafurov, 1990; Gafurov and Firsov, 1992; Nechaeva, 2008).

The practical part of the abovementioned studies mostly implies a typological approach to the assessment of land resources, which suggests allocation of typical areas of land for agricultural purposes. Mostly agrochemical and agrophysical properties of soils and some trace elements essential for growing agricultural plants were under study in these areas. Investigations of the alluvial soils' physical and chemical properties were performed in accordance with standard practices (Sokolov, 1975). These papers examined the distribution of soils and their properties over the elements of the floodplain and their connection with the productivity of natural lands, and soil maps were also produced.

There are no specific peculiarities in the soils of the Ob floodplain. The patterns of vertical soil differentiation are typical of all types of floodplains. The physical and chemical properties of these soils reflect the peculiarities of alluvial soils and the degree of soil-forming process development. Therefore, according to humus content and its distribution, the soil profile consists of two parts: the upper humus and the lower one, where the amount of humus is 8-20 times lower. Detailed profile separation, which is less pronounced, can be viewed in the content of main biogenic elements. The vegetation was studied for agricultural purposes. The vegetation in the Ob floodplain has been well studied. The main papers on vegetation provide full information on the patterns in the vegetation cover in the Ob floodplain (Shibanov, 2009; Baryshnikov, 1961; Vyltsan, 1964; Vyltsan, 1968; Lvov, 1963; Taran and Dymina, 1990; Taran, 1993; Dyachenko and Taran, 2011).

For the same reasons, the meadows in the floodplain were also well studied. The structure of floodplain meadows in the Middle Ob is of significant complexity. Mesophilous phytocenoses, which are connected with the most favourable habitats, are characterised by the largest eco-biological diversity; they are highly fertile and rich in species composition. In the communities of highly dry xeromesophilic and highly moist aerohydrophilic meadows, simplification of the structure occurs to the extent of monodominance, while their productivity decreases (Shepeleva, 1998). Figure 3 shows the meadows of the Ob floodplain.

Figure 3A. Mesophilous meadow in the Ob floodplain.

Figure 3B. Monodominant meadow in the Ob floodplain.

Genetic aspects of soil formation in the Ob floodplain are studied using the example of the Middle Ob. The authors of papers on the soil cover of the Ob floodplain point out that the influence of the factors on the formation and distribution of soils is different within the floodplain. In the soils of the riverine part, there is a poorly pronounced humic horizon and slight gley process. In this part of the floodplain, primitive young soil is found. The humic horizon is well-pronounced in the soils of the central part, where highly fertile wet meadow soils are common. On the nearterrace part on sediments of fine granulometric composition, under the influence of water from the terraces there formed rather moist sod-gley, boggy soil, peat bogs. Our results confirm the previously established variability in the properties of soil. Clearly visible differences in the basic properties of soils across elements of the floodplain (riverine, central, nearterrace) are shown in Table 1 and Figure 4.

In the soils of the riverine part of the Ob floodplain, there is a lower concentration of organic carbon, Ca, and Mg. Statistically significant accumulation of argillaceous deposits and the main elements occurs in the soils of the central and nearterrace parts of the floodplain.

The change in the Ob floodplain soil properties from south to north has a tendency towards a decrease in main nutrients concentration, an increase in acidity, an increase in silt fraction in soil texture and an increase in the coverage of gley and bogged soils in the soil cover (Slavnina et al., 1981; Afanas`eva et al., 1984; Izerskaya and Vorob'eva, 2000).

Table 1. Physico-chemical parameters of the Ob River soils

	Valid N	Mean	Min	Max	Std Dev
Riverine					
% particle, diameter <0.01 mm	75	26.91	3.60	63.30	12.15
Organic carbon (%)	54	1.71	0.11	5.90	1.06
pH (H2O)	77	5.70	3.80	8.00	1.26
Hydrolytic acidity (meq/100g)	67	2.84	0.20	21.78	3.55
Exchangeable Ca (meq/100g)	71	11.42	3.00	39.60	7.13
Exchangeable Mg (meq/100g)	71	3.24	0.40	19.90	3.46
base saturation	71	0.82	0.26	1.0000	0.20
Base saturation (%)	41	25.41	3.00	79.6000	15.60
Middle floodplain					
% particle, diameter <0.01 mm	185	40.62	4.70	79.20	17.02
Organic carbon (%)	138	2.82	0.21	13.53	2.22
pH (H_2O)	197	5.29	3.80	7.60	0.96
Hydrolytic acidity (meq/100g)	166	3.99	0.35	15.18	3.30
Exchangeable Ca (meq/100g)	187	16.51	1.60	56.40	8.71
Exchangeable Mg (meq/100g)	187	4.45	0.16	23.30	3.33
base saturation	187	0.83	0.29	1.00	0.16
Base saturation (%)	100	21.38	2.00	83.80	15.27
Nearterrace floodplain					
% particle, diameter <0.01 mm	36	50.85	33.80	74.29	9.66
Organic carbon (%)	28	3.59	0.41	12.31	2.87
pH (H_2O)	36	5.30	4.00	7.30	0.98
Hydrolytic acidity (meq/100g)	31	4.87	0.70	8.30	2.01
Exchangeable Ca (meq/100g)	35	19.95	10.10	57.59	8.96
Exchangeable Mg (meq/100g)	35	4.74	0.92	27.90	4.59
base saturation	35	0.84	0.66	1.00	0.10
Base saturation (%)	23	11.77	0.10	28.80	6.66

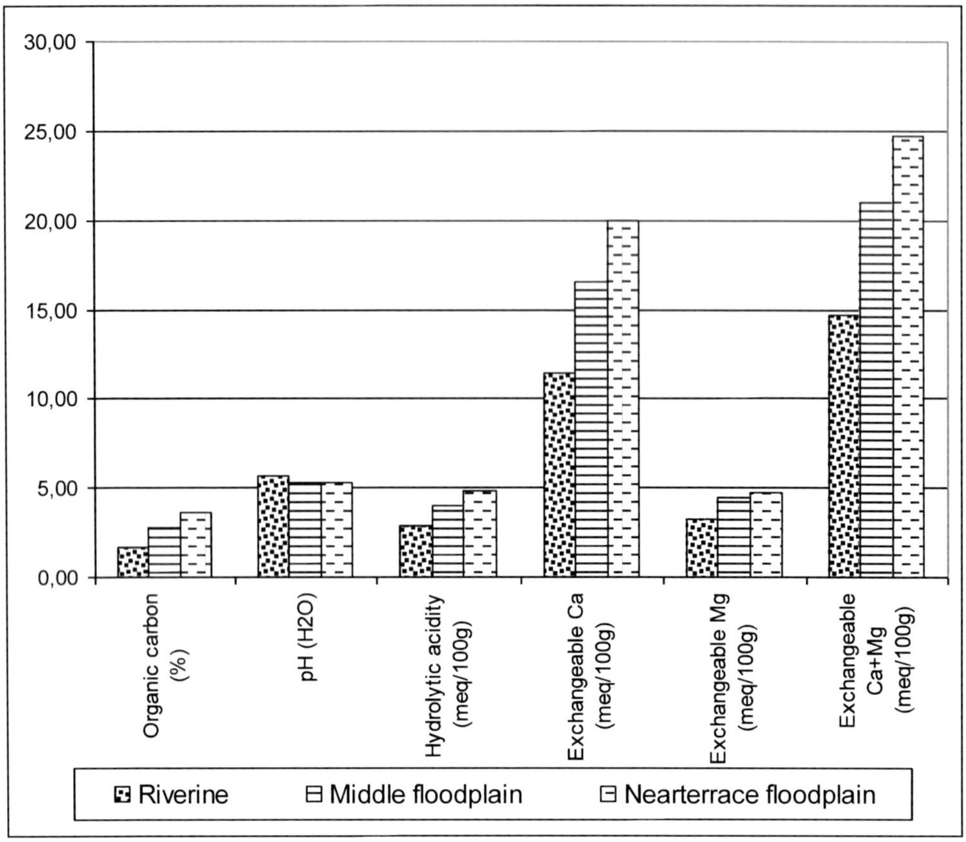

Figure 4. The distribution of the main soil properties according to the Ob floodplain elements (riverine, central, and nearterrace).

In spite of the wealth of information on the Ob floodplain soil cover, a comparison of soils from different parts of the floodplain is somewhat complex. This is due to the high irregularity of the floodplain. It is difficult to choose similar conjugate series of soils from the nearterrace to the riverine parts in different areas of the Ob floodplain. The spatial inhomogeneity of floodplain soil cover and high dispersion of soil properties make it difficult to compare different areas of the floodplain because of the complexity in the organisation of statistical sampling.

To compare different parts of the floodplain and test the hypothesis of changes in properties of floodplain soils from south to north, we used materials from the soil science laboratory of the Scientific Research Institute of Biology and Biophysics at Tomsk State University and materials from the Design Institute of Agriculture Zapsibgiprozem. In the floodplain from 56° to 60° northern latitude, we determined 14 sites of floodplain soils for agricultural purposes, where humus horizons were analysed (Table 2).

The results demonstrated that the method of determining the stations in studying geographical patterns of floodplain soils is suitable and effective. Actually, we can notice changes in some properties of floodplain soils from south to north. The most obvious one is the correlation between the latitude and pH, hydrolytic acidity, exchangeable caution percentage, Ca+Mg and organic carbon (Figure 5).

Table 2. Humus horizon properties of the floodplain soils (average values)

Latitude (degrees)	N	Organic carbon %	pH	Hydrolytic acidity meq/100g	Ca meq/100g	Mg meq/100g	P_2O_5	K_2O
56.22	30	4.88	7.2		20.5	4.1	-	-
56.86	6	4.70	6.1	1.4	21.0	4.8	23.0	8.8
56.94	39	4.20	6.4	1.8	13.7	3.9	26.9	23.9
57.03	44	4.65	5.9	1.5	23.0	8.1	35.9	10.1
57.59	43	4.06	5.3	4.2	16.2	5.2	13.5	12.9
57.82	19	3.98	4.8	4.9	15.6	3.2	10.7	10.8
58.08	21	4.66	4.8	4.4	11.8	3.0	7.1	8.6
58.34	19	4.05	4.9	2.8	9.7	2.0	23.7	9.6
58.53	20	4.04	5.2	4.9	22.2	6.8	-	-
58.89	15	3.17	4.3	4.0	5.0	0.8	11.1	9.5
58.97	19	2.48	4.6	4.0	16.8	1.8	21.0	9.4
59.15	9	2.33	4.3	5.7	6.3	0.7	17.0	7.7
59.47	16	2.15	4.3	6.4	5.4	0.6	19.5	8.2
59.76	15	2.92	3.7	12.4	6.6	2.0	-	-

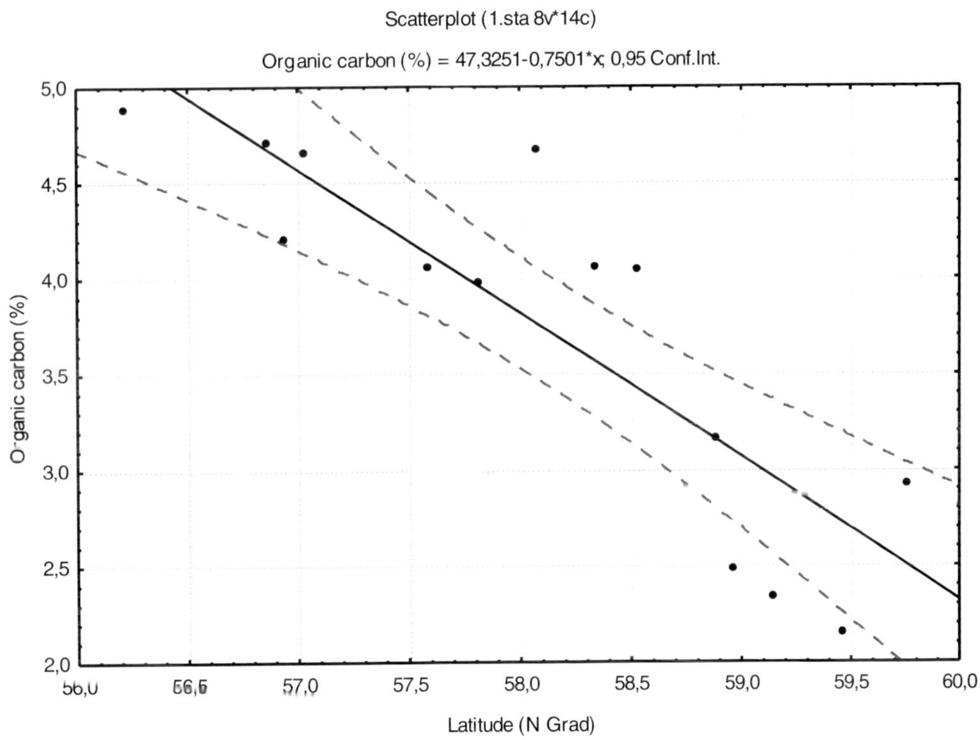

Figure 5. (Continued).

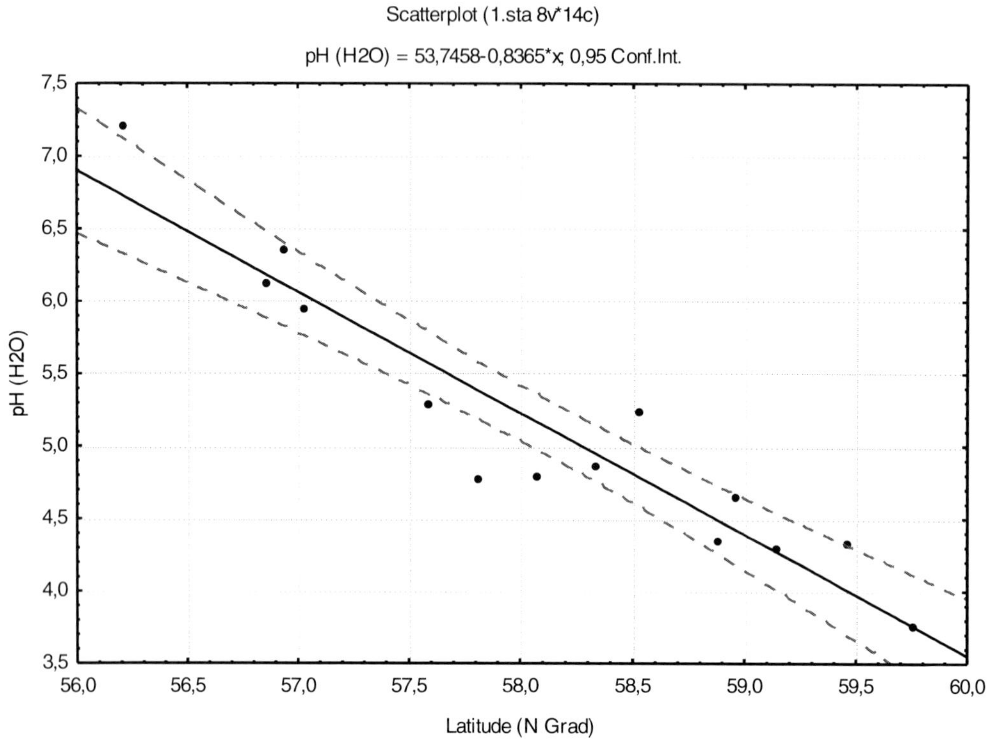
Scatterplot (1.sta 8v*14c)
pH (H2O) = 53,7458-0,8365*x; 0,95 Conf.Int.
pH (H2O)
7,5
7,0
6,5
6,0
5,5
5,0
4,5
4,0
3,5
56,0
56,5
57,0
57,5
58,0
58,5
59,0
59,5
60,0
Latitude (N Grad)

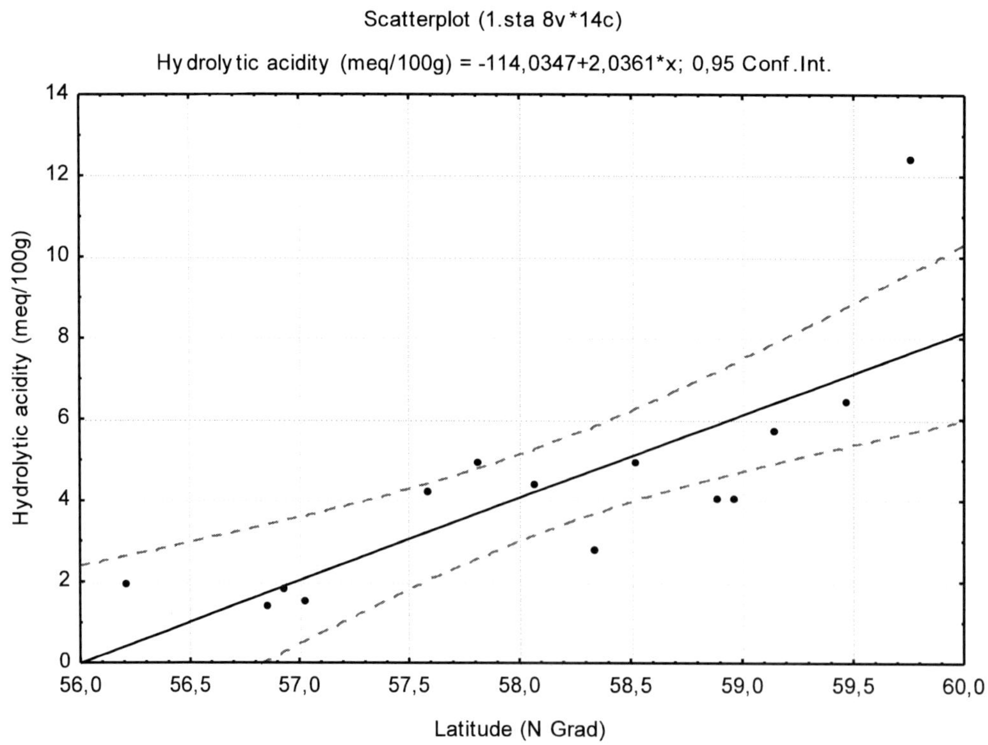
Scatterplot (1.sta 8v*14c)
Hydrolytic acidity (meq/100g) = -114,0347+2,0361*x; 0,95 Conf.Int.
Hydrolytic acidity (meq/100g)
14
12
10
8
6
4
2
0
56,0
56,5
57,0
57,5
58,0
58,5
59,0
59,5
60,0
Latitude (N Grad)

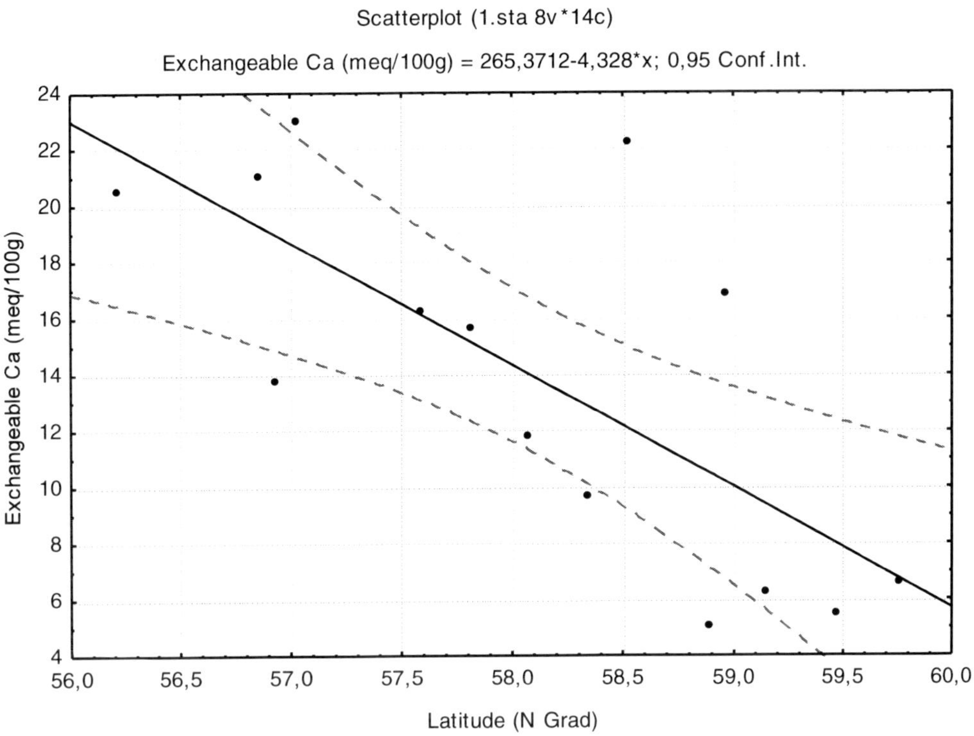

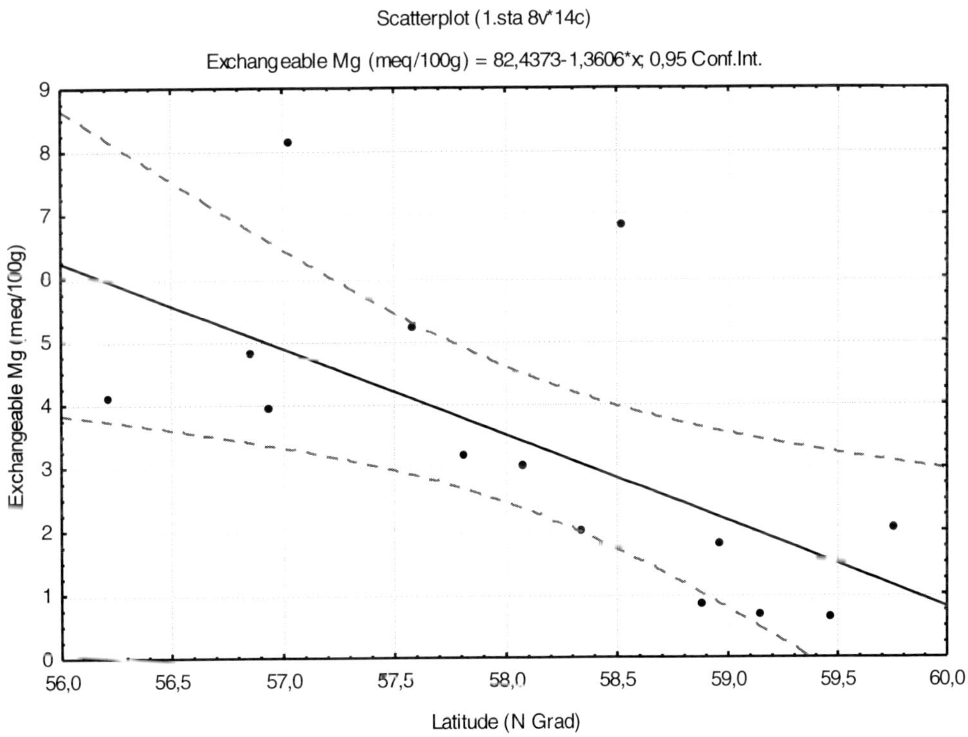

Figure 5. (Continued).

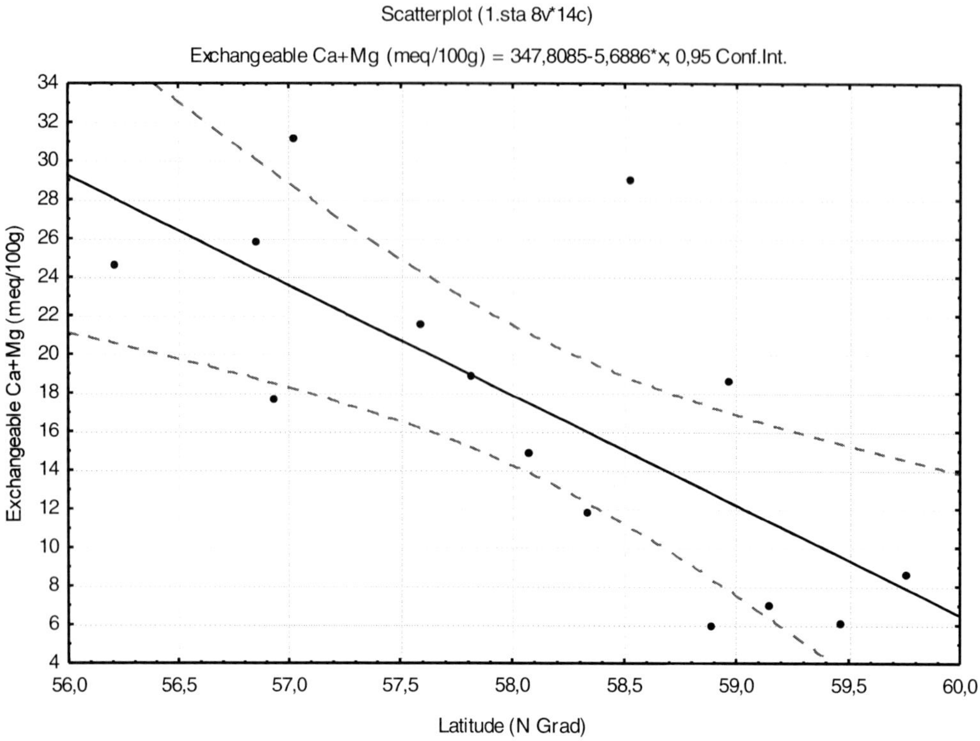

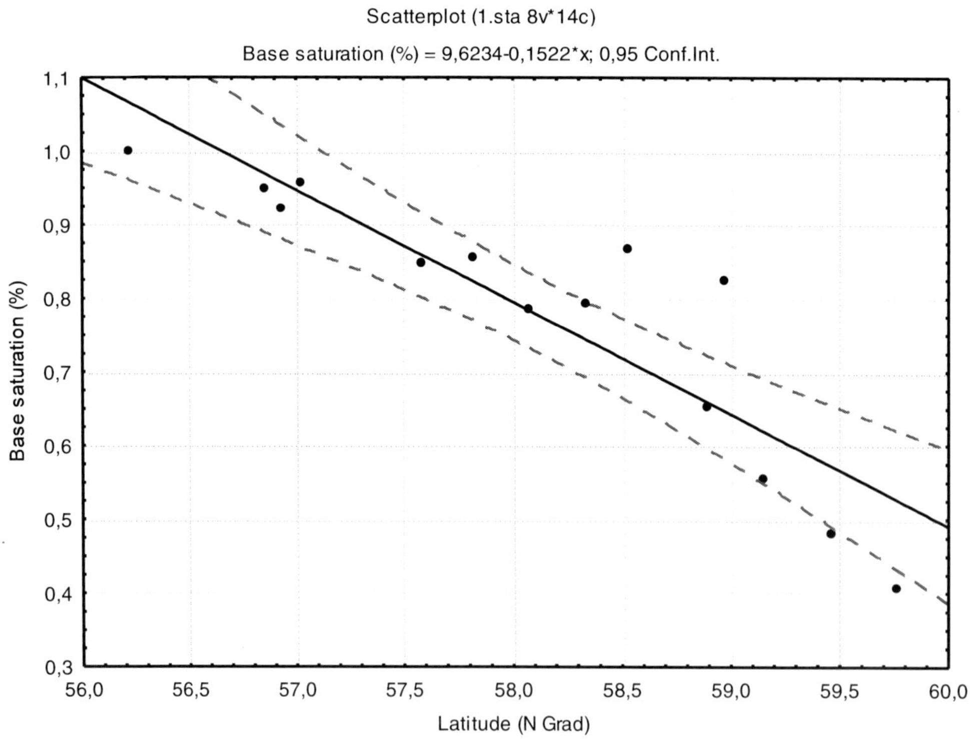

Figure 5. The correlation of soil properties in the Ob floodplain with the geographic latitude of the area.

Such a high degree of correlation between organic carbon and geographical latitude is very significant. Maybe it is because the sampling provided reflects only specific parts of the floodplain used in agriculture and does not contain all the diversity of floodplain soils. Currently we cannot definitively state what affects this distribution of properties in the soils of the Ob floodplain. Perhaps this distribution is conditioned by shifting of sediments according to the reduction of particle size along the river. However, our plots did not show such patterns in changes in the grain texture of the Ob floodplain soils from south to north. A plausible explanation could be the increased amount of limestone in bedrock to the south of the Ob watershed. Dissolution of limestones releases an increased amount of Ca which, in turn, binds humus and accumulates it in the soil profile in the form of Ca humates.

Organic Matter of Alluvial Soils

The organic matter of the floodplain soils and nitrogen are well studied with traditional methods (Afanas`eva et al., 1979; Kakhatkina, 1981; Slavnina et al., 1981; Slavnina et al., 1986; Ivanova and Slavnina, 1981). Nowadays publications give information on the distribution of organic matter in soils in different elements of the floodplain. There are more humic acids than fulvic acids in the composition of organic matter of the soils. Organic matter and other properties of the soils were investigated in the parts of the floodplain, which are promising for agriculture. Perchenko's (2005) comparison of floodplain soils and watershed soils shows that there is more organic matter in the soils of the Ob floodplain, and mobility of humic compounds increases along with an increase in the hydromorphic features of the floodplain soils. Wet meadows in the floodplains store large amounts of carbon in the soil, but drying can lead to a loss of soil organic carbon of up to 5% (Jay et al., 2014). Unfortunately, there is very little research on the composition and molecular structure of humic acids in the Ob floodplain soils (Sartakov, 2006).

Owing to the abovementioned papers, we have some ideas of what the organic matter of the Ob floodplain soils is, but today studies of organic matter in the soils and the interaction of the system 'plant – soil – organic carbon – non-organic carbon' require a different approach. The idea that molecular structure on its own can create stable organic matters and humic substances (biotic or abiotic condensation compounds) which are then stored in the soil has served its purpose. The most recent research allowed us to form a new prospective on the interaction of soils, organic matter and the dynamic of carbon. The preservation of soil organic carbon is determined not by molecular property, but by the property of the ecosystem (Michael et al., 2011). The composition and function of microbial communities and the possibility of (soil organic matter) SOM and their significance for the organic matter in wetlands is also an unexplored problem (Luster et al., 2014). We need different methods of analysis and long-term experiments using permanent plots and lysimeters to study organic matter. Moreover, there is a need for more intense efforts to be put into empirical modeling. For example, modeling of processes of carbon migration in the ecosystem and studying the relevant mechanism using molecular and isotopic tracers. Clearly, there is a need for new approaches and technology already used in related sciences (Michael et al., 2011).

TRACE ELEMENTS IN ALLUVIAL SOILS

Despite the large amount of soil-geographical, soil-geobotanical and soil-agrochemical research that has been conducted, the trace element composition of the Ob floodplain soils is underexplored. There is information in publications which give information only on some agrochemical aspects of trace elements composition of the Ob floodplain soils (Pashneva et al., 1967; Il'in, 1973; Izerskaya, 1984; Palechek, 1984; Izerskaya, 1986; Solovyev et al., 1989; Izerskaya and Vorobyeva, 2000; Ilyin and Syso, 2001; Nechaeva, 2008; Izerskaia et al., 2014).

Generally, the concentration of trace elements in soils of the Ob floodplain complies with a classic geographical pattern. The concentration of trace elements changes from south to north (Izerskaia et al., 2014). We cannot consider that there has been enough exploration of the trace elements speciation in soils, their solubility, mobility and accessibility for the plants. Thus far, only 1 paper has been published on the forms of trace elements in the Middle Ob floodplain and their stock (Izerskaya and Vorobyeva, 2000).

There we find forms of trace elements retrieved by acetate-ammoniacal buffer solution with 4.8 pH, Tamm's extract, 1 n HCl, Grim's extract and other treatment, but unfortunately water soluble forms of trace elements were not studied. Studies of forms of trace elements, stability of their bonds with organic matter, and mineral parts of soils are extremely necessary to understand the interrelation of floodplain soils and floodwater.

There is some information on the general condition of trace elements and heavy metals in the Ob floodplain soils in papers devoted to the pollution of the environment. Generally, the Ob floodplain has been affected by human impact to a very small extent and its soils can be considered environmentally pristine. As a rule, pollution of the floodplain soils is caused by the pollutant input from the floodwater (Nechaeva, 2007). However, the interrelation of floodwater and soils has not been studied. The input of pollutants into the soil is not restricted by pollutant settling from water onto the surface of waterlogged soils. Watering during floods is so large, and the ability of the floodplain soils to process pollutants is so great, that it is hard to detect substances from the water in the floodplain soils. Nonetheless, soil pollution monitoring is always relevant.

In Western Siberia, there is a need for papers devoted to soil pollution because of oil and gas production. There are several oil deposits in the Ob floodplain, and the issues of floodplain soils pollution is highly relevant (Kravchenko, 2013; Shepelev and Mazitov, 2006). Relevant papers are devoted to soil pollution with salt Cenomanian water, which gets to the surface during oil production, as well as oil-salt pollution (Shepelev and Mazitov, 2006).

In this chapter, we do not consider pollution from cities. The only exception is wastewater from the Siberian Chemical Factory (Nikitin et al., 2010). In this paper, data are presented and discussed on the content of ^{90}Sr, ^{137}Cs, $^{23<9240}Pu$ and other artificial radionuclides in water, bottom sediments and alluvial soils of the Tom and Ob rivers from Tomsk to the region of their confluence (through which the radioactive effluents of the Siberian Chemical Combine are transported). Radionuclides are visible artificial tracers, whose source is known and which are necessary to be used in further research of the diverse processes in the floodplain. Floodplain soil deposits (Westrich and Förstner, 2005) may be used as an indicator of pollutants in the basin. The sediments can be dated using a combination of historical and stratigraphic data which provide useful tools for the study of historical records of pollution carried by rivers (Bábek et al., 2008).

CHANGES IN THE SOILS DURING FLOODING OF THE FLOODPLAIN

The specific character of floodplain soil-formation includes two processes occurring in the floodplain soils, i.e., flooding and alluvial. According to Petrov (1979), in the Ob and Irtysh floodplain when compared to other rivers' floodplains, the ecological value of inundation (i.e., the duration of the flood) is larger than that of the alluvial type (Petrov, 1979). The duration of flooding in the taiga zone lasts up to 2-3 months. In the Lower Ob near Salekhard, the snowmelt flood begins during complete freezing of surrounding soil and lasts for 4.5–5 months. First, the lasting snowmelt flood influences vegetation cover, which contains species resistant to flooding. The differences in vegetation caused by alluvial processes are less significant.

Despite the great influence of the flooding period, studies of the impact of floodplain processes in the Ob floodplain on the floodplain soils formation are scarce (Bolotnov, 2007; Nechaeva, 2008; Sartakov and Chumak, 2013; Shepelev and Shepeleva, 1995). This is probably because the floodplain and alluvial processes are difficult to separate from one another. In spring, water from catchments fills the Ob floodplain. Water with products of soil erosion reaches not only the mainstream, but also the floodplain (Bobrovitskaya et al., 1996).

During the flood, part of the flow accumulates on the floodplain soils in the form of silt and other alluvial deposits; some part of the elements and substances is adsorbed by soils from the water. During the flood, the floodplain soils become bottom deposits and their interaction with water can be discussed in terms of patterns established for bottom deposits. For example, sediment cores from lakes in the Ob floodplain contain information over the decades about the source and dynamics of organic carbon coming from the catchment (Dickens et al., 2011). This information could assist in understanding what is deposited in the soils when they are submerged.

The most common drivers of metabolic processes in the system 'bottom deposits - pore solution' is the gradient of redox conditions (Frohne et al., 2015) and the values of pH, as well as the concentration of dissolved organic matter. The mobility of metals depends on changes in the acid-base and redox conditions of bottom deposits. If the conditions are neutral or alkaline, Ca, Na, Mg have high mobility, and Mo, V, U, Se have very high mobility. Under these conditions, the mobility of Al, Cr, Hg and Cu is very low, and the mobility of Pb, Fe, Zn and Cd is low, and they should be accumulated. In the oxidizing atmosphere Ca, Na, Mg, Sr, Mo, V, U and Se are mobile. In a reducing environment, the mobility of Al, Cr, Mo, V, U, Se, Hg, Cu, Cd, Pb Ni, Zn, Co and Fe is low (Plant and Raiswell, 1983). However, the redox buffer system 'bottom deposits – pore solution' of rivers and lakes is far below similar buffering system 'rock (soil) – pore water' of groundwater, swamps and soils (Bourg and Loch, 1995). If the buffer system 'rock (soil) – pore water' is maintained, then during the flood the soil becomes bottom deposit.

An even more complex transformation of deposits precipitated from the water should be expected after water clearance when the oxygen content changes in the soils. For example, it was found that aerobic conditions in the bottom deposits of the Middle and Lower Ob and the distribution of As, Cd, Co, Hg, Mn, Pb and Zn between fine and coarse fractions is controlled by iron content. The effect of iron can be so significant that there is a reverse correlation between the particle size of bottom deposits grain texture and the concentration of metals (Papina, 1999).

After water clearance, during periods of excess moisture of the upper humus horizons, there is an increase in the mobility of iron. If the excess moisture changes to the optimum, the amount of mobile forms of iron decreases by 2 - 4 times (Abramova and Pashneva, 1996). Unfortunately, it is not known in what form the correlation of trace elements and iron occur in the soil profile after water clearance. The fate of the organic matter in soil is also unknown in terms of whether it accumulates in the soil profile or is washed out by flood waters.

Changing of Soils When Moving Alluvial Soil with Flood Waters and Its Accumulation

Despite the fact that the floodplain soils are the result of two specific processes, i.e., flooding and alluvial deposits, soil formation in the floodplain is called alluvial, and floodplain soils are called alluvial soils. The basic properties of floodplain soils are mostly controlled by the alluvial process.

Indeed, the amount of alluvium deposited during snowmelt floods in the Ob floodplain through erosion of watershed soils, as well as the erosion of its banks and the banks of the tributaries is several times higher than the volume of the river suspended material. Approximate calculations show that this process has a huge scale (Makkaveev, 2003). The average rate of erosion of the Ob bank is typically 1-2 meters per year, but sometimes this rate reaches up to 10 meters annually. The Ob has a marked intensification of erosion of banks from south to north (Alexeevsky et al., 2013).

Analysis of cartographic material from 1900 to 2001 and direct measurements made in 2007-2008 by D. V. Kiselev and D. A. Vershinin made it possible to establish the average and maximum rate of erosion of banks during the channel deformation of the Middle Ob. In different sites, the average rates range from 3.6 to 13.3 meters annually, while maximum rates range from 6.4 to 24.2 meters annually (Kiselev and Vershinin, 2010). In periods of high and lasting floods, the bottom channel deposits reach the floodplain. Velikanova and Yarnykh (1970) and Baryshnikov and Samuseva (1999) observed this process in the Ob floodplain.

From the perspective of alluvial soil formation, the composition and patterns of distribution of alluvial sediments in the Ob floodplain have been sufficiently studied. Basic patterns of the alluvial process are controlled by the speed of the water flow. In the Ob floodplain, there are sorted sediments demonstrating reduced particle size along the thalweg and from the channel across the river floodplain (Mizerov, 1953; Gantimurov et al., 1979; Chalov et al., 2004; Kazantseva, 1998; Kazantseva, 2007).

Large rivers have some of the processes and patterns that make them different from small rivers. These processes and patterns are considered by the example of river models: the Ob, Jamuna and Paraná (Ashworth and Lewin, 2012) and help to evaluate the Ob floodplain in the context of general patterns of the formation of alluvial sediments and soils in floodplains. Such general geographical patterns are characteristic of lowland river floodplains in the annex to the Ob floodplain soils systematized in Dobrovolsky et al. (1971) and Balabko (1990). The authors distinguish the zonal-geographical pattern due to the differences in bioclimatic conditions in the floodplains of different natural zones, subzones and provinces; the hydrological and geomorphological pattern due to the process of longitudinal development of the river valley; and the alluvial-geomorphological pattern caused by the transverse differentiation of the floodplain (Figure 6, showing the structure of the soil profile in the riverine, central and nearterrace floodplains).

Figure 6 A. Ob River floodplain Dystric Fluvisols (FAO UNESCO).

Figure 6 B. Ob River floodplain Umbric Fluvisols (FAO UNESCO).

Figure 6 C. Ob River floodplain primitive soil.

Modelling of Generation, Transfer and Discharge of Methane, Organic Carbon in the Floodplain

The existing models of the generation, transfer and emissions of methane are created for wetland ecosystems (for example, Wania, 2007; Stepanenko et al., 2011) and require adaptation for use in floodplain ecosystems. Unlike the floodplain, when modelling wetland ecosystems it is not necessary to take into account the specificity of the organic matter, interaction of methane with the roots of plants, and photosynthetic role of plants. In winter, aquatic ecosystems of the Ob floodplain also experience an acute shortage of oxygen becoming the sources of methane. Methane gas is released into the water bodies of the Ob floodplain. The effects of methane bubbling, similar to those described by Walter and Zimov (2006) in Northern Siberia, are observed. There is still an unresolved question on the fate of oxygen and methane during floods. A better understanding of methane formation by river sediments is also required (Köthe and Gröngröft, 2006).

So far, in the waterlogged riparian soil subjected to floods, there have been no studies of the release of oxygen by plants mentioned by Blom (1999). This mechanism is very important because, during flooding, plants can provide oxygen to methanotrophic and nitrifying bacteria. Oxygen is necessary for methanotrophic bacteria to oxidate the methane, and for nitrifying to ensure the processes of oxidation of nitrogen compounds. Thus far, there has been no research on the processes occurring when substances precipitated from alluvium are mixed with organic matter in humus horizons of flooded soils.

During modelling of geochemical processes in floodplain ecosystems, we should take into account another feature of the soil cover of the Ob floodplain due to the ability of water flow, especially during snowmelt floods, to transport the heat. The Ob river water flows from south to north, from the warm-steppe region to the cold Arctic. Because of this difference, the valley of the River Ob accumulates more heat and the duration of the warm period is greater in the floodplain than in the basin (Malik, 1987). This means that the floodplain soils are warmer and there are longer active biological processes than in soils on watersheds. Research by Agafonov and Gurskaya (2013) showed that in the floodplain of the Lower Ob, heat accumulates because the flow creates a special regime of air temperature. For example, this provides an additional radial growth of tree rings in tree species like spruce, Siberian pine and larch. The Ob and its floodplain impact on the climate of Western Siberia and the global climate, but this effect remains to be explored.

CONCLUSION

Thus, the river floodplain is a huge geochemical barrier regulating the input of substances from the catchment into the World Ocean. The soil cover of the Ob floodplain and its tributaries play an important role in the conceptual model of the hydrochemical flow. However, the data found in available papers on the floodplain soils are not used for these models. This is because the research on the soil cover of the Ob floodplain was conducted mostly for agricultural purposes. There was no research on the most significant function of the soil cover of the Ob floodplain as a source of methane, organic carbon, sedimentator, and huge geochemical barrier regulating the input of dissolved and suspended material from the catchment of Western Siberia to the World Ocean. We need to develop the ideas of the global functions of the Ob floodplain. Moreover, there is a need for greater efforts in hydrological and biogeochemical modelling.

The Ob floodplain is the best site for developing approaches to the research of global processes in the floodplains of rivers in the Northern hemisphere. Firstly, the Ob has a developed floodplain, which lies within several soil-geographical zones, including permanently frozen soils. Secondly, the Ob floodplain today is the least exposed to anthropogenic transformation out of all rivers' floodplains with north-south orientation. Thirdly, there are about 150,000 rivers which are tributaries of various types, and more than 400,000 lakes in the Ob watershed. Their catchments are natural model aquatic reservoirs of varying size suitable for studying factors affecting hydrological and biogeochemical processes.

REFERENCES

Abramova MD, Adam AM, Golubykh OS, Kravchenko LB, Moskwitina NS, Neznamova EG, Nesvetailo VD, Nefediev PS, Parshina NV, Pashneva GE, Popkov VK, Popkova LA, Ruzanova AI, Strekozov VA, Shepelev AI, Shepeleva LF, Cicareva LK. Biological resources of the floodplain of the Middle Ob: dynamics and prognosis. Tomsk: Biology and Biophysics Research Institute. 1996;212. (In Russian).

Abramova MD, Pashneva GE. The soils of the floodplain of the Middle Ob andmeliorative condition. In: Tanzibaev M. G. (Ed) The soils of the floodplain of the Middle Ob, their meliorative condition and agrochemical characteristics. Tomsk: Tomsk State University. 1996;43–109. (In Russian).

Afanas`eva TV, Balabko PN, Sumerin MB, Tereshina TB. The soils of the floodplain of the upper and middle Ob. In: Kovalev R. V. (Ed.) *Problems of use and protection soil in Siberia and the Far East.* Novosibirsk: Nauka. 1984;62 – 65. (In Russian).

Afanas`eva TV, Tereshina TB, Bilik AM. Group and fractional composition of humus excessively moist soils of the floodplain of the river Ob. Vestnik Moskowskogo gosudarstvennogo universiteta, *Seria* 17, Pochvovedenie. 1979;1:55–61. (In Russian).

Agafonov LI and Gurskaya MA. The Influence of the Lower Ob River Runoff on Radial Growth of Trees. *Contemporary problems of ecology*, 2013;6:7:779–787. (in Russian).

Agafonov LI. Flow Lower Ob and its changes in the XX century. News of Academy of Sciences. *Geographical Series.* 2010; 4: 68–76. (In Russian).

Alekin OA, Brajnikova LB. Flow solutes from the territory of the USSR. Moscow: *Nauka.* 1964; 144 pp.

Alexeevsky NI, Chalov RS, Berkovich KM, Chalov SR. Channel changes in largest Russian rivers: Natural and anthropogenic effects. *International Journal of River Basin Management.* 2013;11:2:175-191. (in Russian).

Amon RMW, Rinehart AJ, Duan S, Louchouarn P, Prokushkin A, Guggenberger G, Bauch D, Stedmon C, Raymond P. A, Holmes RM, McClelland JW, Peterson BJ, Walker SA, Zhulidov AV. Dissolved organic matter sources in large Arctic rivers. *Geochimica et Cosmochimica Acta.* 2012;94:1:217-237.

Ashworth PJ.a, Lewin Jb. How do big rivers come to be different? *Earth-Science Reviews.* 2012;114:1-2:84-107.

Atavin AA, Belonenko NB, Buligina OP, Voronkov GB, Gering MA, Kalinin BM, Kirillov BB, Liubimov SA, Popova NB, Prokhorova NB, Savkin VM, Cherniaev AM. Ob basin. In: Cherniaev, A. M. (Ed.), *Russia: River basins,* Ekaterinburg: Aqua-Press. 2000;536. (In Russian).

Avetov NA, Bulgakov DS, Shishkonakova EA. Agroecological Characteristic of the Middle Ob Soils. Plodorodie, 4, Moscow: Russian Scientific Research Institute of Agricultural Chemistry named D.N. Pryanishnikov. 2008;39-41. (In Russian).

Bábek Ondrej, Hilscherová Klára, Nehyba Slavomír, Zeman Josef, Famera Martin, Francu Juraj, Holoubek Ivan, Machát Jirí, Klánová Jana. Contamination history of suspended river sediments accumulated in oxbow lakes over the last 25 years. *Journal of Soils and Sediments.* 2008;8:3:165-176.

Balabko PN. The Development of the floodplain soil formation and classification problems floodplain soils. *Pochvovedenie.* 1990;9:28–33. (In Russian).

Baryshnikov MK. Meadows Lower Ob, their characteristics and prospects of use. Proceedings of the Research Institute of Agriculture of the Far North, 10, Norilsk. 1961;115 – 158. (In Russian).

Baryshnikov NB, Samuseva EA. Human impact on the self-regulating system – basin - river flow - channel. St. Petersburg, RGGMU. 1999;220. (In Russian).

Black PE. Watershed function. *Journal of the American Water Resources Association.* 1997; 33:1:1-11.

Blom CWPM. Adaptations to Flooding Stress: From Plant Community to Molecule. *Plant Biology*. 1999;1:3:261–273.

Bobrovitskaya NM, Zubkova C. and Meade RH. Discharges and yields of suspended sediment in the Ob' and Yenisey Rivers of Siberia. In; DE. Walling and B. W Webb (Eds.) *Erosion and Sediment Yield: Global and Regional Perspectives*. Wallingford, UK, IAHS. 1996.

Bolotnov VP. Application of flood index in monitoring flood-plain ecosystems (by the example of the middle Ob flood-plain). Bulletin of the Tomsk Polytechnic University. 2007;310:3:26-30.

Bourg ACM, Loch JPG. Mobilization of heavy metals as affected by pH and redoxconditions. In: W. Salomons and W. M. Stigliani (Eds) *Biogeodynamics of pollutants in soils and sediments*. Berlin: Springer-Verlag. 1995;87–102.

Burlakova LM, Kazanceva LG. The soils of the floodplain of the Upper Ob within the Altai Territory. In: Sevastiyanov V. K. (Ed.), Soil and agronomic research in Siberia. Collection of Scientific Papers for the 100th anniversary of Professor N. V. Orlovsky. Barnaul: AGAU. 1999;1:38–40. (In Russian).

Chalov RS, Zavadsky AS, Panin AV. River meanders. Moscow: Moscow University. 2004;371. (In Russian).

Dickens AF, Baldock J, Kenna TC, Eglinton TI. A depositional history of particulate organic carbon in a floodplain lake from the lower Ob' River, Siberia. *Geochimica et Cosmochimica Acta*. 2011;75:17:4796-4815.

Dobrovol'skiy GV, Afanas'eva TV, Balabko PN, Vostokova LB, Tereshina TV. Land resources of the Middle Ob floodplain and its rational use. In: Kovalev RV. (Ed.) *Problems of use and protection soil in Siberia and the Far East*. Novosibirsk: Nauka. 1984;161-166. (in Russian).

Dobrovolsky GV, Afanasyeva TV, Remezova GL, Balabko PN. Types of floodplains of the Ob River in the southern taiga subzone. In: Kovalev R. V, (Ed.) *Land resources of Siberia*. Novosibirsk: Nauka. 1974;29-34. (In Russian).

Dobrovolsky GV, Afanasyeva TV, Remezova GL, Stroganov MN, Palechek LA, Balabko PN. Types floodplain of the Ob River. Biologicheskie nauki. 1971;4;117-121. (In Russian)

Dobrovolsky GV, Afanasyeva TV, Remezova GL. Typology of the floodplain of the River Ob. In: Popov, A. I. (Ed.), *Natural conditions of Western Siberia*. Moscow: Moscow State University. 1973;3:107-126.

Dyachenko AP, Taran GS. By bryoflora floodplain forests of the Ob River in the southern taiga subzone. Vestnik Tomskogo gosudarstvennogo universiteta. *Biologia*. 2011;3:15:75 – 91. (In Russian).

Dyukarev AG. Landscape and dynamic aspects of the taiga soil formation in Western Siberia. Tomsk: NTL. 2005; 284. (In Russian).

Frohne T, Diaz-Bone RA, Du Laing G, Rinklebe J. Impact of systematic change of redox potential on the leaching of Ba, Cr, Sr, and V from a riverine soil into water. *Journal of Soils and Sediments*. 2015;15:623-633. DOI 10.1007/s11368-014-1036-8.

Gafurov FG, Firsov VP. Pedogenesis in long flooded landscapes high latitudes. Sverdlovsk: USSR Academy of Sciences, *Ural Branch*. 1992; 147. (In Russian).

Gafurov FG. Agro recommendations and ways of rational use of floodplain soils of the lower Ob. In: Gorchakovsky, P. L, Firsova, V. P, (Eds.) Ecological aspects of the food problem. Sverdlovsk, USSR Academy of Sciences, *Ural Branch*. 1990;66-72. (In Russian).

Gantimurov II, Zaitseva TF, Fedorov LM. Materials for the characterization of the soils of the floodplain of the Ob River within the Novosibirsk region of Siberia. In: Kovalev R. V, (Ed.) Land resources of Siberia. Novosibirsk: *Nauka*. 1974;34-45. (In Russian).

Gantimurov II, Zaitseva TF, Siuhina MS, Fedorova LM. Terms and conditions of soil formation and soil floodplain Ob River at the upper boundary region of the side erosion. In: Kovalev R. V. (Ed.) Specificity soil in Siberia. Novosibirsk: *Nauka*. 1979;192-202. (In Russian).

Il'in VB. Biogeochemistry and agricultural chemistry of Mn, Cu, Mo, B in the southern part of Western Siberia. Novosibirsk: *Nauka*. 1973;389. (in Russian).

Ilyin VB, Syso AI. Trace elements and heavy metals in soils and plants of the Novosibirsk region. Novosibirsk: *Nauka*. 2001;229. (In Russian).

Ioganzen BG. Ob River floodplain (nature, development, reclamation). Novosibirsk: West-Siberian Publishing House. 1968;157. (In Russian).

Ivanova RG, Slavnina TP. Fractional composition and dynamics of nitrogen in floodplain soils. In: Tanzibaev MG. (Ed.) The soils of the floodplain of the Middle Ob, their meliorative condition and agrochemical characteristics. Tomsk: Tomsk State University. 1981; 123-154. (In Russian).

Izerskaia LA, Vorobyev SN, Vorobyeva TE, Kolesnichenko LG, Zakharchenko AV. The concentration of Mn, Cu, Zn, Co, B, Sr, Cd and Pb in alluvial soils of the Ob River (forest-steppe, southern taiga and central taiga). International *Journal of Environmental Studies*. 2014;71:5:691–697.

Izerskaia LA. Trace element (Mn, Cu, Co, Zn) resources, their concentration and distribution regularities in floodplain soils. Tanzybaev M.G. (Ed.), Soils of the Middle Ob floodplain, their ameliorative condition and agrochemical characteristics. Tomsk: Tomsk State University. 1981;194-205 (in Russian).

Izerskaya LA, Vorob'eva TE. Heavy metal compounds in alluvial soils of the middle Ob valley. Eurasian Soil Science. 2000;33:1:49-55.

Izerskaya LA. Content of molybdenum in soils floodplain of the Ob River and its availability to plants. In: Izerskaya LA. (Ed.), Soil management and soil cover Western Siberia, Tomsk: Tomsk State University. 1986;82-88. (In Russian).

Izerskaya LA. Reserves of manganese, copper, cobalt, zinc in soils of the floodplain of the Ob and their availability to plants. In: Burlakova LM, Broadcasting VA, Kolesnik GI, (Eds) The problems of soil fertility improvement in the conditions of the Altai Territory. Novosibirsk: Nauka. 1984;16-24. (In Russian).

Jay B. Norton, Hayley R. Olsen, Laura J. Jungst, David E. Legg, William R. Horwath. Soil carbon and nitrogen storage in alluvial wet meadows of the Southern Sierra Nevada Mountains, USA. *Journal of Soils and Sediments* Vol. 2014;14:1:34-43.

Kakhatkina MI. Features humus genesis of floodplain soils. In: Tanzibaev, MG. (Ed.) The soils of the floodplain of the Middle Ob, their meliorative condition and agrochemical characteristics. Tomsk: Tomsk State University. 1981;111-123. (In Russian).

Karen E. Frey, Donald I. Siegel and Laurence C. Smith Geochemistry of west Siberian streams and their potential response to permafrost degradation. *Water resources research*. 2007;43:W03406:1–15.

Kazantseva LG. Landscape-geochemical zoning of the floodplain of the Upper Ob. Barnaul: Altai State University. 2007;175. (In Russian).

Kazantseva LG. Spacial differentiation of the light soluble salts in the alluvial soils of the Upper Ob. Izvestia Altayskogo gosudarstvennogo Universiteta. Barnaul: Altai State University. 1998;113-115. (In Russian).

Kiselev DV, Vershinin DA. Channel deformation in the middle reaches of the Ob River. Vestnik Tomskogo gosudarstvennogo universiteta. 2010;337:185–188. (In Russian).

Köthe Harald, Gröngröft Alexander. Prognosis of Methane Formation by River Sediments (9 pp) Journal of Soils and Sediments. 2006;6:2:75-83.

Kravchenko IV, Shepeleva LF, Shepelev AI. Trace elements structure of soils and some species of plants of the petropolluted territories of flood plain of Central Ob. Problemy regionalnoy ecologi. 3. Moscow: Ltd. Camerton. 2013;23-28. (In Russian).

Lopatin GV. Sediment of rivers of the USSR (education and transport). Moscow: Geografgiz. 1952;368. (In Russian).

Luster J, Kalbitz K, Lennartz B, Rinklebe J. Properties, processes and ecological functions of floodplain, peatland, and paddy soils. Geoderma. 2014;228-229:1-4.

Lvov Yu. A. To the characteristic the Ob River floodplain vegetation. *Trudy Tomskogo gosudarstvennogo universiteta.* 1963;152:258–267. (In Russian).

Makkaveev NI. Riverbed and erosion in the basin. Moscow: MSU Faculty of Geography. 2003;355. (In Russian).

Malik LK. Ecological-reclamation aspects of agricultural development of floodplains West Siberian rivers. In: Syroechkovsky EE. (Ed.), *Problems of flood plains of the northern rivers.* Moscow: Agropromizdat. 1987;160-166. (In Russian).

Michael WI. Schmidt Margaret S. Torn Samuel Abiven, Thorsten Dittmar, Georg Guggenberger, Ivan A. Janssens, Markus Kleber, Ingrid Kögel-Knabner, Johannes Lehmann, David A. C. Manning, Paolo Nannipieri, Daniel P. Rasse, Steve Weiner, Susan E. Trumbore. Persistence of soil organic matter as an ecosystem property. *Nature.* 2011;6:487:49 – 56.

Mizerov BV. To the materials in the structure of the West Siberian Plain floodplains (for example, the Ob river and some of its tributaries). Trudy Tomskogo gosudarstvennogo universiteta. Tomsk: Tomsk State University. 1953;124.159–170. (In Russian).

Nechaeva EG. Differentiation of the soil cover in the lower Ob river valley. *Eurasian Soil Science*. 2008;41:11:1156-1161.

Nechaeva EG. Impact of oil mining on soils in the middle Ob' River reaches. In: GV Motuzova, (Ed.), Modern Problems of Soil Pollution, Moscow, Moscow State University. 2007;171–174. (in Russian).

Nechayev EG, Scrubs Y. Agrogeochemical aspects alluvial process in the Ob-Irtysh valley. Geografia I prirodnye resursy, 1, Novosibirsk: Geo. 1993;87-94. (In Russian).

Nepryakhin EM. Floodplain soils of the southern and southeastern regions of Tomsk Oblast. In: Ioganzen, B.G. (Ed.), Nature of the floodplain of the Ob River and its economic development. Trudy Tomskogo gosudarstvennogo universiteta, 152, Tomsk: Tomsk State University. 1963;240-257. (In Russian).

Nikitin AI, Kryshev II, Bashkirov NI, Valetova NK, Dunaev GE, Kabanov AI, Katrich I.Yu, Krutovsky AO, Nikitin VA, Petrenko GI, Polukhina AM, Selivanova GN, Chumichev VB, Shkuro VN. Up-to-date Content of Long-Lived Artificial Radionuclides in the Area of the Tom and Ob Rivers Impacted by the Siberian Chemical Combine Discharges.

Izvestia vyshikh uchebnykh zavedeniy. Yadernaya energetika, 3, Obninsk: Obninsky institut yadernoy energetiki. 2010;66-76. (In Russian).

Palechek LA. Trace elements in soils of the floodplain of the Middle Ob. In: Dobrovolsky GV. (Ed.) The soils of river valleys and deltas, their rational use and protection. Abstracts Union Conference 25-27 December. 1984. Moscow: Moscow State University. 1984;71-72 (In Russian).

Papina TS, Tretyakova EI. Eyrikh AN. Influence factors on the distribution of heavy metals on the abiotic components of aquatic ecosystems Middle and Lower Ob. *Khimia v interesah ustoichivogo razvitia*. 1999;7:553 - 564. (In Russian).

Pashneva GE, Serebrennikov VV, Slavnina TP. The content of some trace elements in soils of the Tomsk region. In: Filippov VR. (Ed.), Trace elements in the biosphere and their application in agriculture and medicine Siberia and the Far East. Ulan-Ude. 1967;164-171. (In Russian).

Pashneva GE. The structure of the soil cover floodplain. Ob within the Tomsk region. In. Tanzybaev M.G. (Ed.) The soils of the floodplain of the Middle Ob, their meliorative condition and agrochemical characteristics. Tomsk: Tomsk State University. 1981;17-42. (In Russian).

Perchenko NA. The gumus composition of the Ob meadow soils. Vestnik Novosibirskogo gosudarstvennogo Agrarnogo Universiteta. Novosibirsk: Novosibirsky gosudarstveny Agrarny Universitet. 2005:22–26. (In Russian).

Petrov IB. The Ob–Irtysh Floodplain: Classification and Evaluation of Lands. Novosibirsk: *Nauka.* 1979;136. (In Russian).

Plant JA, Raiswell R. Principles of environmental geochemistry. In: Thornton I, (Ed.), *Applied environmental geochemistry*. London: Academic Press. 1983;1-39.

Puzanov AV, Balykin SN, Balykin DN, Saltykov AV. Trace elements and radionuclides in mountain-forest soils of Upper Ob river basin. Mir nauki kultury obrazovania 2. Gornoaltaysk: Mir nauki kultury obrazovania. 2007; 8-12. (In Russian).

Rybalskyi NG, Omelyanenko VA, Dumnov AD, Leopards AR, Doronichev NA, Koriakin AS, Miroshnichenko NA, Muraveva EV, Samotesov ED, Nikolaev IV, Chernogayev GM, Pugach SL, Cherepansky MM. State Report On the status and use of water resources of the Russian Federation in 2007. Moscow: NIA-Priroda. 2007; 408. (In Russian).

Sartakov MP, Chumak VA. Infrared spectra of alluvial soil humic acid absorption in the Ob-Irtysh floodplain. *Bulletin of the Krasnoyarsk State Agrarian University,* 8, Krasnoyarsk: Krasnoyarsk State Agrarian University. 2013;53–56. (In Russian).

Sartakov MP. X-ray characterization of humic acid soils Ob-Irtysh floodplain formed in different moisture conditions. *Bulletin of the Krasnoyarsk State Agrarian University,* 5, Krasnoyarsk: Krasnoyarsk State Agrarian University. 2006;62–65. (In Russian).

SAS.Planet. Project SASGIS SAS, 2015, Group share information cartographic character, published on the Internet. Available online at: http://sasgis.org.

Savichev OG, Ivanov AO. Atmospheric deposition in the Middle Ob river basin and their impact on the hydrochemical river flow. Izvestia RAN, seria geograficheskaya, 1, Moscow: Moscow State University. 2010; 63-70. (In Russian).

Savichev OG. Hydrochemical drain in the middle Ob river basin. In: Savichev OG. (Ed.) *Bulletin of the Томск Polytechnic University*. 2007;310:1:27-31.

Schipper A, Loos S. Hydrological and hydrochemical system analysis of the Ob valley mires. MSc thesis. University Utrecht. 2003; 50.

Semenova Natalya M, Vorobyov Sergey N, Kolesnichenko Larisa G, Ruzanova Albina I. Geoecological estimation of White Lakes system in Vasyugan Landscape Reserve (Tomsk Region). Vestnik Tomskogo gosudastvennogo universiteta, 365, Tomsk: Tomsk State University. 2012;194-200. (in Russian).

Seredina VP, Andreeva TA, Alekseeva TP, Burmistrova TI, Tereshchenko NN. Oil-contaminated soils: properties and reclamation. Tomsk: TPU. 2006;270. (In Russian).

Shepelev AI, Mazitov RG. The impact of oil pollution on the salt properties of soils of the floodplain of the Middle Ob. Interexpo Geo-Siberia. 3 (1), Novosibirsk: Siberian State Academy of Geodesy. 2006;144-149. (In Russian).

Shepelev AI, Shepeleva LF. Principles of ecologically domestic evaluation of floodplain soils: soil-genetic aspects. In: Vorobyev V.N.(Ed.) *Problems of regional ecology*. Tomsk: Krasnoye znamya. 1995;5:152. (in Russian).

Shepelev AI. Alluvial soil formation in the floodplains of the taiga zone of Western Siberia (for example floodplains Middle Ob). PhD thesis. Novosibirsk. 1999;35. (In Russian).

Shepeleva LF. Organization floodplain meadow communities of the Middle Ob. PhD thesis. Novosibirsk. 1998;33. (In Russian).

Shibanov AA. The vegetation cover of the floodplain of the Upper Ob. PhD thesis. Barnaul. 2009;260. (In Russian).

Shvartsev SL. Hydrogeochemistry supergene zone. Moscow: Nedra. 1998; 366.

Slavnina TP, Kahatkina MI, Ivanova RG, Seredina VP, Izerskaya LA. Food regime of soils of the central part of the floodplain of the Middle Ob. In: Ioganzen B.G. (Ed.), *Ways of rational use of soil, plant and animal resources of Siberia*. Tomsk: Tomsk State University. 1986;48-50. (In Russian).

Slavnina TP, Pashneva GE, Kakhatkina MI, Ivanova RG, Abramova MD, Seredina VP, Izerskaya LA. The soils of the floodplain of the Middle Ob, their meliorative condition and agrochemical characteristics. Tomsk: Tomsk State University. 1981;226. (In Russian).

Smetanin IS. The soils of the floodplain of the river Ob. Nature of the floodplain of the Ob River and its economic development. *Trudy Tomskogo Gosudarstvennogo universiteta*. Tomsk: Tomsk State University. 1963;152:17–42. (In Russian).

Sokolov AV. (Ed.) Methods for soil agrochemical properties investigation. Moscow: Moscow State University. 1975;656. (in Russian).

Solovyev GA, Ustyak SA, Popov VV. Influence of ecological and genetic characteristics of soils in the nature of income and accumulation of zinc, cadmium and lead in the southern taiga subzone of Western Siberia. In: Bobovnikovoy, TS.I. and Malakhov, S.G. (Eds.) *Migration of contaminants in soils and adjacent environments*. Tr. 5 All-Union. soveshch, Obninsk, 12-15 January. 1987. Gidrometeoizdat. 1989;183-189. (In Russian).

Stepanenko VM, Machulskaya EE, Glagolev MV, Lykosov VN, Modeling emission of methane from permafrost lakes. Izvestiya RAN. *Atmospheric and Oceanic Physics*. 2011;47:2:275-288. (In Russian).

Taran GS, Dymina GD. Different one-year variability of grass communities in the floodplain of middle Ob. Izvestia Akademii Nauk SSSR, Seria Biologicheskskie Nauki, 2.Moscow: USSR Academy of Sciences. 1990;85-92.

Taran GS. Sintaksonomichesky review floodplain forest vegetation middle Ob (Alexander segment). *Sibirsky biologichesky jurnal*, 6. Irkutsk: East-Siberian State University. 1993;85-91. (In Russian).

Temerev SV. Assessment of the status of aquatic ecosystems to chemical stress. *Polzunovsky Vestnik* 2, Barnaul: Altai State Technical University. 2006; 185 - 190. (In Russian).

Velikanova ZM, Yarnykh NA. Field investigations floodplain hydraulics array in high water. *Trudy GGI*, Moscow: Nauka. 1970;33–53. (In Russian).

Vyltsan NF. Materials for the Study of floodplain meadows Ob River. Uchenye zapiski Tomskogo Universiteta. 1964;49:158-167. (In Russian).

Vyltsan NF. Meadows River valley Ob River northern districts of the Tomsk region. In: Zemcov, A.A. (Ed.) *Nature and Economics Alexander's oil region* (Tomsk region). Tomsk: Publishing House of Tomsk University. 1968;212-225. (In Russian).

Walter KM, Zimov SA, Chanton JP, Verbula D, Chapin FS. Methane bubbling from Siberian thaw lakes as a positive feedback to climate warming. Nature. 2006,;443:71–75.

Wania R. Modelling northern peatland land surface processes, vegetation dynamics and methane emission. PhD thesis. University of Bristol. 2007;122.

Westrich Bernhard, Förstner Ulrich. Sediment Dynamics and Pollutant Mobility in Rivers (SEDYMO): Assessing Catchment-Wide Emission-Immission Relationships from Sediment Studies. BMBF Coordinated Research Project SEDYMO (2002–2006) *Journal of Soils and Sediments*. 2005;5:4:197-200.

In: Riparian Zones
Editor: Oleg S. Pokrovsky
ISBN: 978-1-63484-613-4

Chapter 12

SPATIAL STRUCTURE AND DYNAMICS OF TOM RIVER FLOODPLAIN LANDSCAPES BASED ON GIS, DIGITAL ELEVATION MODEL AND REMOTE SENSING

Vadim Khromykh** and Oxana Khromykh
Geography Department, Tomsk State University,
Tomsk, Russia

ABSTRACT

The Tom river valley is characterized by very complex landscape structure. It is determined by its location in the transition zone from Altay Mountain Region to the Western Siberian Platform. The landscape systems of the floodplain have high rate of the natural transformations. Besides, a lot of landscape systems have undergone serious anthropo-genic modification during XX-XXI centuries.

The purpose of this research is the analysis of the spatial structure of Tom valley landscapes and their natural and anthropogenic changes from the end of XIX century. The object of this research is the Tom river valley within the limits of the Tomsk Region. Here we focus on the description of floodplain landscape systems, their modern status, natural dynamics and anthropogenic modification over last 100 years. For this, we used both the traditional methods of the research in physical geography (including field observations) and the newest methods of GIS-mapping and complex spatial analysis based on remote sensing data and digital elevation models using software ArcGIS 10 (ESRI Inc.), ERDAS Imagine (ERDAS Inc.).

The results allowed creating large geodatabase and GIS "Tom river valley" (including the set of digital landscape and other thematic maps within the scale 1:10000–1:25000), digital elevation model (DEM) of valley in TIN and GRID formats, morphometric indexes of floodplain landscape systems (based on DEM), and characterize main trends of landscape dynamics in the different parts of the valley.

Keywords: floodplain landscapes, Tom River, GIS, spatial analysis, remote sensing, landscape dynamics

---

* geo@mail.tomsknet.ru.

INTRODUCTION

The study of the natural complexes in large river valleys and riparian zones has both theoretical and practical importance. In these valleys, which are often crossing different landscape zones, there is an expressive specific nature of the landscapes structure. On the other hand, river valleys are the most cultivated areas. The Tom river valley is characterized by complex landscape structure, which is conditioned by its location in a transition zone from the Altay Mountain Region to the Western-Siberian Platform (Figure 1). The landscape systems of the valley have high rate of the natural transformations. Besides, a lot of landscape systems have undergone serious anthropogenic modification. Particular intensive changes of the geosystems in the Tom River valley are observed from the middle of the XX century due to the increased industrial and agricultural activity such as sand and gravel extraction from the bed of the river, draining and melioration as well as plough-lands. Other important factors modifying the landscape are transport construction and expansion of the urban and rural areas.

Figure 1. General map of the study site (red frame).

STUDY SITES, SAMPLING, SOURCES AND METHODS

The length of the Tom River is 827 km. The Tom River joins the Ob Rive some 2677 km above Ob's estuary. The width of the river Tom bed is 300–800 m (Figure 2 and 3). The catchment area is more than 62,000 km^2. The average water flow is 1110 m^3/s, the highest flow is 13600 m^3/s [1]. The Tom River is fed by the snow falls (50-65%), rain (14-32%) and underground water (20-29%).

Figure 2. The Tom River near Yarskoe. The width of the riverbed is near 400 m.

Figure 3. The Tom River near Orlovka (not far from the confluence with the Ob River). The width of the riverbed is more than 700 m.

In this work were used different map sources, remote sensing data and field research materials (landscape profiles, geobotanical test areas) as described below. The following maps were used:

- topographic maps (scale 1:25000) and plans (scale 1:10000);
- maps of forest survey (scale 1:50000);

- soil map of Tomsk Region (scale 1:100000);
- maps of geological survey (scale 1:200000);
- topographic plans dated 1896 (scale 1:10000);
- topographic plan of Tomsk dated 1933 (scale 1:10000);
- soil and vegetation maps of Tomsk Region dated 1928-1929.

The following remote sensing data were used:

- aerial photos dated 1944-1973 (scale 1:7000–1:30000);
- multi-spectral space satellite images from Terra (Aster) dated 2002-2006 with the spatial resolution 15 m;
- multi-spectral space satellite images from QuickBird II dated 2005-2009 with the spatial resolution 2.4 m.

Field observations of the floodplain and riparian zone were performed by authors during 2001-2015 years. They included more than 200 points (including geobotanical test areas and soil descriptions). All points were geo-referenced with the GPS. The accuracy of the geo-referencing was between 4 and10 m.

In addition to complex physical geography, field and remote sensing methods, we used the newest GIS-techniques, including building geodatabases, digital thematic mapping and morphometric analysis based on digital elevation models. For this, we used ArcGIS 10 (ESRI Inc.), ERDAS Imagine (ERDAS Inc.), Easy Trace 8.3 Pro (EasyTrace Group).

Our approach to landscape GIS-mapping included next stages:

- digitizing of all map sources and remote sensing data with the geo-referencing to the Gauss-Kruger cartographic projection (15th zone, Pulkovo, 1942);
- construction of the general geodatabase, which unites all spatial and attributive information (including field research materials);
- elaboration of the digital elevation model (DEM) based on the elevation data from topographic maps (1:25000);
- morphometric analysis of the valley based on DEM and "map algebra";
- differentiation of large geomorphologic elements of the valley based on the results of the morphometric analysis, geological maps, multi-spectral space satellite images Terra (Aster) and field research materials; mid-scale landscape mapping;
- detail differentiation of vegetation and soils inside these large geomorphologic elements of the valley based on topographic maps (1:25000), topographic plans (1:10000), soil and vegetation maps, aerial photos, multi-spectral high-resolution space satellite images QuickBird II and field research materials; large-scale landscape mapping;
- spatial and geostatistical analysis of geosystems (mean area, slopes, aspects etc.) and their arrangement (predominant neighbors with common boundaries);
- analysis of dynamics and evolution of valley geosystems based on the comparison of different dated maps, remote sensing data and field research materials.

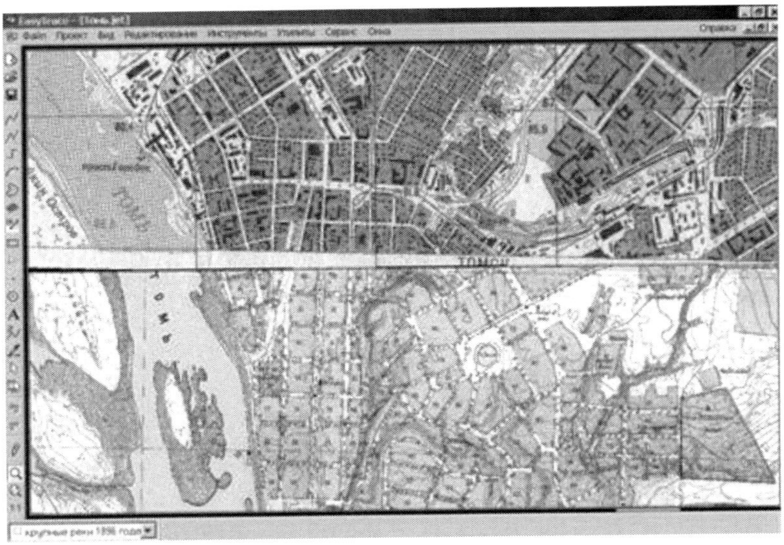

Figure 4. Geometric correction of the 1930 topographic plan (below) with the modern topographic map in EasyTrace.

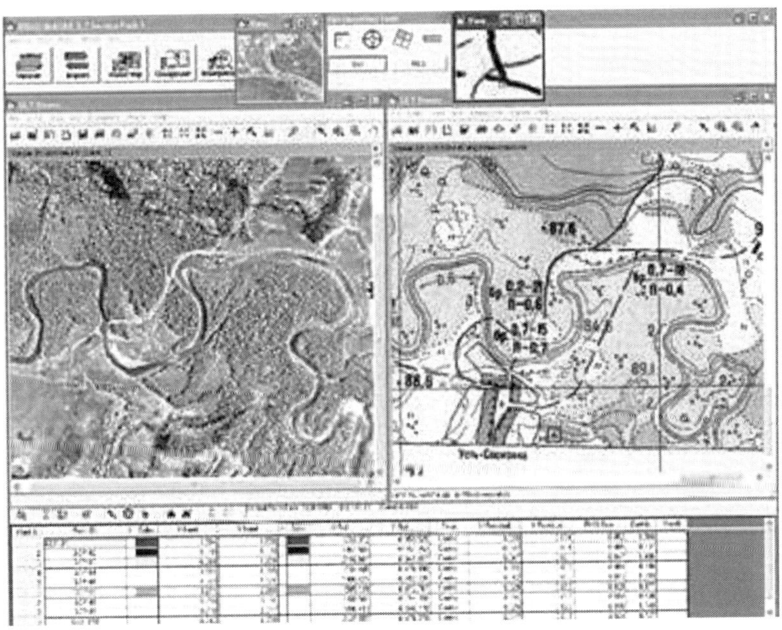

Figure 5. Geometric correction of the aerial photo with the modern topographic map in ERDAS.

As a GIS software we used ArcGIS 10 (ESRI Inc.), ERDAS Imagine (ERDAS) and EasyTrace (Easy Trace Group). Topographic maps were digitized by the scanning (300 dpi, 8-bit/pixel format, TIFF) and the subsequent vectorization in EasyTrace. These sources became the topographic base for the rest of the materials. Afterwards, raster topographic and thematic maps dated 1896, 1930, 1928, and 1929 have been added to the project in EasyTrace. The geometric correction was made based on an irregular network reference points using affine transformations (Figure 4).

The findings were exported to ArcGIS Geodatabase. On the basis of exported files we created the following feature classes in the ArcGIS: relief (linear and point objects), major rivers and lakes (polygons), small rivers (linear objects), vegetation (polygons), urban and industrial areas (polygons), pipelines, power transmission lines (linear objects). As a result, all the data were transferred into a single geodatabase. The table with descriptions of points of field observations was also added to the geodatabase. Space images were originally in digital form. The geometric correction of Quick Bird images and Aerial photos was made based on an irregular network of reference points with affine transformations in ERDAS Imagine (Figure 5).

RESULTS

Digital elevation model of the Tom valley was created using ArcGIS 3D Analyst module by Delaunay triangulation. As the raw data, we used the digitized isohypses (total 3 338 lines), and elevation points, including the edges of water (3 374 points). We also used the polygon and linear objects of the drainage system (a total of 1 310 842 lines and polygons), as well as the contours of the lakes with the known water's edge (185). The objects of the drainage system used in the calculation of DEM and polygons with known lakes water's edge were used for approximation of the flat surfaces. This allowed building triangulation irregular network (TIN), consisting of 1,042,373 triangles with a range of altitudes from 67.8 to 195 m. For this, we used the TIN which is a computer database (34.7 MB) for the relief of the Tom valley. It includes its floodplain and riparian zone. In addition to the heights for each triangle, the network stores information about the angle of inclination of the slope and the exposure.

For the first time, we built the series of large-scale maps of key indicators of relief: hypsometric map, the maps of steepness and the slope exposure, allowing in-depth morphometric analysis. For this, the card obtained were converted to raster format, GRID, using a regular grid with a pitch of 10 m. As a result, the raster maps “algebra” and detailed analysis of the topography of the cells of 10x10 m became available. The analysis showed that the largest areas in the valley (26%) are occupied by the areas with altitudes of less than 80 meters. Chiefly, these areas belong to the flood and riparian zones. This indicates the predominance of the floodplain over the other elements of the relief of the Tom river valley.

More than half the area of the lower valley (57%) is represented by almost flat areas with slopes less than 0.3°. The proportion of areas with slopes greater than 60° was very small (0.005%). Depending on the slope exposure, all cells DEM were classified into eight categories. In the valley, the surface is dominated by areas without severe exposure (35.2% of the area) and areas with north-west slopes (10.4%) and western (10.2%) exposure. This is caused by the overall slope of the valley oriented towards the northwest. The greater area of the right bank exhibits western exposure [5].

Figure 6. 3D-Model of the Tom Valley and its riparian zone, which is based on Delaunay triangulation using ArcScene ArcGIS.

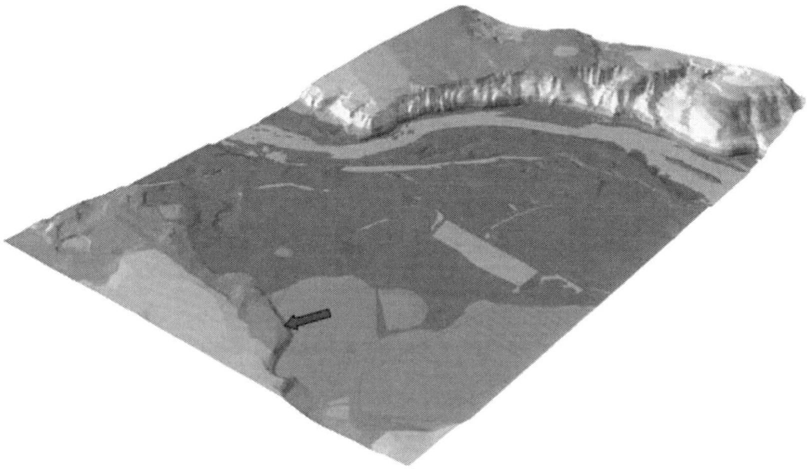

Figure 7. The fragment of the Tom Valley 3D-Model near Tomsk (red arrow indicates the second terrace).

The program ArcScene ArcGIS GIS package based on a digital terrain model allowed creating three-dimensional model of the Tom River valley bottom. After the increase of the vertical scale of five to ten times, the horizontal features on this model are clearly visible (Figures 6 and 7). Based on aerial photographs and satellite images of Terra (Aster), a three-dimensional topographic model was constituted (Figure 8).

A large-scale landscape mapping was allocated to 5,147 tracts circuits, divided into 112 types. The GIS spatial analysis was conducted based on tracks and calculated metrics. The zonal statistics for stows map steepness of the slopes was counted using ArcGIS Spatial Analyst module and the average slope of each tract was determined. This allowed us to estimate the degree of drainage geosystems and reduce subjectivity in the characterization of relief in tracts title. For example, the tract with a mean slope of less than 0.2° was defined as a flat area, 0.2-0.5° as lined sections and more than 0.5° as sloping sites. Also, it was concluded that the best drainage geosystems are located in the upper portion of the valley, where the average slope was 0.92° against 0.58° in geosystems of the lower portion. On the basis of a complex spatial analysis in GIS tracts were identified prevailing "neighbors" of the common borders. We also studied tracts position that helped in some cases (especially in the riverbed floodplain) to trace paragenetic series of geosystems considering genetic cohesion and coupling.

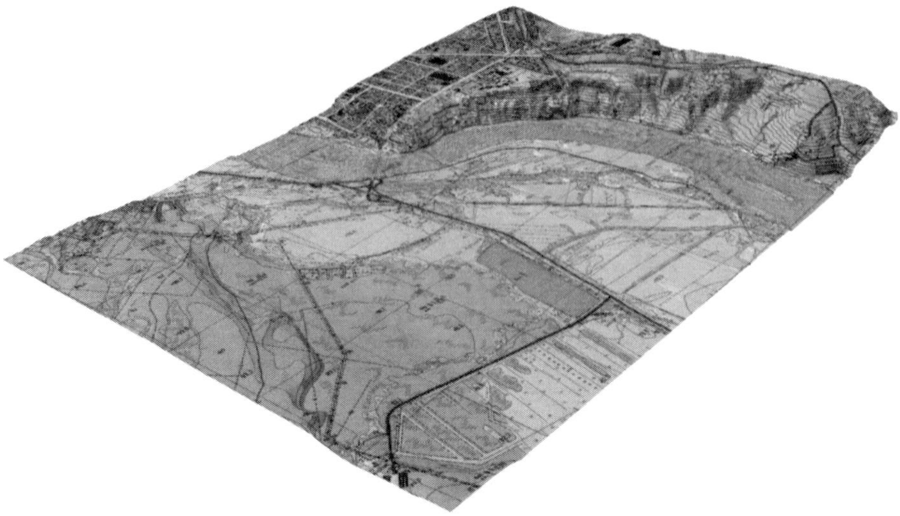

Figure 8. The fragment of the Tom Valley 3D-Model near Tomsk. The model is based on modern topographic map.

The result of spatial analysis confirmed the findings of the preliminary morphometric analysis about the prevalence of floodplain type of terrain, which occupies more than 515 km^2, or 41.3% of the area of the lower part of the Tom valley. The floodplain of the Tom River pulled strip from south to north in the upper section of the valley and from the southeast to the northwest. On the lower portion, the floodplain surface has increased markedly in the area Bogashevo deflection (Baturino – Kolarovo) below and near Tomsk Samus depression. The relative excess of the floodplain surface above the water edge is less than 7 to 8 m (except for the high riverine trees). The floodplain type of terrain has the hollow structure and a large variety of the lower rank geosystems, most of which has a characteristic elongated shape. These geosystems form unique paragenetic series, which, according to V. V. Kozin [7], can be represented as a sequence "the riverbed-flow–bayou lake-fen-swampy meadow-lowland meadow- plain meadow." The floodplain terrain has typical elongated structure of its constituent elements, as can be seen on the satellite images (Figure 9 and 10).

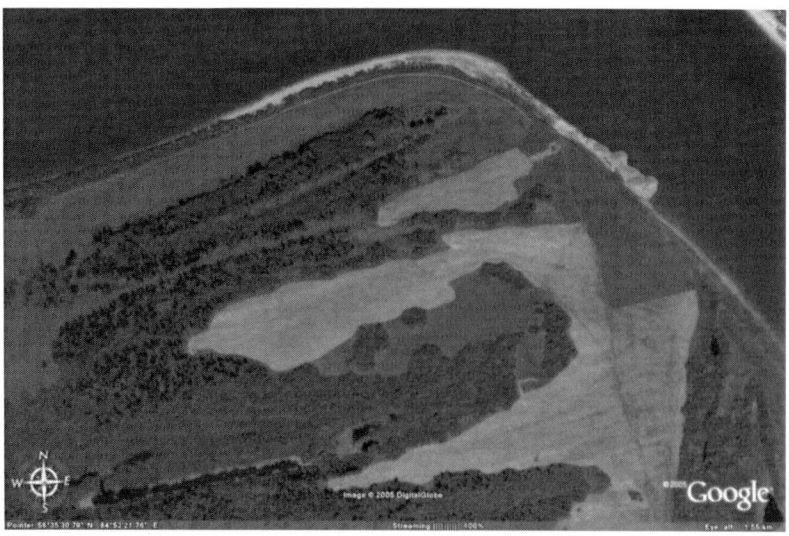

Figure 9. The elongated structure of floodplain geosystems opposite Seversk (space image QuickBird II).

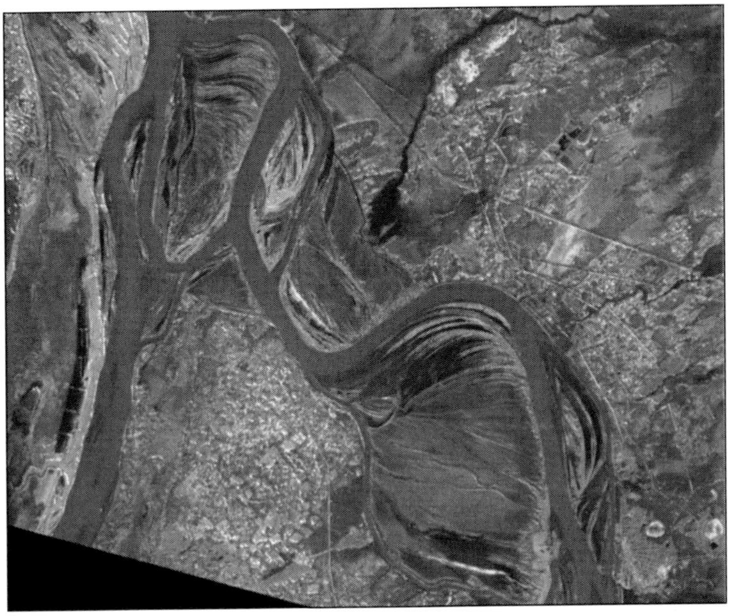

Figure 10. The characteristic elongated arcuate geosystems structure of the segment-ridged floodplain near the estuary of the river Tom on satellite image Terra (Aster), 25.04.2001.

In the flood plain, 31 type tracts were allocated. In the development of floodplain geosystems the leading role, along with the climate, belongs to the hydrological regime of the river, which distinguishes them from other geosystems [8]. The greatest impact on the differentiation tracts have the duration and the degree of flooding which depend on the position relative to the main channel or duct (riverine, urban or terrace floodplain), and the age.

The emerging low floodplain areas of the riverbed represent the first link in the chain of riverine tracts. Here, the elementary paragenetic complex (EPGC) are associated with maximum values of alluvial supply [7]. The gravel, pebble and sand banks with pioneer vegetation on alluvial primitive layered soils ar ewidely distributed in this zone. At the same time, the upper section of the valley (above the mouth of the River Bol. Kirghizka) are dominated by gravel and pebble (see Figure 2). At the bottom of the river valley they are dominated by sand and silt (see Figure 3). The average size of these landscape features is small, less than 0.11 km^2. Of all the "neighbors" in the largest number of cases (27% of the length of borders) the tract borders the low trees with willow alluvial sod layered soils, which are the next link EPGK riverine and floodplain dominating the riverbed (4.5% of the entire floodplain). A lot of these ecosystems occur in the Tom river mouth zone. The height of willows here can reach 7 meters. On the upper reaches, these ecosystems are found mainly in the floodplain between the villages Kaltay and Kolarovo. The height of willows may reach 10 m, and the grass is represented by nettles. In the hollows, the tracts of equisetum-sedge meadows on alluvial meadow soils are developed. The forb-grass meadows predominate on the surfaces of the high plains with the dominance of timothy grass, cocksfoot, and couch grass on sod alluvial soils. The lower part of the valley has riverine trees pine and birch and aspen-birch mixed forests with grass on sod alluvial soils. The undergrowth in such forests is dense and presents various kinds of willows (goat, ash-gray, etc.), bird cherry, pea, rowan and currants. The meadow horsetail, nettle, Kentucky bluegrass, water knotweed, red clover, angelica Siberian, fern, and geranium forest prevail in the herbage.

The central floodplain pronounced hollow morpho-structures (see. Figure 9-10) characteristic for segment-island plains lowland rivers [9]. In the low central floodplain, the oxbow-form hollow EPGK, continuing evolutionary series riverine EPGC [7]. The main physico-geographic process here is filling negative fluvial forms with lithogenic and biogenic sediments. The willow thickets alluvial silty-gley soils (the upper section) dominate in depressions on the periphery of dead channels and oxbow wetlands. Thin peat bogs (the lower portion) occupy 5.1% of the floodplain. The vegetation of these eco-geosystems is very thick and high (>1 m). It comprises goutweed, sorrel horse, tansy, geranium grass, yarrow, lady's bedstraw real and tenacious, nettle, horsetail, meadow sedge (Figure 11). Aligned oxbow lowering occupies lowland sedge bog with birch trees on peat soils. In the depressions, the sedge meadows on alluvial meadow-bog soils are quite common. Shallow depressions are usually filled with tract polydominant meadows and willow shrubs on alluvial meadow soils (3.8% of the area). As the distance from the river and canals increases, the oxbow-troughs and forest-meadow appear. Young central floodplain part is occupied mainly by willow and willow-birch nettle-grass forests on sod alluvial soils (11.7% of the area). Significant number of eco-geosystems is developed on the lower portion and opposite to Seversk (see. Figure 9). The height of the trees in these settings reaches 18-20 m. In the area of the mouth (in the vicinity of Pushkarev) and within the islands of the lower portion of the Tom River (Chernilschikovsky, Fir, Isayevsky) there are zones with poplar and willow, poplar, birch shrub forb forests on alluvial sod soils which occupy quite large area (average 0.57 km^2). On high ridges there are forb-grass meadows with predominance of meadow grass, cocksfoot, couch grass, meadow foxtail (12.8% of the area), and pine and birch and aspen forest shrub forb alluvial sod soils. At the same time, due to the strengthening of zonal factors, the high floodplain meadows are more prevalent in the upper portion, and the forest are predominant on the bottom of the river valley. On the slopes of high ridges, especially on the islands of the lower portion, we observed a proliferation of spruce-fir forests on alluvial shrub sod podzolic soils (Figure 12). Some authors [10] relate this to the presence of dark coniferous taiga in Chernilschikovsky island with outliers of Tom first terrace (see Figure 12). As a unique tract, almost non-flooding very high areas were also found in these settings. They presumably represent ancient riverine banks, south from Kolarovo on Pine and Bolshoi Aydakovo Islands with the birch forests on alluvial sod soils.

Aligned sections of terrace floodplain wetlands are usually occupied by pine and pine-birch forests on sedge peat-gley soils (7.9% of the floodplain) or birch and aspen-birch forests on peat-gley and peat soils (7.8% of the area). In the most downstream on the right bank there are several very large tracts (an average of 2.72 km^2) of wetlands fir forests developed on clayey-peat-gley soils. The flat depressions have spread sedge hypnum pine and birch-sedge-moss bogs on peat soils. In such ecosystems, the peat accumulation in the surface of the marsh may eventually rise to the flood-free elevations, forming a kind of small terrace above the floodplain, for example, Kandinsky peat bog with a capacity of up to 4 m and a swamp in the estuary of Poros [11]. Lowlands and sub-valleys of the small rivers in the terrace floodplain wetlands are often occupied by spruce and fir and birch forests on silty peat soils. There are areas with birch and aspen-pine forests on alluvial sod weakly podzolized soils (3.2% of the area) develop don high areas not subjected to flood. The undergrowth in these woods is presented and rosehip rowan and herbage such as fern, horsetail meadow, sanguisorba, sedge, and reed grass.

Figure 11. A 2-meter goutweed among willows in the depression of the central floodplain (near the mouth of the river Yakunina).

Figure 12. Fir-spruce shrub forest in the central floodplain (Elovy Island).

In addition to these relatively pristine eco-geosystems with a low degree of human modification we distinguished floodplains with a high degree of human modification. The most common agricultural geosystems (Figure 13) are arable land (7.7% of the floodplain) and arable land with artificial irrigation (4.4% of the area) in the modified alluvial sod soils. Common garden areas on the modified alluvial sod and meadow soils. Being modified as a result of transport construction tracts dominated by clearing transmission lines and pipelines through birch and pine and dark coniferous forests with polydominant meadows on alluvial sod, and meadow soils. In the lower reaches on the right there recovering cutting overgrown young pine and birch forests on alluvial sod soils. On the pine bogs terrace floodplain the common Geosystems with bilge drainage canals and peat on reclaimed peat soils are frequently found.

As a result of external influences or under the influence of self-development, in accord with the opinion of Sochava [12], Krauklis [13], and Isachenko [14], three levels of changes can be distinguished: the cyclic changes (both diurnal and annual), dynamics and evolution. The largest natural changes of the landscape structure observed in the floodplain are associated with the transformation of terrain erosion-accumulation activity of rivers. Natural dynamics and evolution of geosystems has been studied in several key areas of Tomsk by comparing the topographic maps in 1896 with a modern landscape map as described below. An intense coastal erosion and significant shift of the Tom River channel (up to 155 m) are observed in these areas (Figure 14, 15).

Figure 13. Arable land and hayfields in the floodplain on the left coast of Tom (south from Tomsk).

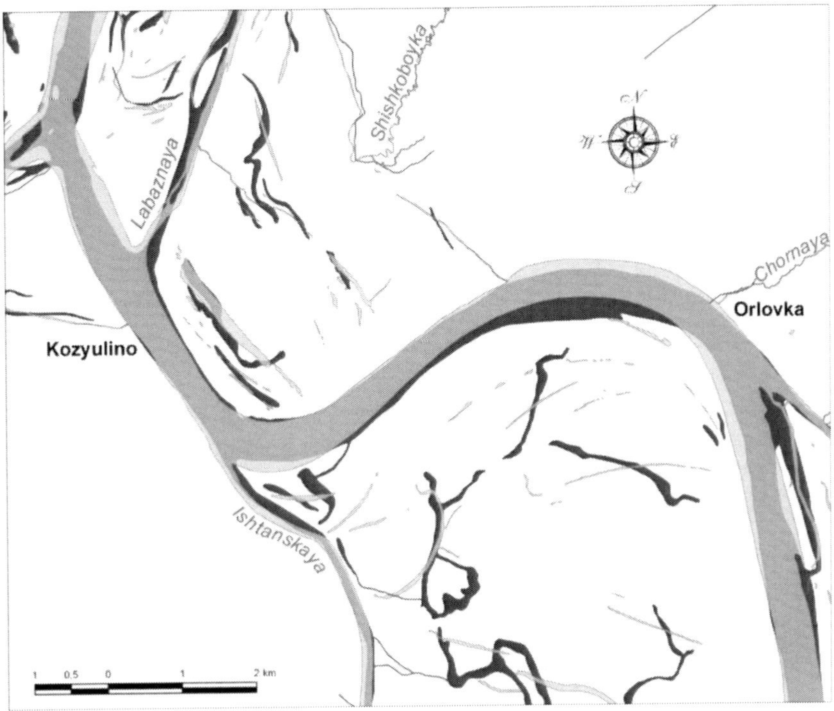

Figure 14. The changes of Tom riverbed position, channels and oxbow lakes from 1896 (lower dark blue layer) to 1998 (upper light blue layer) near Orlovka–Kozyulino.

Figure 15. Fallen trees as a result of Tom shore erosion.

Figure 16 and 17 show changes in natural systems for 100 years in the key area near the mouth of Ishtanskaya Channel.

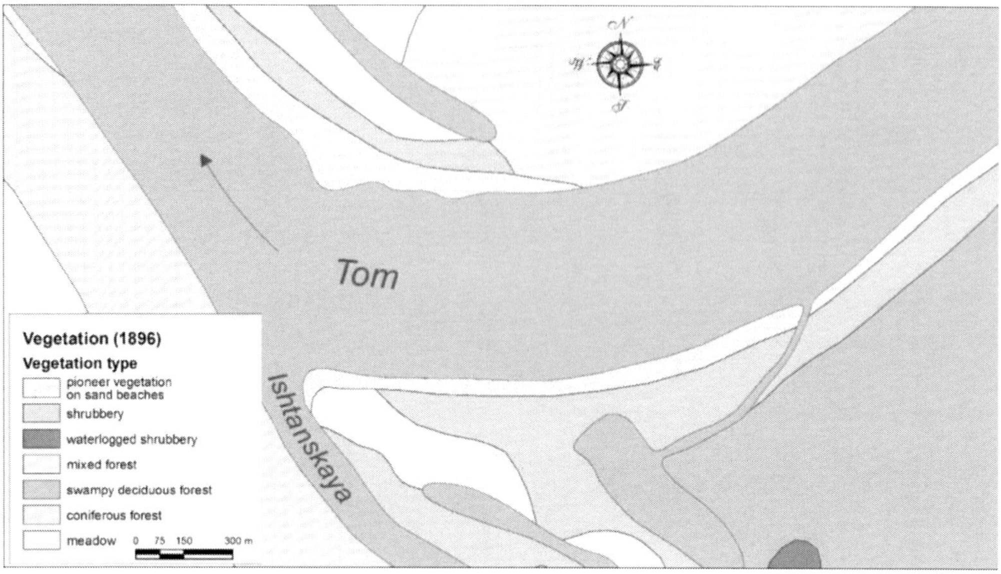

Figure 16. The fragment of the vegetation type map in 1896.

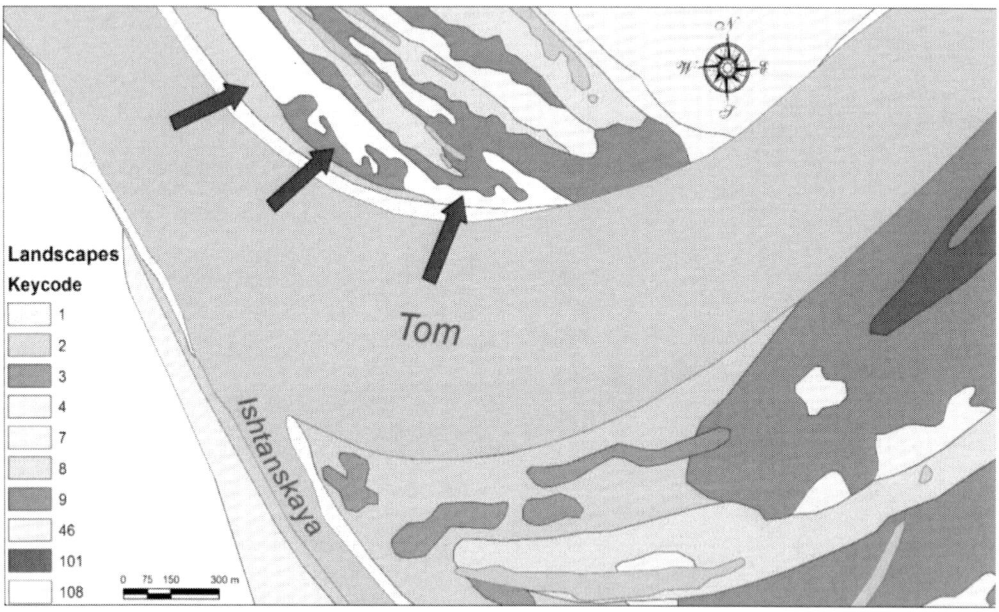

Figure 17. The fragment of the modern landscape map (red arrows show young floodplain geosystems, which appeared during last 100 years). Key codes: 1 – sand beaches with pioneer vegetation on primitive layered soils; 2 - low banks with thickets of willow alluvial sod layered soils; 3 - hollows with equisetum-sedge meadows on alluvial meadow soils underdeveloped; 4 - depression with polydominant meadows on alluvial meadow soils; 7 – surface of riverine banks and the tops of zones with pine and birch forests on sod alluvial soils; 8 - deep depressions with sedge meadows on alluvial meadow marsh soils; 9 - zones with nettle-grass willow forests on alluvial underdeveloped soils; 46 - zones on the 2nd terrace with pine and birch forests on light gray forest soils; 101 - i deep depressions on the periphery of bayou-lakes with swampy thickets of willow on alluvial silty-gley soils and shallow peat bogs; 108 - tops of the high areas with forb-grass meadows on alluvial sod soils.

The appearance of inundated floodplain eco-geosystems is in line with the pattern of sediment accumulation. Upon the change of the water level during the flood, the low water parts of riverine sandbars emerge from the water and can be retained by vegetation (tract with code 1 in Figure 17). The highest of them are formed during the wet years, not flooded in the subsequent low tide. This increases the probability of securing vegetation. The sedge and canary are gradually formed on sandy soils, leading to the increase of the thickness of loamy silt to 30 cm during 5 to 6 years, followed by the development of young willows with sparse herbage [15]. This is due to the fact that in the temperate latitudes of Eurasia, it is a pioneering willow bushes, which require almost no soil and grow directly on the sand [16]. For several consecutive dry years, when high islands are not covered with water, the willow starts to develop their root system on them. According O. N. Kolesnikova [17], in the first 7 to 13 years the riverine sandbars represent the most dynamic stage of development.

The appearance of the bush on the surface of the shallows during the flooding causes an increase in the roughness of the bottom and, consequently, favors the accumulation of thinner material (silt, organic detritus). The latter, in turn, stimulates further development of the vegetation. The precipitated particles of sand or pebble cover the surface of the former side-forming layer of silt floodplain. On its surface, in addition to willow, many species of herbaceous plants can settle, and the riparian soils begin to develop. Subsequently, under the canopy of willow, large representatives grasses appear such as purple loosestrife, loosestrife, crepis Siberian [18]. With increasing thickness of the floodplain loam up to 1.5-2 m in the stand, significant changes occur. First, grasses develop, then other shrubs (usually currants) follow. Finally, the projective cover of grass reaches 50%. On the site of a former sandbank (the lower part of the right bank of Figure 16), an eco-geosystem of low bank with thickets of willow may develop (the tract with code 2 in Figure 17). This completes the process of converting the coastal side to flood banks. A natural evolution of the gravel-pebble and sandy shoals with riverine pioneer vegetation on alluvial primitive layered soils includes low riverine thickets of willow trees with alluvial sod layered soils.

Further development of the floodplain surface is associated with new bank formation. Parts of the rapids, first covered with water, are being gradually filled with sediments. Aqueous vegetation overgrows and becomes dominant downward, forming a natural boundary of hollows riverbed floodplain with sedge meadows on alluvial meadow soils (the tract with code 3 in Figure 17). Over time, these features evolve into the central floodplain where they dominate the meadows and willow shrubs on alluvial meadow soils (tract with code 4 in Figure 17). In deep depressions, we observe numerous sedge meadows on alluvial meadow marsh soils (tract with code 8 in Figure 17. The deformations of the floodplain channel may take decades, and shorten the entire development cycle to hundreds of years [16]. When the bank is at a height of 4-5 m above the water's edge, willow bushes replaced willow and willow-birch forests nettle-grass (tract with code 9 in Figure 17, instead of bush in Figure 16). According to Ryazanov and Surkov [15], this process takes about 50 years. In depressions willow lasts longer. Thus, the transformation of a young sandy central floodplain takes an average of 60-70 years. After several successions of steps in the central flood plain, the wood is converted into meadow (Tracts codes 108, 4, and 8 in Figure 17 in place of the mixed forests in Figure 16). The process of changing forest landscape meadow until now has not been studied in detail, and this issue is debatable. Apparently, this process is defined as the development of forest communities under control of external factors such as changes in the mechanical composition and thickness of alluvial deposits, the depth of the ground water, and the amount of mineralized litter [8].

Along with the natural dynamics of the floodplain landscape, the Tom River in the last 100 years encountered significant anthropogenic modification of the natural systems. The key factors of anthropogenic modification of the geosystems in the study area, in our opinion, are industrial and agricultural activities of man, as well as the expansion of residential areas and heavy transport construction. As a result of extensive mining of sand and gravel from the 1950th, the Tom River formed large pits. Such pits could be created in about 500 years during the natural cycle, since the stock of the benthic sediment is only 200 thousand. m^3 [19]. Since the pit is a trap for sediment from above, an uncompensated removal of material occurs below the pit. This greatly enhances the erosion. Below the pit, the erosion of the river bottom occurs. The size of incision decreases downstream since the river reduces its bias and conveying capacity in line with a decrease in the inflow of bottom sediment on top. For low-flow water levels in Tom for the period of active mining from 1960 to 1987, a drop of water surface changed from 2.7 to 0.7 m, corresponding to the slope of the water surface was 1 cm/km [19].

We conducted with the help of GIS-mapping the comparison of modern drainage system with the drainage system of the year 1896. This comparison revealed that the Public island located downstream the bridge has disappeared (Figure 18). Here, over 100 years, the riverbed of Tom narrowed from 770 m to 180 m. Human activities led to drying up the left side of the bed. The depth of the channel has substantially reduced, in some areas up to 1 m (profile line N - O in Figure 19; Figure 20). Located below the bridge section of the river now is a natural sump largest sediment deposited in the form of numerous islands, composed of sand and gravel (Figure 18).

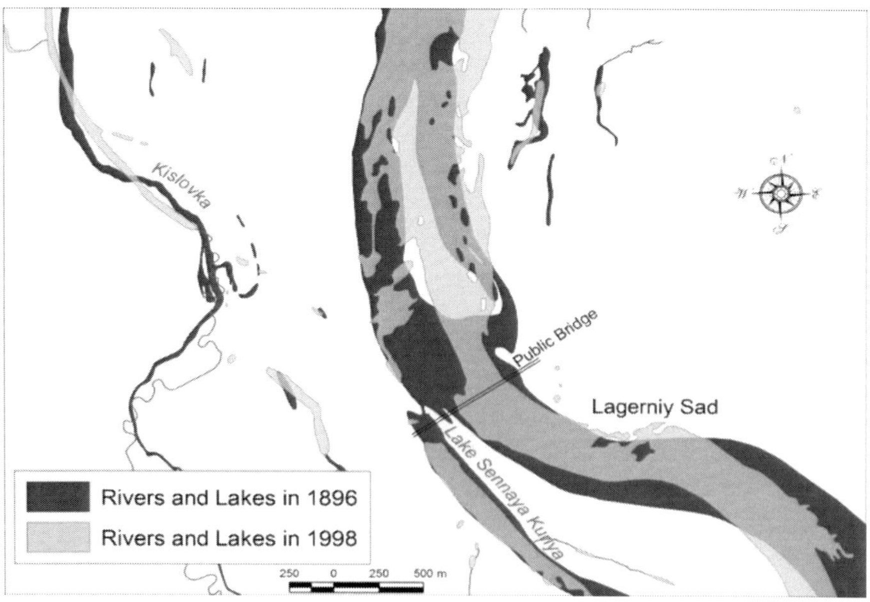

Figure 18. The changes of position of the Tom riverbed, channels and oxbow lakes from 1896 (lower dark blue layer) to 1998 (upper light blue layer) near Public Bridge.

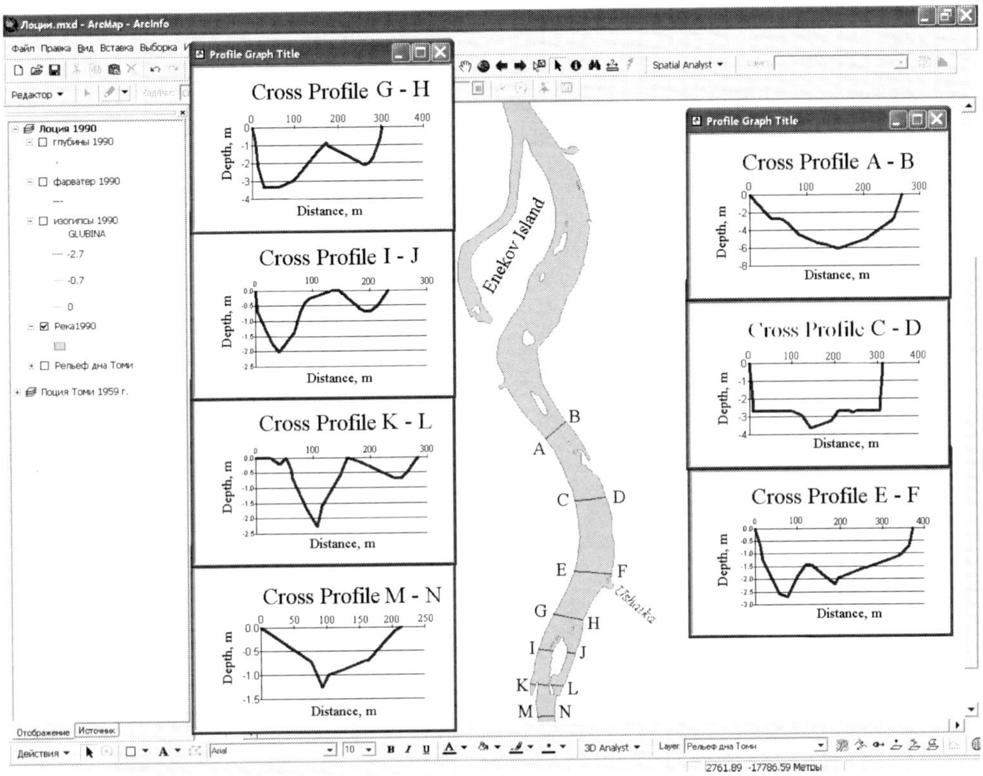

Figure 19. The cross profiles and depth profiles below Public Bridge, built by the authors on the basis of sailing directions in 1990 with the use of ArcGIS 3D Analyst.

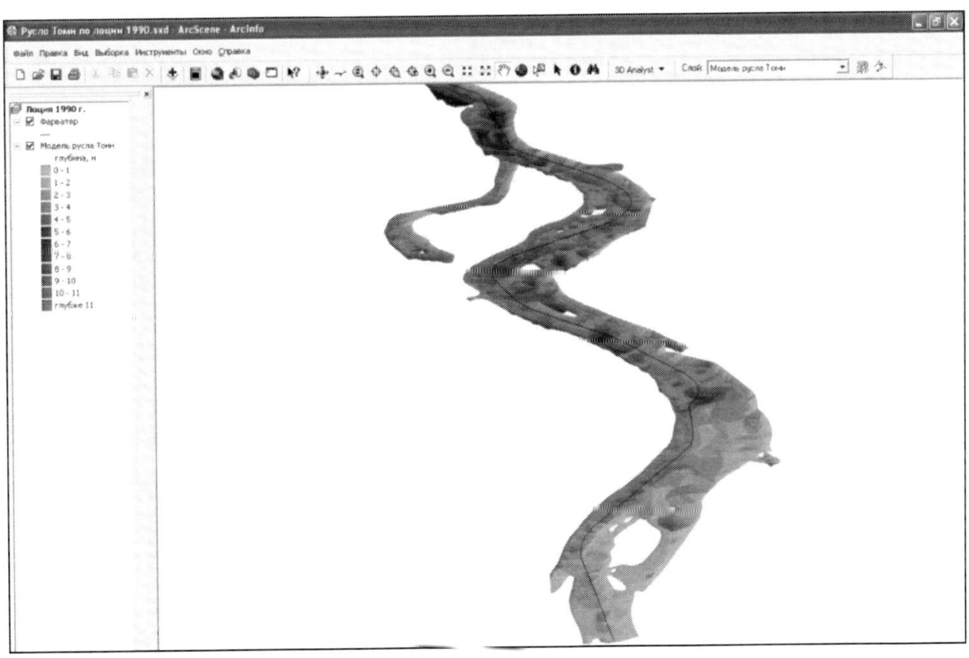

Figure 20. 3D-Model of Tom riverbed, built by the authors on the basis of sailing directions in 1990 with the use of ArcScene ArcGIS.

Severe narrowing of the Tom River channel observed at the site of rock outcrops next to the camp garden. Here, along the right bank of the river, even at moderate levels of the water, the bedrock surface is exposed at the width of about 300 m (Figure 21). Intensive lateral erosion of the river caused the need to strengthen the shore with the help of special structures (Figure 22). The surfaces of the rocks (shales of the Lower Carboniferous) are also subject to physical erosion due to high flow velocities in the low-water period (up to 3 m/s), as evidenced by rounded fragments of the shales detected on the pebble beach.

Figure 21. Rock outcrops (shale), forming threshold in line with Tom near Lagerny Sad.

Figure 22. The bank protection coast from the destruction.

The mode of flooding of the river valley has changed, which led to its restructuring. Most of the newly formed 50-300 meters wide floodplain is gravel and pebble covered with isolated clumps of sedges, horsetails, and other pioneer species. The rear part of the side-width of 25-100 meters in 4-5 years overgrown with willows height of 2-2.5 m, and has more sparse herbage. There is a deposit of sand and loamy silt. This is the embryo of the future geosystem central floodplain. It follows by loamy alluvium accumulation, which develops fields and sod layered soil under nettle, willow reed. In the area of the decreasing river level the willow shrubs shifted to the river bed about 2 m in height. The riverine floodplains are formed on the side of the former riverbed and floodplain, at the central location along the river bed.

Figure 23. The fragment of the Tom Valley 3D-Model near Tomsk. The model is based on the aerial photo dated 1954 (red arrow – bayou-lake Kalmatskoe, green arrow – bog).

Figure 24. The fragment of the Tom Valley 3D-Model near Tomsk. The model is based on the Terra (Aster) space image dated 15.05.2003 during flood (red arrow, pond instead bayou-lake Kalmatskoe; green arrow, the system of melioration channels instead the bog; yellow arrow, new ponds near village Kislovka).

For agricultural use in the study area, the network of the drainage canals and ponds was built on the site Kalmatskoe floodplain lake and the village Kislovka. The changes in the drainage system during last 50 years are shown in Figures 23 and 24. Under the influence of the construction of waterworks, the ecosystem significantly changed. Large ponds changed the microclimate, which becomes softer. The natural forest and grassland complexes were replaced by the ecosystems, resistant to high water table (instead of mixed forests develop willow, instead of dry grasslands - sedge marshes). Below the dam, the floodplain and riparian zone have dried.

As a conclusion, this work demonstrates that, at present, almost all riparian landscapes of the Tomsk River in the vicinity of Tomsk are exposed to anthropogenic modifications to various degree. The vector of the landscape dynamics is directed towards considerable desiccation due to the lowering of the groundwater level, which occurred due to the overlapping of various anthropogenic factors.

REFERENCES

[1] Geographic Collegiate Dictionary: Geographical names.– Moscow, 1983.– 528 p.

[2] Beruchashvili N. L. *Personal Computers in Geography.* – Tbilisi, 1992. – 180 p.

[3] Burrough P. A. Principles of Geographical Information Systems for Land Resources Assessment. – Oxford, 1996. – 194 p.

[4] Cartography/ed. by A. M. Berlyant.– Moscow: Aspect, 2003.– 477 p.

[5] Khromykh O. V., Khromykh V. V. The landscape Analysis of Tom valley based on GIS: natural dynamics of valley geosystems and their changes in consequence of anthropogenic impact. – Tomsk: NTL, 2011. – 160 p.

[6] Khromykh O. V., Khromykh V. V. *Digital Elevation Models.* – Tomsk: NTL, 2011. – 188 p.

[7] Kozin V. V. *Paragenetic landscape analysis of river valleys.* – Tyumen, 1979. – 86 p.

[8] Khromykh V. S. The functioning and dynamics of the floodplain landscapes. – Tomsk, 2008. – 128 p.

[9] Makkaveev N. I., Chalov R. S. *Channel Processes.* – Moscow, 1986. – 264 p.

[10] Shepelev A. I., Shepeleva L. F., Rosnovsky I. N. and oth. Assessing the environmental status of the island Chernilschikovsky//Ecological assessment of Seversk and the 30-km zone SCC.– Tomsk, 2000.– P. 116–124.

[11] Gnuchih V. G. The morphological characteristics and the water regime of flood plains//*Questions of melioration in Tomsk* Ob. Vol. 1. – Tomsk, 1974.– P. 64–79.

[12] Sochava V. B. *Introduction to the study of the geosystems.*– Novosibirsk: *Nauka*, 1978.– 319 p.

[13] Krauklis A. A. Problems of the experimental study of landscapes. – Novosibirsk: *Nauka*, 1979.– 233 p.

[14] Isachenko A. G. *The theory and methodology of geography.* – Moscow: Academy, 2004. – 400 p.

[15] Ryazanov P. N., Surkov V. V. Floodplain natural-territorial complexes of the lower reaches of the river Tom and some of the trends of their change//*Geography and natural resources.* – 1986. № 1. – P. 59–65.

[16] Chernov A. V. River floodplains – their origin, development, and optimum utilization of//*Soros Educational Journal*. – 1999. – №12. – P. 47–54.
[17] Kolesnikova O. N. Structure and dynamics of the floodplain landscape on the example of floodplains southeast taiga zone of West Siberian Plain: PhD Dissertation. – Tomsk, 1988. – 237 p.
[18] Kuminova A. V. *The Vegetation of Kemerovo Region*. – Novosibirsk, 1949. – 167 p.
[19] Kamenskov U. I. *Channel and floodplain processes*. – Tomsk, 1987. – 171 p.

In: Riparian Zones
Editor: Oleg S. Pokrovsky

ISBN: 978-1-63484-613-4

Chapter 13

BENTHIC INVERTEBRATE COMMUNITY FLOODPLAIN-RIVER SYSTEM BASIN VASYUGAN (MIDDLE OB): CONSEQUENCES OF OIL FIELD EXPLORATION

***D. S. Vorobiev*[1,*]*, Y. A. Noskov*[1]*, V. K. Popkov*[1]
and A. I. Ruzanova[1]
[1]Institute of Biology and BIO-GEO-CLIM Laboratory,
Tomsk State University, Tomsk, Russia

ABSTRACT

This study describes the benthic communities of different types of reservoirs in the Vasyugan River floodplain-river system. Twenty-one macroinvertebrate taxonomic groups of animals were found in the Vasyugan basin during the period of study (1992-2002). The most frequently encountered groups were Chironomidae (90%), Oligochaeta (72%) and Mollusca (57%). In the floodplain of the Vasyugan river basin the most productive floodplain ponds had maximum values of abundance and biomass of benthic organisms (2527ind./m^2; 11.74g/m^2); whereas the smallest quantitative indexes were in the river bed (927ind./m^2; 2.87g/m^2). Among the different types of sediment, the largest community development was observed in the silt (2468ind./m^2; 9.31g/m^2) and detrital-silty areas (2621ind./m^2; 9.12g/m^2); The lowest abundance and biomass of benthic organisms accounted for sediments of sand (672ind./m^2, 1.58g/m^2) and clay (433ind./m^2; 1.79g/m^2).

The sediments of water bodies in the territory of oil fields were significantly different ($p < 0.05$) on concentrations of petroleum products. detrital (370mg/kg), silt (89mg/kg) and silty-sand (37mg/kg). A strong correlation ($r = 0.94$) between the concentration of oil products and the content of total organic matter in the sediments was found. Productivity of bottom cenoses in subordinate and flood waters below the oil fields decreased more than twofold as a result of the development of hydrocarbon deposits in the Vasyugan basin.

[*] E-mail: danilvorobiev@yandex.ru.

Keywords: Western Siberia, Vasyugan River, Benthic invertebrate community, oil contamination

Introduction

Intensive pollution of rivers in the northern territory of Tomsk Region and the lower part of the Ob River is connected with the activities of oil producing companies. The main source of pollution of the river systems is oil spills. The Vasyugan river basin is experiencing strong anthropogenic pressure, which is expressed in oil pollution of soil and water systems (Vorobiev and Noskov 2015a). Many water bodies polluted with oil are not cleaned since there are no standards for the oil and hydrocarbon content in bottom sediments (Vorobiev and Noskov 2015b).

Due to the limited lability, long life cycle and accumulation of oil in the bottom layers of water and sediment, benthic organisms are at a disadvantage when there is a contamination at the water bodies (Mikhailova 1987, Belitskaya et al. 1999, Mochalova et al. 2002). Quality criteria of bottom sediments should include indicators that take into account the quantitative development of zoobenthos (Vinogradov et al. 2002).

Ecological monitoring of bottom sediments, using indices of abundance and biomass of major taxonomic groups of benthic communities, is essential in the period of intensified monitoring activities in the areas of oil production. It should be noted that benthic invertebrates, especially aquatic oligochaetes, actively participate in the self-purification of bottom sediments from oil (Vorob'ev et al. 2010, Vorobiev 2013, Vorobiev et al. 2014).

The aim of this work was to study the distribution and spatial-temporal dynamics of indices of abundance and biomass of benthic communities of the reservoirs of the Vasyugan basin and its change as a result of oil pollution for use in environmental monitoring, notably its floodplain and riparian zone.

The objectives were:

1. To study the distribution of benthic organisms on the types of water bodies and habitats in the floodplain of the Vasyugan river basin system;
2. To investigate the dynamics of abundance and biomass of benthos of the Vasyugan floodplain-river system and identify the factors influencing them;
3. To study the effects of oil pollution on indicators of abundance and biomass of benthic communities of the floodplain-river basin and the riparian zone of the Vasyugan river system.

To this end, we describe the distribution of groups of macrozoobenthos in different types of water bodies and habitats in the floodplain of the Vasyugan river basin system. A correlation between the abiotic factors of aquatic environment (water and temperature regimes) and quantitative indicators of benthic communities in the floodplain-river system of the lower portion of the Vasyugan River was detected.

Using analysis of variance the effect of oil pollution, we revealed the quantitative characteristics of certain groups of benthic communities in the Vasyugan River floodplain-river system.

1. Study Sites, Sampling and Analyses

The materials referred to are the result of hydro-ecological research of the Vasyugan river basin from 1992 to 2002. Over the entire period of the investigation 529 samples of macrozoobenthos in heterogeneous reservoirs of the Vasyugan river basin (Table 1.1; Figure 1.1) were collected. Each sample included the contents of the bottom grab 2-4.

To assess the impact of oil on benthic communities, 37 sediment samples for the determination of petroleum products and total organic matter were collected. The samples were processed in the hydrochemical laboratory of Tomsk Oil and Gas Research and Design Institute using IR spectrophotometry. To assess the degree of contamination of sediment in petroleum products used grading proposed by Uvarova (1989):

up to 5.4 mg/kg - clean;
5.5 - 25.5 mg/kg - lightly polluted;
25.6 - 55.5 mg/kg - moderately polluted;
55.6 - 205.5 mg/kg - polluted;
205.6 - 500 mg/kg - dirty;
more than 500 mg/kg - very dirty.

The types of water bodies were determined according to the typology proposed by Ioganzen (1951) for the Middle Ob basin: river waters subordinate and flood waters.

Observational data from hydrometeorological posts located in the basin of the Vasyugan River were provided by the Kolpashevo zonal hydrometeorological observatory.

For sampling of zoobenthos we used a Petersen grab with an area of capture $1/80m^2$. The primary analysis of the samples was carried out immediately with a magnifying glass and tweezers. Organisms of zoobenthos were fixed with 70% of alcohol, labeled and packed for transportation. Further processing of the material was carried out in laboratory conditions. The characteristics of benthic animals were quantified as to the abundance and biomass in $1m^2$, and the number of environmental groups (S).

To determine the occurrence of organisms of zoobenthos we used the expression: V = v/n, where v is the number of samples where the Group was registered; n is the total number of samples. "Oligohaeta" Goodnight-Whytley index (Goodnight and Whitley 1961) was used as an indicator of bioindication.

Table 1.1. Number of selected hydro-biological samples in flood-basin of the Vasyugan river system

Type of reservoir	Number of sample
The bed of the river	186
Ducts (secondary channels)	34
Tributaries I, II, III, and other orders (up to 58*)	210
Subordinate water bodies (inlets, bays)	23
Floodplain lakes	76
Total	529

Note: * - number of investigated tributaries I, II, III, and other orders.

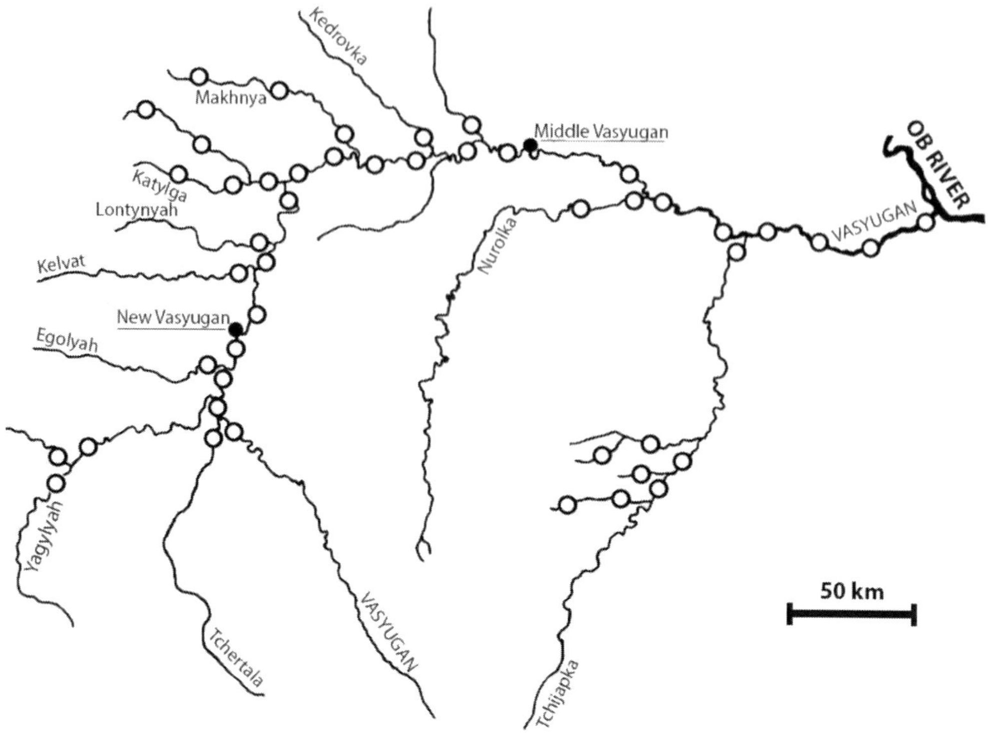

Figure 1. Location of hydrobiological sampling in the floodplain and riparian zone of the Vasyugan river system.

Processing of the material was carried out with the optics: MBS - 9. The organisms were weighed on the torsion balance "BT."

The statistical comparison t-test, variance and correlation analysis was performed using the software package STATISTICA® for Windows, pre-processing of the data and descriptive statistics - in the tabular processor of Microsoft ® EXCEL.

RESULTS

2. Benthic Communities and Their Distribution by Types of Reservoirs

Quantitative indicators of benthic organisms' development in the flood plain and river systems depend on many different factors prevailing in the particular type of reservoir. The main features of river water are a constant water flow, homothermia, the presence of dense sediments (sandy or clay-various extent of silting). A duct is the first stage of a parting of a reservoir flood plane, an intermediate link between the riverbed and the subordinate water body. The ducts are shallower (compared to the riverbed), and are on the initial stages of silting of sediments due to the decrease of the flow rate. Subordinate waters are waters that are always associated with the river flow; they represent the transitional stage of the river flow to the floodplain lakes. These reservoirs have the features of the two systems, approaching the river (during the flood), or the lake (in the autumn). The main features are absence of flow, intensive processes of sedimentation, and presence of aquatic vegetation. Floodplain lakes are reservoirs which interact with the river for a short period, mainly during the spring flood. They are presented at various levels of flooding reservoirs.

2.1. River Waters

Throughout the Vasyugan River we encountered 13 taxonomic groups of benthic animals (Table. 2.1). On the upper site of the river the smallest variety of sediment types (habitat) were observed, which is typical for lowland rivers of the Middle Ob basin, where the upper sites of large rivers' erosion prevails over accumulation. The main biotope here is clean sands areas, where the number of organisms of macrozoobenthos was 847 (100–2108)ind./m^2 and biomass was minimum – 0.9551 (0.0300–2.2000)g/m^2. Chironomid larvae dominated quantitatively (77%) and biomass dominated by bivalve mollusks (74%). The biotope of pure clay is interesting because at a minimum number of benthic animals – 240ind./m^2, it has a maximum biomass – 5.584g/m^2 due to the predominance of large mollusks in the biotope (95%). In the areas of silted sand the quantitative characteristic increased up to 1213 (1000–1360) ind./m^2 and 1.0227 (0.8600–1.2760)g/m^2 on average, the base of the benthos here, both in number (74%), and biomass (40%) were Chironomid larvae.

At the middle site of the river Vasyugan 9 taxonomic groups were found (Table 2.1). The greatest development, both in abundance and biomass reached the benthos in the areas of pure silt – 1999 (0–6595) ind./m^2 and 8.1721 (0–64.1870)g/m^2, which was dominated by the number of Oligochaetes (45%) and Chironomids larvae (33%). The base of biomass were mollusks (67%), subdominant group were Oligochaetes (24%). The second highest productivity areas were silted sand. The average abundance and biomass amounted to 1235 (40–3080)ind./m^2 and 1.3830 (0.0200–5.1200)g/m^2. It is necessary to note the high occurrence of Chironomid larvae (94%) in this biotope, which hold the lead in terms of quantity in the bottom community, accounting for 65% of the population and 39% of the biomass. In this biotope, as in the biotope of pure silt, can be noted the high diversity of benthic animals – 9 taxonomic groups. In areas of sand massively developed Chironomid larvae, making 82% of population and 40% of the biomass of bottom animals. The poorest abundance and biomass of benthos was presented in the middle site of Vasyugan in the biotope of clay, where the quantitative characteristic reached the minimum value – 240ind./m^2 and 0,050g/m^2 and only two taxonomic groups of benthic organisms were found – Chironomidae and Heleidae.

The lower site of the Vasyugan River had a very diverse benthic population, where 12 groups of animals were encountered (Table 2.1). Less productive areas were silted sands, where quantitative characteristics of zoobenthos were 959 (80–3280)ind./m^2 and 2.4349 (0.1000–11.7880)g/m^2. Chironomids dominated in numbers, accounting for 66% of the total benthos abundance. Biomass was dominated by Shellfishes (53%) and Chironomid larvae (28%). Relatively large size of the biomass at low abundance was in areas of sand. Here the communities with the numerical dominance of chironomids (46%) and mollusks, which occupied 83% of the biomass of benthos developed. The average abundance of benthic animals accounted for 334 (0–4160)ind./m^2 and 1.8126 (0–60.4800)g/m^2.

2.2. The Ducts (Secondary Channels)

In the ducts of the lower site of the Vasyugan River 11 taxonomic groups of benthic animals were found (Table. 2.2).

Table 2.1. Quantitative measures of macrozoobenthos in the Vasyugan River. The dominants are given in bold

Group	Abundance, ind./m^2		Biomass, g/m^2		W, mg	V, %
	N ± m	%	B ± m	%		
			Upper site			
Nematoda	3 ± 3	0.35	0.0009 ± 0.0008	0.06	0.300	7
Oligochaeta	46 ± 16	5.32	0.1040 ± 0.0413	6.50	2.261	**53**
Hirudinea	3 ± 3	0.35	0.0015 ± 0.0013	0.09	0.500	7
Acarina	4 ± 3	0.46	0.0033 ± 0.0027	0.21	0.825	13
Ephemeroptera	8 ± 6	0.92	0.0428 ± 0.0350	2.68	5.350	13
Trichoptera	4 ± 3	0.46	0.0152 ± 0.0133	0.95	3.800	13
Chironomidae	561 ± 141	**64.85**	0.1830 ± 0.0438	**11.45**	0.326	**100**
Heleidae	126 ± 44	**14.57**	0.0322 ± 0.0115	2.01	0.256	**67**
other Diptera	3 ± 3	0.35	0.0027 ± 0.0027	0.17	0.900	7
Mollusca	107 ± 34	**12.37**	1.2132 ± 0.4120	**75.88**	11.338	**73**
Total	**865** ± 157		**1.5988** ± 0.4309			
			Middle site			
Nematoda	2 ± 1	0.15	0.0018 ± 0.0012	0.06	0.900	5
Oligochaeta	353 ± 127	**26.58**	0.7350 ± 0.2898	**23.08**	2.082	**60**
Hirudinea	1.29 ± 0.90	0.10	0.0040 ± 0.0034	0.13	3.101	3
Ephemeroptera	1.90 ± 1.43	0.14	0.0030 ± 0.0022	0.09	1.579	3
Trichoptera	15 ± 6	1.13	0.0800 ± 0.0548	2.51	5.333	13
Chironomidae	713 ± 142	**53.69**	0.3775 ± 0.1048	**11.85**	0.529	**94**
Heleidae	140 ± 23	**10.54**	0.0390 ± 0.0064	1.22	0.279	**73**
Chaoborinae	2.81± 1.24	0.21	0.0105 ± 0.0054	0.33	3.737	8
Mollusca	99 ± 38	7.45	1.9344 ± 0.9988	**60.73**	19.539	34
Total	**1328** ± 198		**3.1852** ± 1.1821			
			Lower site			
Nematoda	1.59 ± 1.15	0.22	0.0010 ± 0.0008	0.03	0.629	3
Oligochaeta	215 ± 52	**30.41**	0.5080 ± 0.1331	**17.78**	2.363	**57**
Hirudinea	1.57 ± 0.78	0.22	0.0065 ± 0.0042	0.23	4.140	4
Acarina	0.59 ± 0.44	0.08	0.0002 ± 0.0001	0.01	0.339	2
Coleoptera	0.36 ± 0.36	0.05	0.0007 ± 0.0007	0.02	1.944	1
Ephemeroptera	1.07 ± 0.82	0.15	0.0022 ± 0.0016	0.08	2.056	2
Simullidae	1.82 ± 0.81	0.26	0.0025 ± 0.0011	0.09	1.374	5
Trichoptera	23 ± 9	3.25	0.0520 ± 0.0260	1.82	2.261	14
Chironomidae	330 ± 51	**46.68**	0.5660 ± 0.2163	**19.81**	1.715	**84**
Heleidae	51 ± 10	7.21	0.0147 ± 0.0025	0.51	0.288	**41**
Chaoborinae	4 ± 2	0.57	0.0130 ± 0.0072	0.45	3.250	4
Mollusca	77 ± 15	**10.89**	1.6906 ± 0.5739	**59.17**	21.956	**44**
Total	**707** ± 96		**2.8574** ± 0.6849			

Group	Abundance, ind./m²		Biomass, g/m²		W, mg	V, %
	N ± m	%	B ± m	%		
Vasyugan river (all sites)						
Nematoda	1.84 ± 0.79	0.20	0.0011 ± 0.0006	0.04	0.598	4
Oligochaeta	247 ± 52	**26.65**	0.5513 ± 0.1243	**19.24**	2.232	**58**
Hirudinea	1.67 ± 0.58	0.18	0.0052 ± 0.0027	0.18	3.114	4
Acarina	0.73 ± 0.35	0.08	0.0003 ± 0.0002	0.01	0.411	2
Coleoptera	0.27 ± 0.22	0.03	0.0004 ± 0.0004	0.01	1.481	1
Ephemeroptera	1.94 ± 0.82	0.21	0.0056 ± 0.0031	0.20	2.887	3
Simullidae	1 ± 0.48	0.11	0.0014 ± 0.0007	0.05	1.400	3
Trichoptera	19 ± 6	2.05	0.0584 ± 0.0237	2.04	3.074	13
Chironomidae	477 ± 58	**51.46**	0.4725 ± 0.1315	**16.49**	0.991	**89**
Heleidae	86 ± 11	9.28	0.0242 ± 0.0029	0.84	0.281	**54**
Chaoborinae	3.28 ± 1.38	0.35	0.0112 ± 0.0046	0.39	3.415	5
other Diptera	0.27 ± 0.22	0.03	0.0002 ± 0.0002	0.01	0.741	1
Mollusca	87 ± 16	9.39	1.7333 ± 0.4728	**60.50**	19.923	**43**
Total	**927 ± 90**		**2.8651 ± 0.5622**			

Note. N - (ind./m²); B - (g/m²); m - a mistake; W - the average individual weight of the body; V - occurrence.

Table 2.2. Quantitative measures of macrozoobenthos in the ducts of the Vasyugan River. The dominants are given in bold

Group	Abundance, ind./m²		Biomass, g/m²		W, mg	V, %
	N ± m	%	B ± m	%		
Nematoda	8 ± 4	0.47	0.0015 ± 0.0008	0.02	0.188	14
Oligochaeta	968 ± 368	**56.81**	2.6313 ± 1.0420	**36.64**	2.718	**76**
Hirudinea	4.52 ± 2.30	0.27	0.0138 ± 0.0100	0.19	3.053	10
Acarina	1.48 ± 1.48	0.09	0.0021 ± 0.0021	0.03	1.419	3
Simullidae	11 ± 8	0.65	0.0152 ± 0.0102	0.21	1.382	10
Trichoptera	17 ± 10	1.00	0.2207 ± 0.1610	3.07	12.982	17
Chironomidae	394 ± 79	**23.12**	0.5673 ± 0.1559	7.90	1.440	**93**
Heleidae	43 ± 17	2.52	0.0155 ± 0.0057	0.22	0.360	**38**
Chaoborinae	16 ± 12	0.94	0.0536 ± 0.0397	0.75	3.350	10
other *Diptera*	3 ± 3	0.18	0.0014 ± 0.0014	0.02	0.467	3
Mollusca	238 ± 72	**13.97**	3.6584 ± 1.0547	**50.95**	15.371	**79**
Total	**1704** ± 394		**7.1808** ± 1.4665			

The richest benthic communities of the ducts were presented in muddy areas where the abundance and biomass of organisms in average was 2623 (120–10440)ind./m² and 11, 9302 (0.52–26.4)g/m², respectively. Development of the benthos in sands had a minimum value of 417 (27–1600)ind./m² (biomass – 1.175 (0.0080–5.5120)g/m²), where the dominant benthic organisms were Oligochaeta (43%), not very typical for this type of biotope. The biomass formed three groups: Chironomids (40%), Shellfishes (32%) and Oligochaeta (28%). Other types of sediment habitat are very rare.

2.3. The Tributaries of the Vasyugan River

Over the entire investigation period in 58 of the tributaries of the Vasyugan River 21 taxonomic groups of macrozoobenthos were found (Table. 2.3). In the tributaries the biggest variety of habitats was observed, which was allocated to 11 types of sediments. The most frequent sediments were sand, silt and silty-sandy.

A major role in the tributaries of habitats with the contents of detrital fractions (detrital, detrital-sandy, muddy-detrital, detrital-clay). The greatest quantitative development benthic communities reached on detrital-muddy areas where the maximum average abundance (4623 (20–22080)ind./m^2) and biomass (11.6090 (0.01–37.42)g/m^2) of bottom animals in the tributaries were registered.

In the areas of silty sands the number of benthic organisms was 2307 (80–12000) ind./m^2 and 5.306 (0.44–28.34)g/m^2. The basis of the population was formed by Chironomids (67%) and the biomass was dominated by Mollusks (50%), subdominant were Chironomids (19%) and Oligochaeta (17%). On detrital sediments the number of benthic organisms amounted to 2358 (80–8880)ind./m^2 and 2.1444 (0.32–4.32)g/m^2. On the sands the abundance of zoobenthos was 1024 (0–4200)ind./m^2 and 2.06 (0–8.37)g/m^2. The high number of organisms formed by Chironomid larvae (46%) and Oligochaeta (29%) and biomass – by Molluscs (53%) and Oligochaeta (28%). The lowest quantitative indicators in the tributaries of the Vasyugan River were on sediments of clay.

Here, the number of benthic organisms was 614 (133–1600)ind./m^2 and 1.1502 (0.08–4.62)g/m^2.

2.4. Subordinate Reservoirs

In the subordinate reservoirs of the Vasyugan River we found 11 taxonomic groups of benthic organisms (Table 2.4). The sediments were mostly silts.

2.5. Floodplain Lakes

The most developed system of flood waters is in the middle and lower reaches of the Vasyugan River. In the flood waters of the middle site of the Vasyugan we found 6 taxonomic groups of zoobenthos (Table 2.5). In the floodplain lakes only muddy sediments were registered. Higher quantitative indicators were in the flood waters of the lower site, where the benthos consisted of 12 groups (Table 2.6). In this area, as well as in the middle site, only silt sediments were registered.

3. Interannual Dynamics of Benthic Communities in the Floodplain of River System of Lower Vasyugan

3.1. The Vasyugan Riverbed (Lower Site)

In connection with a clear predominance of sandy soils in the lower site of the Vasyugan, the interannual dynamics of benthos in the sandy sediments have been considered.

Table 2.3. Quantitative measures of macrozoobenthos in the tributaries of the Vasyugan River

Group	Abundance, ind./m²		Biomass, g/m²		W, mg	V, %
	N ± m	%	B ± m	%		
Nematoda	2.21 ± 0.71	0.09	0.0010 ± 0.0004	0.02	0.452	5
Oligochaeta	635 ± 96	**25.94**	1.7055 ± 0.2934	**26.44**	2.686	**78**
Hirudinea	12 ± 3	0.49	0.0817 ± 0.0238	1.27	6.808	13
Amphipoda	8.56 ± 7.21	0.35	0.0646 ± 0.0548	1.00	7.547	2
Araneina	0.35 ± 0.25	0.01	0.0002 ± 0.0002	0.00	0.571	>1
Acarina	8.59 ± 4.30	0.35	0.0022 ± 0.0011	0.03	0.256	6
Odonata	0.56 ± 0.44	0.02	0.1856 ± 0.1811	2.88	331.429	>1
Ephemeroptera	23 ± 9	0.94	0.0244 ± 0.0073	0.38	1.061	12
Trichoptera	17 ± 8	0.69	0.0434 ± 0.0145	0.67	2.553	9
Coleoptera	3.45 ± 1.68	0.14	0.0969 ± 0.0922	1.50	28.087	4
Chironomidae	1443 ± 221	**58.95**	1.4391 ± 0.3343	**22.31**	0.997	**91**
Heleidae	81 ± 11	3.31	0.0374 ± 0.0062	0.58	0.462	**51**
Chaoborinae	3.16 ± 1.74	0.13	0.0093 ± 0.0053	0.14	2.943	2
Limoniidae	9 ± 8	0.37	0.0336 ± 0.0300	0.52	3.733	3
Simulidae	0.52 ± 0.52	0.02	0.0006 ± 0.0006	0.01	1.154	1
Tabanidae	3.48 ± 1.18	0.14	0.0245 ± 0.0101	0.38	7.040	6
Lepidoptera	0.56 ± 0.33	0.02	0.0628 ± 0.0576	0.97	112.143	2
Plecoptera	0.35 ± 0.25	0.01	0.0003 ± 0.0002	0.00	0.857	>1
other Diptera	0.21 ± 0.21	0.01	0.0027 ± 0.0027	0.04	12.857	>1
Megaloptera	14 ± 5	0.57	0.0753 ± 0.0265	1.17	5.379	8
Mollusca	182 ± 22	7.43	2.5603 ± 0.5243	**39.69**	14.068	**62**
Total	**2448** ± 271		**6.4514** ± 0.8009			

Note. See note for table 2.1.

Table 2.4. Quantitative measures of zoobenthos of subordinate reservoirs of the Vasyugan river basin (lower site)

Group	Abundance, ind./m²		Biomass, g/m²		W, mg	V, %
	N ± m	%	B ± m	%		
Nematoda	7.4 ± 4.1	0.56	0.0030 ± 0.0022	0.05	0.405	17
Oligochaeta	380 ± 122	**28.99**	0.9641 ±0.3520	17.14	2.537	**94**
Hirudinea	2.7 ± 1.8	0.21	0.0644 ± 0.0456	1.14	23.852	11
Acarina	4.5 ± 4.5	0.34	0.0089 ± 0.0089	0.16	1.978	6
Isopoda	118 ± 118	9.00	0.1600 ± 0.1600	2.84	1.356	6
Trichoptera	4.5 ± 4.5	0.34	0.0178 ± 0.0178	0.32	3.956	6
Chironomidae	547 ± 158	**41.72**	1.3211 ± 0.4031	23.48	2.415	**94**
Heleidae	50 ± 22	3.81	0.0182 ± 0.0076	0.32	0.364	39
Chaoborinae	15.6 ± 9.8	1.19	0.0711 ± 0.0509	1.26	4.558	17
other Diptera	2.3 ± 2.3	0.18	0.0011 ± 0.0011	0.02	0.478	6
Mollusca	179 ± 59	13.65	2.9960 ± 1.3390	**53.26**	16.737	**89**
Total	**1311** ± 239		**5.6257** ± 1.5098			

Note. See note for Table 2.1.

Table 2.5. Quantitative measures of zoobenthos flood waters of the middle site of the Vasyugan River

Group	Abundance, ind./m^2		Biomass, g/m^2		W, mg	V, %
	N ± m	%	B ± m	%		
Nematoda	17 ± 17	0.85	0.0122 ± 0.0122	0.14	0.718	9
Oligochaeta	858 ± 594	**42.88**	2.2238 ± 1.5379	26.36	2.592	**91**
Chironomidae	778 ± 400	38.88	4.5167 ± 2.8338	**53.54**	5.806	**91**
Heleidae	14 ± 10	0.70	0.0439 ± 0.0246	0.52	3.136	27
Chaoborinae	304 ± 153	15.19	0.9407 ± 0.3794	11.15	3.094	**64**
Mollusca	30 ± 27	1.50	0.6995 ± 0.5664	8.29	23.317	18
Total	**2001** ± 803		**8.4368** ± 4.2313			

Note. See note for Table 2.1.

Table 2.6. Quantitative measures of zoobenthos flood waters of the lower site of the Vasyugan River

Group	Abundance, ind./m^2		Biomass, g/m^2		W, mg	V, %
	N ± m	%	B ± m	%		
Nematoda	2.1 ± 1.10	0.08	0.0010 ± 0.0006	0.01	0.476	7
Oligochaeta	1623 ± 378	**58.49**	5.6656 ± 1.3725	**46.03**	3.491	**89**
Hirudinea	6.67 ± 2.85	0.24	0.0693 ± 0.0397	0.56	10.390	14
Acarina	2.19 ± 1.14	0.08	0.0026 ± 0.0016	0.02	1.187	7
Ephemeroptera	0.32 ± 0.32	0.01	0.0107 ± 0.0107	0.09	33.438	2
Trichoptera	0.43 ± 0.43	0.02	0.0062 ± 0.0062	0.05	14.419	2
Coleoptera	3.48 ± 2.52	0.13	0.2136 ± 0.1524	1.74	61.379	4
Chironomidae	692 ± 135	24.94	3.51 ± 0.698	28.55	5.079	**91**
Heleidae	87 ± 30	3.14	0.0590 ± 0.0157	0.48	0.678	**59**
Chaoborinae	68 ± 19	2.45	0.1696 ± 0.0444	1.38	2.494	34
Megaloptera	0.81 ± 0.81	0.03	0.0071 ± 0.0071	0.06	8.765	2
Mollusca	289 ± 50	10.41	2.59 ± 0.384	21.04	8.963	77
Total	**2775** ± 408		**12.31** ± 1.42			

Note. See note for Table 2.1.

Comparing the quantitative indices of zoobenthos of lower site of the Vasyugan from samples of different years (1996 – 2001), revealed that abundance and biomass of the benthos subjected to significant annual changes (Table 3.1). These changes are associated with the development of the dominant core of bottom communities (oligochaetes, mollusks and larvae of chironomids).

The peaks of the benthos were caused by the development of oligochaetes and chironomid larvae, which formed the basis of the population in those years (1996 and 2000).

In order to study the relationship between quantitative indices of benthos in sandy sediments with abiotic factors (temperature, maximum and annual average water level, and the maximum water level of the previous year) a correlation analysis were performed (Table 3.2).

ANOVA showed a significant nonlinear effect of the maximum and average annual water level on the number of benthic organisms according to the years 1996-2001 ($p < 0.001$). A significant influence of the factors on the total biomass of benthos was not found.

Table 3.1. The interannual changes in abundance and biomass of benthos in sandy sediments of the lower site of the Vasyugan

1996		1997		1998		1999		2000		2001	
N	B	N	B	N	B	N	B	N	B	N	B
784	2.14	190	0.32	80	0.10	96	0.09	621	0.75	102	0.88

Note. N – abundance (ind./m^2); B – biomass (g/m^2).

Table 3.2. Correlation of some abiotic factors with quantitative indicators of zoobenthos on sandy sediments of the Vasyugan River (data of 1996-2001.)

Characteristic	Total benthos		Oligochaetes		Chironomids		Mollusks	
	N	**B**	**N**	**B**	**N**	**B**	**N**	**B**
The average annual water level	-0.26	0.15	-0.02	-0.03	-0.28	-0.07	-0.11	0.19
Maximum water level	-0.24	-0.14	-0.02	-0.01	-0.24	0.01	-0.09	-0.14
Average water temperature (May-July)	0.13	-0.09	-0.19	-0.19	**0.32**	-0.24	0.01	0.02
Average water temperature of previous year (May-October)	0.27	**0.36**	**0.32**	0.29	0.12	**0.38**	0.02	0.19
Maximum water level of previous year	-0.05	-0.28	**-0.37**	**-0.34**	0.22	**-0.45**	-0.04	-0.08

Note to Table 3.2: N is the abundance, ind./m^2; B is the biomass, g/m^2;
* significant correlation coefficients ($p<0.05$) are highlighted in bold.

3.2. Long-Term Dynamics

Research into the benthic communities of the Vasyugan basin was conducted in the 1970s (1971-1973, 1977) before the exploitation of hydrocarbons in the basin served as "background" data for the study and analysis of successional processes occurring in aquatic ecosystems of basin under influence of hydrocarbons. During the study period (1993-2001) in terms of abundance and biomass of benthic organisms the most productive waters were floodplain ones (Figure 3.1).

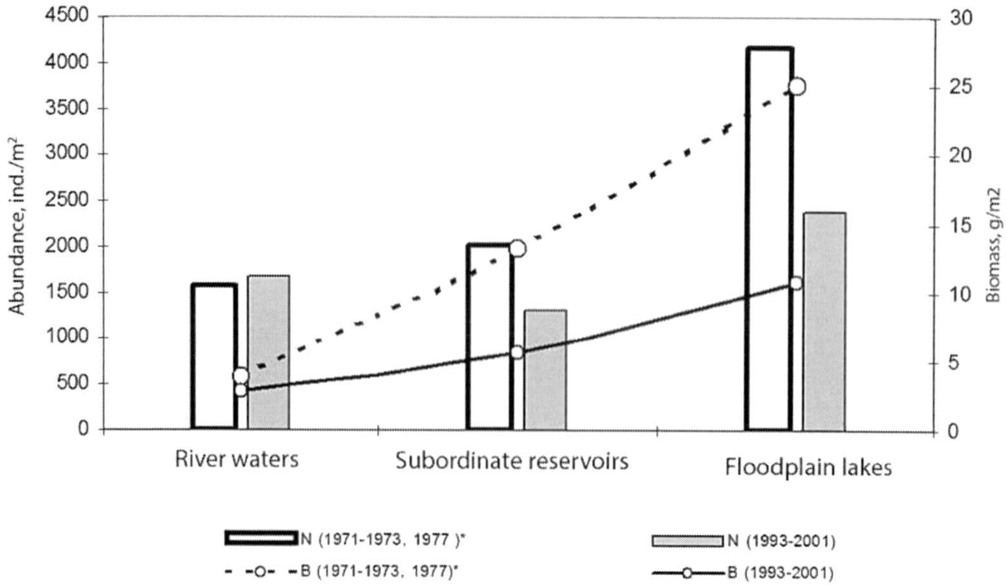

Figure. 3.1. Indicators number (N) and biomass (B) of bottom community of different water bodies of the Vasyugan floodplain-river system; * - Before the beginning of the exploitation of oil deposits.

The lowest biomass of zoobenthos in 1993-2001 noted in the subordinate system of reservoirs. In the studied period, when the oil fields in the floodplain of the river have already been exploited, in the benthic communities of flood waters the productivity of cenoses sharply decreased, and the diversity of the population decreased. In place of mollusks, previously leading in the Vasyugan floodplain, domination passed to oligochaetes, capable of withstanding significant oil pollution. Depressed benthic communities of Vasyugan floodplain waters show a significant accumulation of oil components in sediments that suppress the development of benthic organisms.

4. Benthic Communities under Oil Contamination

4.1. Oil Products in the Bottom Sediments of the Vasyugan River Basin

The development of oil fields in the Vasyugan basin started in the 1970s (Oil and Gas of Tomsk Region, 1988). During the decades of their exploitation a large number of rivers and streams of the basin were polluted. In the summer of 2000 samples of bottom sediments from the Vasyugan and its tributaries flowing through the territory of oil fields were taken for analysis of oil content (Figure 4.1).

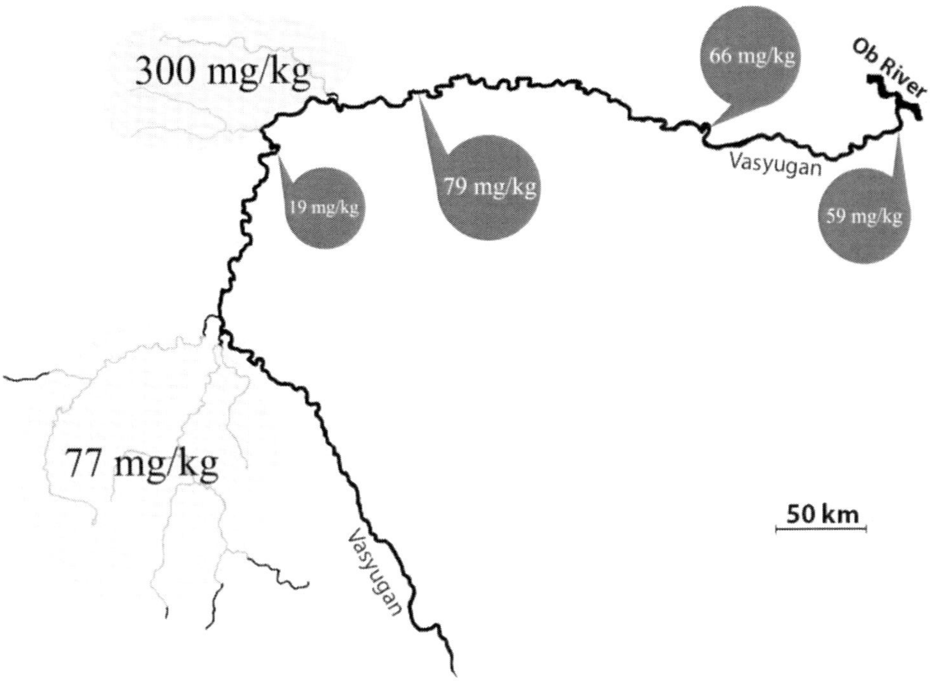

Figure. 4.1. The content of oil in the bottom sediments of the Vasyugan river basin (based on sampling of 2000). The gray background marks the territory of oil fields.

Data analysis showed that different parts of the basin are quite different in content of oil products in bottom sediments.

Streams flowing through the territory of the oil fields on average have significantly higher oil content in the sediments. In the basins of the tributaries of the middle Vasyugan are the oldest deposits (catchment areas of the rivers Katylga, Makhnya, Cheremshanka), the development and exploration of which was carried out over 30 years. These streams are the main carriers of petroleum hydrocarbons that pollute downstream areas of rivers. The concentration of oil in the sediments of these areas is on average 300mg/kg. According to the classification of Uvarova (1989) the sediments of the area should be classified as "dirty."

Oil exploration in the river basins of Yagylyah and Chertala began relatively recently, and the oil concentration in the samples of bottom sediments was 4 times lower (77mg/kg (classified as contaminated)). Oil content in the bottom sediments of the rivers flowing through the territory of the field mainly depends on the duration of their exploration. Other authors also note that "*the highest degree of contamination is registered on the fields, the exploration life of which exceeds 15-20 years*" (Remorov et al. 1998, p.106). Contamination of lands and waters occurs due to the high accident rate of "old" deposits. In 1996 on the territory of Pervomaisky oil field NGDU "Vasyuganneft," which was put into active operation in 1984, in one year 127 accidents occurred, and in 1994 – 195 (Bagautdinov et al. 1998).

There are three main sources of oil pollution of lands: (1) oilfield bushes; (2) infield pipelines and (3) main pipelines. The most intense flushing oil into the rivers occurs during the spring flood.

Above the mouth of the Katylga River the concentration of oil in the sediments was only 19mg/kg (classified as slightly polluted soils). Below the influx of the rivers Katylga, Makhna, Cheremshanka, Kedrovka the concentration of petroleum hydrocarbons increased fourfold – up to 79mg/kg (contaminated). Obviously, such an increase in a relatively small part of the river (within 100km) was due to entering the waters above the rivers. Furthermore, in the lower part of the river (in 250-400km below the influx of polluted tributaries), the oil content in the sediments had a mildly declining trend. In the area of the mouth of the Nyurolka – 66mg/kg (moderately polluted) and Staroyugino village – 59mg/kg (moderately polluted). According to Turov et al. (2002), in 2000, at the mouth of the Vasyugan River the concentration of oil products amounted to 55mg/kg, which is consistent with our data. According to Uvarova (2000), the oil content in the sediments of the Ob River in 1995-1998 was on average: near Nizhnevartovsk – 28mg/kg, in the district of Surgut – 33mg/kg. These concentrations are 2-3 times lower than in the Vasyugan River in 2000. Analyzing the data on the content of oil products in the water and bottom sediments of the water bodies of the Vasyugan basin, it should be noted that petroleum contamination spots continue to exist and contaminate the Vasyugan floodplain river systems. The main amount of oil is concentrated in the sediments of oil-producing areas, and has a marked tendency to decrease slightly downstream.

4.2. Benthic Communities of the Vasyugan River and the Influence of Oil Fields Exploration

In order to study the effect of exploration of oil fields on quantitative indicators of the Vasyugan's benthic communities we divided the Vasyugan into 3 sections (Figure 4.2): 1 – the section above the mouth of the Katylga River (relatively clean); 2 – from the mouth of the Katylga to N. Tevriz village and 3 – the lower part of the river to the mouth of Nyurolka river; Secondly, we chose and divided samples according to the type of bottom sediments: sandy, silty-sandy and silty. To determine the effect of oil pollution on quantitative and bioindicative ("oligochaetes" index) indicators of benthic organisms ANOVA was used.

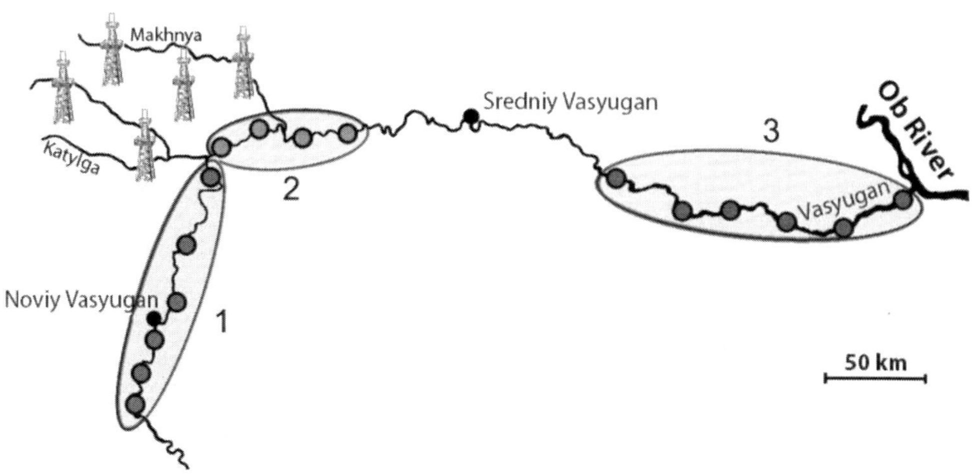

Figure. 4.2. The parts of the Vasyugan River with different anthropogenic impacts: 1 – above the mouth of the Katylga River (conditionally clean); 2 – from the Katylga River to the village of N. Tevriz (100 km section near the sources of pollution); 3 – from Nyurolka River to the mouth of the Vasyugan (the part away from the pollution sources for 300-450km).

A significant difference of three sites on sandy soils was found only in terms of the total number of benthic species ($n = 72$; $p < 0.05$) and chironomid larvae ($n = 72$; $p < 0.05$).

It clearly shows a reduced number of chironomid larvae, and, because of their dominance, the total number of benthos. The structure of benthic community was also affected. The role of chironomids in the communities was noted to reduce. In a relatively clean area (above the river Katylga), the share of the population was 86%, which then declined to 74%, and in the lower site declined to 46%. An increase of the role of Oligochaets in the total number of benthos should be noted (up to 3%, 6% and 20%, respectively). The main role of biomass in sandy sediments was played by shellfish, increasing its value to the mouths of the river up to 85%.

Changes in the benthic community structure of the Vasyugan River certainly occur due to the influence of contaminated tributaries. Chironomid larvae reduced in abundance and biomass in sandy soils nearer the mouth of the river. Sandy habitats were dominated by a grazing chain: small organisms are little and decompound chains are feeble. Similar changes are taking place as a result of oil pollution of river systems. Oil can be considered as an additional organic – close to sources of pollution the abundance and biomass of benthic organisms is greater on average. There is a change in the structure of benthic communities and a decrease of zoobenthos abundance along with the distance from the source of pollution. The significant role in benthos here is played by the organisms using a filtration type of feeding (shellfish). Oligochaetes are relative resistance organisms to pollutants and also increase their role in the community. Using analysis of variance, any significant effect of oil on the quantitative and structural indeces of the benthos in the areas containing silted sand and silt on three above-mentioned areas of river Vasyugan has not been identified. However, finding no effect of oil pollution on the quantitative indicators of benthic and its constituent groups, the presence of pollutants influenced the bioindicative indicator – "Oligochaet's index." The share of oligochaetes in the number of benthic organisms increased along with the distance from the source of pollution ($n = 40$; $p < 0.01$). This fact explains the great endurance of oligochaetes to pollutants derived from petroleum compared with other groups of benthic organisms.

It should be noted that of all the types of the water bodies of the Vasyugan basin most affected by oil pollution are flood waters, which occurs due to the hydrological characteristics of the accumulation of oil, suppressing the development of the benthic fauna. The most negative impact of oil pollution is on quantitative zoobenthos manifested in sediments with a low organic content (sand sediments). From quantitative indices of bottom biota of the Vasyugan, the total abundance and dominant largest group of chironomids on sandy soils and oligochaetes on silt sediments significantly reflect the degree of anthropogenic load. There is a reduction in the quantitative characteristics of benthos in the sites of the Vasyugan from pollution sources to the mouth, which indicates an increase in the toxic effects of oil components along with the distance from pollution focuses. According to GOST 17.1.4.01-80 in determining the oil, low polar and non-polar hydrocarbons determine the most characteristic components of oil and its processing. Indeed, these fractional components are contained in large amounts in crude oil (from 70% to 90%; Hydrochemical Dictionary, 1988). However, this technique does not consider the contents of the most toxic oil components to aquatic organisms, such as polycyclic aromatic hydrocarbons. According to Belitskaya et al. (1999), studies in the waters of the middle Ob showed accumulation in sediments predominantly PAHs and heavy paraffins (> C20). It is obvious that for environmental monitoring sediment contamination by oil products, it is necessary to take into account the most toxic components of oil.

Conclusion

1. Twenty-one taxonomic groups of macrozoobenthos in the Vasyugan basin were found. The most abundant taxa in the Basin were Chironomidae (90%), Oligochaeta (72%) and Mollusca (57%).
2. Of all the types of water bodies of the Vasyugan floodplain-river system investigated (rivers, ponds clauses, floodplain lakes), the most productive bottom cenosis were floodplain lakes (2527ind./m^2; 11.74g/m^2), and the least – the riverbed (927ind./m^2, 2.87 g/m^2). The largest communities development was observed in silt (2468ind./m^2, 9.31g/m^2) and detrital-muddy areas (2621ind./m^2, 9.12g/m^2). The lowest abundance and biomass was recorded in sand (672ind./m^2, 1.58g/m^2) and clay (433ind./m^2, 1.79g/m^2).
3. The concentration of oil in bottom sediments on the territory of oil fields significantly differed ($p < 0.05$) on: detrital (370mg/kg), silt (89mg/kg) and silty-sandy (37mg/kg). A strong correlation ($r = 0.94$) between the concentration of oil and the total content of organic matter in sediments was revealed.
4. As a result of oil pollution a reduction in the productivity of river floodplain bottom cenosis along with the distance from the source of contamination downstream occur. The Vasyugan riverbed revealed a decrease in the amount of benthic organisms in the sandy sediments from the sources of pollution to the mouth using analysis of variance (1256ind./m^2 (upper site – relatively clean), 605ind./m^2 (middle site) 343ind./m^2 (lower site)); on silty sediments – increasing the “Goodnight-Whitley” oligohaete’s index (0.05 – 0.25 - 0.37, respectively).
5. As a result of the development of hydrocarbon deposits in the Vasyugan basin over two decades, the productivity of bottom cenosis decreased in secondary channels (to 2028ind./m^2 and 13.24g/m^2 to 1310 copies./m2 and 5.63g/m^2) and flood zone (with 4183ind./m^2 and 25.00g/m^2 to 2379ind./m2 and 10.80g/m^2).

Acknowledgments

The investigation was performed as part of research topics of the Department of Ichthyology and Hydrobiology and the BIO-GEO-CLIM Laboratory at the Institute of Biology, Tomsk State University.

References

Bagautdinov, A. K., Mangazeev, V. P., Rastrogin, A. E., 1998. The objectives for improving the development of deposits in connection with the stabilization of oil production at OAO «Tomskneft» VNK. *Bulletin of VNK*, 1, 34–39 (in Russian).

Belitskaya, E. A., Guznyaeva, M. Yu., Kadychagov, P. B., Russkikh I. V., Turov Y. P., 1999. Organic contaminants in the waters of the middle Ob. Ecology of the Siberian rivers and flood plains of the Arctic/Ed. V. V. Zueva. - Novosibirsk: Publishing House of SB RAS, 122–129 (in Russian).

Goodnight, C. J., Whitley, L. S., 1961. Oligochaetes as indicator of pollution. Proc. 15 th Annual Ind. Waste Conf. Pardue, 139–142.

GOST 17.1.4.01-80, 1983. Conservancy. Hydrosphere. General requirements for methods of determination of oil products in natural and waste waters. – Moscow: Publishing House of standarts, p. 4.

Hydrochemical Dictionary/Comp. A. A. Zenin, N. V. Belousova. – L.: Gidrometeoizdat, 1988. – p. 240.

Ioganzen, B. G., 1951. Some of the questions of the productivity of reservoirs. Proceedings of Tomsk University, 115, 371–386.

Mikhailova, L. V., 1987. Features of river pollution with oil products. Proceedings of XXVIII Union hydrochemical meeting. Leningrad, 113–114 (in Russian).

Mochalova, O. S., Antonova, N. M., Gurvich, L. M., 2002. The role of dispersants in the processes of oil transformation and oxidation in aquatic environment. *Water Resources*, Vol.29, pp. 202-205. doi: 10.1023/A:1014961506446.

Oil and Gas of Tomsk Region, 1988. A collection of documents and materials.–Tomsk, p. 359.

Remorov, V. V., Sidorenko, T. N., Schwab, N. A. Impacts of land contamination of areas of oil and gas deposits on the condition of ecosystems, 1998. *Bulletin of VNK*, 1, 106–107.

Uvarova, V. I., 1989, Current status of the level of contamination of water and soil of the Ob-Irtysh basin. *Collection of scientific works GosNIORH*, 305, 23–33 (In Russian).

Vinogradov, G. A., Berezina, N. A., Lapteva, N. A., Zharikov, G. P., 2002. Use of structural characteristics of bacterio- and zoobenthos for assessing the quality of bottom deposits: Case study of water bodies in the Upper Volga basin. *Water Resources*, 29, 299–305. doi: 10.1023/A:1015680329937.

Vorob'ev, D. S., Frank, Yu. A., Lushnikov, S. V., Zaloznyi, N. A., Noskov, Yu. A., 2010. Oil Decontamination of Bottom Sediments Using Limnodrilus hoffmeisteri (Oligochaeta: Tubificidae). *Contemporary Problems of Ecology*, 1, 15–18. doi:10.1134/S1995425510010042.

Vorobiev, D. S. Biological basis for cleaning of bottom sediments of water bodies from oil and oil products, 2013. Unpublished doctoral thesis. Tomsk: Tomsk State University. 385 p.

Vorobiev, D. S., Merzlyakov, O. E., Noskov, Yu. A., 2014. Oil decontamination of bottom sediments: past, present and future. *Procedia Chemistry*, 10, 158–161. doi:10.1016/j.proche.2014.10.027.

Vorobiev, D. S., Noskov, Yu. A., 2015a. Oil Contamination of the Ob Basin. *Internat. J. Environ. Studies*, doi.10.1080/00207233.2015.1027591.

Vorobiev, D. S., Noskov, Yu. A., 2015b. Norm setting for oil and petroleum products in bottom sediments and their quality assessment according to Russian hydro chemical parameters. *Internat. J. Environ. Studies*, doi:10.1080/00207233.2015.1033967.

In: Riparian Zones
Editor: Oleg S. Pokrovsky

ISBN: 978-1-63484-613-4

Chapter 14

DYNAMICS OF FLOODPLAIN LANDSCAPES

V. S. Khromykh*

Department of Geography, Tomsk State University, Tomsk, Russia

ABSTRACT

Floodplain landscapes are unique due to their variability and stability. Their structure is constantly changing under the influence of external factors. At the same time, the mechanism of self-regulation of floodplain geosystems strives to maintain its functioning. One can recognize the floodplain geosystem as steady ones for a period of more than a year. The direction of the floodplain landscape evolution is defined by the watercourse variations, and the overall evolution is defined by the perennial cycle of the floods. The main role in floodplain dynamics is played by hydrodynamical factors such as erosion-accumulative activity of the river and flood cycles. The dynamics of the level of underground waters and the role of soils in landscape dynamics are also discussed. Additionally, the changes of the floodplain green are considered, and the dynamics of floodplain landscapes is described. Successions of phytocenosis can be a reliable instrument for detecting of directional landscapes dynamics. The fluctuation of vegetation coverage is clearly expressed in rivers' floodplains. The catena includes zones of forestation, grassland- and bog-formation. The evolution of vegetation in the seral series reflects the processes of floodplain landscapes' self-development. The consistent patterns of this dynamics should be taken into account when conducting economic activity in floodplains and reclamation of the floodplain landscapes.

Keywords: landscape, floodplain, erosion, alluvium, flood, dynamics, development, water

---

* Corresponding author: V. S. Khromykh. Department of Geography, Tomsk State University, Russia. E-mail: valery_khromykh@mail.ru.

Figure 1. An aerial view of the Middle Ob floodplain.

INTRODUCTION

Floodplains are unique landscapes. In the figurative expression of R. A. Yelenevsky (1936), floodplains are "the natural coin box." Indeed, having adequate water resources, a favorable climate and high levels of soil fertility, flood lands possess rich biological resources (Figure 1).

Floodplain geosystems extend to all landscape-climatic zones and regions. They occupy more than 4% of the land surface and 2.6% of the territory of the former USSR (570,000 km^2) (Voropai, Kunitsa and Levitsky, 1974). The biological productivity of floodplain landscapes is unusually high. According to L. E. Rodin and N. I. Bazilevich (1965), despite floodplains' small area they are responsible for almost 10% of annual worldwide production.

Floodplain landscapes have the greatest dynamism and smallest age compared to any other watershed landscapes. This is primarily due to variability in water levels – a leading factor in the formation of floodplains. Overall, hydrodynamic factors such as the erosive-accumulative activity of the river, the level and duration of saturation of the floodplain and the deposition of alluvium and local and provincial features of the territory such as climate and tectonics play an important role in the development and formation of floodplain landscapes and their dynamics. All other components are influenced by these factors and play a subordinate role. Changing of the floodplain landscapes includes their self-development from the stage of a young river stream to the subclimax state.

1. EROSIVE-ACCUMULATIVE ACTIVITY OF THE RIVER

All the diversity of floodplain landscapes is inextricably linked with riverbed processes and the erosive-accumulative activity of the water flow in the riverbed. The river, washing away its banks, destroys some landscapes, and by depositing the products of accumulation, creates other complexes. River erosion and accumulation define the foundations of the topography of floodplain, have a decisive importance in the deposition of alluvial sediments, affect the groundwater regime, etc. The floodplain is a derivative of the modern riverbed, since it is formed by overgrowing of the highest parts of the accumulative forms of riverbed relief under conditions of low water and lack of water.

The erosive-accumulative activity of rivers is actually the only factor in the formation of water stows of the riverbed, which constitute a special class of river landscapes. They are characterized by dynamics associated with seasonal changes in depths and flow velocity and with continuously flowing dynamic processes in the riverbed flow.

"Due to the prevailing traditional typological approach to the study of the nature of floodplains, oriented in an explicit or latent form, mainly in ecology, the patterns of riverbed process, part of which is the formation of floodplain, are insufficiently taken into account" (Petrov, 1979, p. 3).

Riverbed processes are changes in the external appearance of the riverbed, and its horizontal and vertical position. In the riverbed islands, shallows, etc. can periodically appear and disappear, for a long time the river might crash into the underlying rocks, or, conversely, rise, lining the bottom with material which is moved with it or its tributaries above this point. Riverbeds also move horizontally, while the riverbed flow erodes one bank and lays deposits on the opposite one.

Such dynamics of processes are caused by the mutual influence of river flow and the riverbed containing it. The riverbed controls the flow and shapes its velocity field. The flow, together with the velocity distribution, affects the shape of its riverbed, itself creating the riverbed which corresponds to its velocity field. This process (deformation of the riverbed) occurs as a result of movement with the water flow of different riverbed formations by redeposition of sediments.

Consequently, the riverbed will change constantly, including horizontal displacements - meandering, or the forming of bends. On the postponed sediments (convex) bank will be formed, growing both in height and horizontally, the riverine shallow, (beach, "sand") - the thickness of alluvial deposits, whose surface will eventually break free from the water at low levels, and at high it will be flooded. In the vertical section of thickness of alluvial deposits, there is a decrease in their size bottom-up, due to a decrease in the depth and velocity of flow upon an increase in the shallows and waste of the river to the side. Opposite, the concave bank (the reach area) will erode and the washed away material will move into the category of river sediments. Reaches, the deepest parts of the riverbed, are interspersed with shoals – the straightened sections of the riverbed, having a lesser depth and large speeds of flow.

Initially, a straight or slightly sinuous riverbed increases gradually its development of bends. In the initial stages of this process (bends at angles up to about 100°) the bend moves across the valley. Meanders are also able to move slowly downstream. This is due to the predominance of bank erosion in the upper (downstream) part of the bend, and depositing of the products of erosion at its lower border. This process is continuous, when the meandering occurs in a limited fashion (the borders of the valley of the river are the limiters).

During a free meandering process there is a change in the form of a bend from slight meandering, sinusoidal, to loop-shaped (Figure 2). The continuing curvature of the riverbed leads to a weakening of the erosive-transporting ability of the flow. Increasing the diameter of the bends determines over time the convergence eroded shores of adjacent meanders. The result is (more often during the period of high water) a natural gap between meanders, which leads to a straightening of the riverbed. After the formation of the new riverbed the development of a new bend begins, and the entrance to the old bend overlaps with sediment deposits for several years. After the closing of the upper mouth the channel turns into a backwater, which is connected with the riverbed through its lower mouth. A shallowed backwater, when its lower mouth closes, turns into a horseshoe-shaped oxbow lake (Figure 3).

Figure 2. Loop shape bend channel.

Figure 3. Disconnected meanders turned into horseshoe-shaped oxbow lakes.

The incomplete meandering of rivers is possible. In this case, straightening channels are formed at earlier stages in the development of bends. In free meandering the bend is usually straightened, when its length exceeds the length of the straight line connecting the inflection points of the riverbed, 2-4 times, and in an incomplete meandering only 1.6 times. The riverbed is divided into two sleeves, which can function simultaneously for tens of years, and be divided by an island which is displaced across a current as a result of erosion and accumulation. Over time, the main riverbed is closed by sediments or turns into an oxbow lake.

The process of developing and dying of bends repeats constantly. A plot formed during the horizontal riverbed deformation of one bend is called a segment of the floodplain (Figure 4). Part of the bottom of the river valley, enclosed between the lines connecting the vertices of the bends on both banks of the river, is called the belt of meandering. The riverbed migrates to the bottom of the valley and at the approach to its slopes erodes them, broadening into a floodplain and giving bow-shaped outlines to its borders.

Thus, the process of river floodplain formation and its morphological features (topography and structure) is the consequence of the redepositing of sediments by the river, of the horizontal riverbed deformations, i.e., of a certain course of riverbed processes. In the most general case the floodplain is understood as the thick layer of sedimentary deposits, whose surface is periodically flooded by water flow, paving its riverbed to this thickness. In the layer of alluvial deposits, two groups of phases can be identified: 1) riverbed alluvium composed of bottom sediments usually of coarse mechanical structure; 2) floodplain alluvium composed of suspended sediments. The dynamics of the formation of each floodplain segment lead to the following patterns in its relief and structure:

1. floodplain deposits are characterized by a decrease in particle size from the base of thickness to its covering layer;
2. the height of the floodplain decreases from the upper to the lower (downstream) portion of the segment, the capacity of the floodplain phases of alluvium in this direction is increasing; this is because the upper part of the segment is an area of slowdown of river flow passed along the overlying concave bank and replete with the products of its erosion, i.e., the greatest number of tractional sediments is laid here;
3. the surface relief of a segment is an alternation of arcuate ridges and inter-ridge depressions repeating the contours of the convex bank with sediment deposition; each ridge undergoes the main stages of development in the phase of its riparian arrangement. The generation of the ridge begins after its preceding ridge reaches a height approximately equal to the average of the maximum water level. The base of the future ridge is located in the zone of the greatest accumulation of sediments - at a certain distance from the edge of the convex shore. The distance between adjacent ridges and their size depends on the average annual volume of accumulation of deposits and speed of the displacement of the eroded (concave) bank. "Every next floodplain ridge is formed further downstream than the previous one" (Chernov, 1983, p. 39). Furthermore, the oldest ridges of floodplain segments have a smaller height than that formed in the subsequent stages. This is due to the reduction in the speed of a current, and, therefore, the increase of accumulation in the process of development of a bend. With further increases in the steepness of the bend and lowering of the energy of flow, yet lower and narrower ridges begin to form. Thus,

the profile through the floodplain segment will be symmetric and the highest ridges will be in the central part of the segment. An increase in the width of inter-ridge depressions to the bottom part of the floodplain segment is also noted.

Consider the dynamics of erosive and accumulative processes, which floodplain segment may be subjected to, once off the belt of meandering through straightening of the riverbed, but as a result of the wandering of the river it was again under the influence of riverbed processes. This effect manifests itself primarily in the formation of sediment phases of superimposed riparian floodplains. Riparian shafts are formed on the eroded bank of the floodplain segment in the period of water overflow over the edge, when the sediments are deposited along the riverbed. While the larger particles settle closer to the shore, shallow ones are meted out into the floodplain. There is a vertical sorting of sediments. Deposits of the superimposed riparian floodplain always overlap the humified soil layer and have a reverse order of strata for floodplain sediments - lower layers are formed by smaller particles, and the upper layer by larger ones. A superimposed riparian floodplain is usually short-lived, because is constantly destroyed by lateral erosion. In the Middle Ob, it usually has a width of 0.8-1.2 km and moves together with the movement of the riverbed (Khromykh, 1975).

Thus, the erosion and accumulation of meandering rivers creates floodplains whose structural unit is the segment or its surviving part. Such a floodplain is characterized by considerable diversity in height and structure and, as a consequence, soil and vegetation cover, i.e., landscapes.

Besides the above-described meandering of rivers, there are other types of riverbed processes (Popov, 1968, 1969): belt-ridge type, side type, many-sleeve channel and floodplain types (Figure 5).

Figure 4. Floodplain segment. There is a lake in the lower central part.

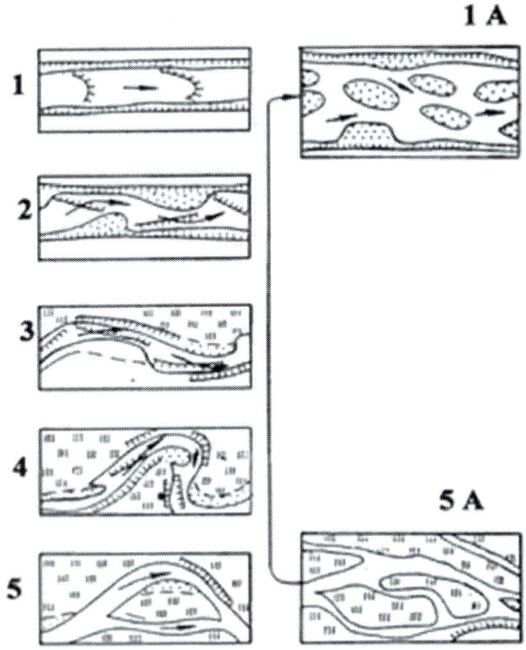

Kondratiev and Popov, 1973): 1 - belt ridge type; 2 - side type; 3 – type of limited meandering; 4 - type of free meandering; 5 - type of unfinished meandering; 1A - side type or the many-sleeve channel type; 5A - many-sleeve floodplain type.

Figure 5. Typing of riverbed process.

The first two types of processes are not accompanied by a formation of the floodplain of rivers and, therefore, for the undertaken research are not of particular interest. Characteristic features of the many-sleeve channel are the presence of one wide channel in the period of high water and crushing it with ridges on separate sleeves in the summer low-water period. The many-sleeve channel type has two varieties: 1) middle shoal type; 2) island type. In the first case, the riverbed shoals - middle shoals and side shoals, formed with the bottom sediments - are mobile and the speed of their movement depends on the flow of the river and the size of sediments which compose them. When the recession of a flood is stretched, there has been stagnant water above the surface of the shallow, and a deposit of a sufficient amount of suspended solids, the bottom sediments overlap by small deposits. Such a surface of the shallow in low-flow period covered by vegetation, which in its turn strengthens the process of deposition of suspended particles. In this way the middle shoal is transformed into an island, and the second type of a many-sleeve channel is formed.

A many-sleeve floodplain develops when by the type of unfinished meandering, the rate of the dying of the straightened bends lags behind the rate of formation of new ducts. The floodplain is dismembered by lots of flow, each of which develops its own type of riverbed reformations.

If by different types of meandering we can predict certain scheme redeposition alluvial deposits, in the case of many-sleeve types the forecast is quite complicated, because its own type of riverbed process is formed in each sleeve, and the deposition of sediments is complicated by the interaction of the individual channels in the places of their fusion and fission. The intensity of erosive-accumulative process by other equal things is greater than the degree of kinetics of the riverbed flow, defined by the slopes of the valley floor. The latter are determined by geological structures of high-order and their tectonic regime. Consequently, its contribution to the development of riverbed and floodplain processes contribute to the vertical deformations of the riverbed, which affect the longitudinal slope of the riverbed and the character of the transverse profile of the floodplain. On the plots of raising of the earth's crust the rivers deepen their riverbeds, i.e., produce erosion, and a stepwise floodplain is formed, a trend towards narrowing of the floodplain is noticeable. The marsh natural territorial complexes are dated for near-terrace floodplain. The natural territorial complexes, close to the zonal types, are widespread. The lakes have a crescent shape.

In areas of subsidence of the crust there occurs accumulation and the superimposed or diked floodplain is formed. Thus, according to L. V. Yakovleva (1978), the power of the silt, deposited by the formation of the superimposed floodplain in the middle Ob and covered the 50-70 cm horizon of the soils, close in structure to zonal, is in the riparian zone of 240-275 cm, and to the rear is reduced to 5-30 cm. The morphological structure of zones of accumulation has a broad floodplain, with a predominance of landscapes with a hydromorphic structure and lakes. The horizontal deformations of the riverbed reach the greatest value. It is here that the process of waterlogging progresses.

Thus, the vertical deformations of the riverbed determine the direction of development of geocomplexes - either by way of formation of zonal landscapes with an increase in the role of climate or hydromorphic landscapes with increased waterlogging (Surkov, 1997).

Irreversible deformation is also confined to the upper (erosion) and lower (accumulation) links of the hydrographic network. But in most cases, the intensity of these processes is measured in millimeters per year, while the deformations of the riverbed caused by the transport of the sediments with the river are expressed in meters per year. Consequently, the volumes of unidirectional deformations of the riverbed in modern rivers are thousands of times smaller than the volumes of reversible deformations by the redeposition of sediments with the river.

Summing up the consideration of river erosion and accumulation, we can conclude that the form and dynamics of the riverbed process, the direction and speed of the horizontal deformation of riverbed are decisive in formation of floodplain, and, in turn, depend on a number of factors. According to A. I. Voeikov, "the rivers can be seen as a product of the climate" (1884, p. 98). "However, the hydrological regime depends not only on the climate but also on the tectonic features of the area: by the border of the crossing of tectonic structures the rate of flow of the river, the degree of meandering of riverbed, etc. are changing" (Annenskaya, 1982, p. 45). Consequently, the well-known formula of A. I. Voeikov "is too narrow and did not fairly represent reality" (Solntsev et al., 1976, p. 76).

The main components of riverbed processes are the costs of water and sediments, i.e., rivers - not just the flows of water, but also the masses of solid particles drifting with the water. Consequently, patterns of riverbed processes, and hence formation of floodplain depend on a whole range of factors (climatic, geological, geomorphological, soil-botanical). The main ones are the size and mode of water delivery to the surface of the catchment area, the geological structure and topography of the basin, the state of its surface (the degree of coating with the turf, vegetation cover, the degree of soil freezing and so on), the mechanical properties of soils of the bed of the river (the bottom and the shores). Peculiarities of floodplains can be affected by secondary factors, such as processes of the formation of bogs, the karst and the thermokarst, the deluvial takeaways, etc. Thus, in their morphological appearance and in the water regime the rivers bear features of the basic landscape where their flow is formed.

It should also be emphasized that river erosion and accumulation are processes that depend not only on external factors, the riverbed dynamics acts as a self-developing and self-regulating process. It is the cyclical process in both space and time.

In Western Siberia, the totality of these factors determines the prevailing in the dynamics of riverbed processes of the majority of rivers with the type of free meandering and the development of the segment-ridged floodplains corresponding to it. So Ob, flowing "surrounded by sandy loams and loams of fluvial, fluvioglacial and lake-alluvial origin," has formed a floodplain of the large width (Chernov, 1983, p. 20).

Dynamics of riverbed processes, deformations of riverbed and formation of floodplain can be studied best using aerial and satellite images of different times. Their comparison allows us to determine most clearly and objectively the direction and magnitude of planned changes of the riverbed and floodplain forms.

2. Flood Cycles

The specific properties of floodplains are formed due to the water regime of the rivers. This allows us to consider their territories as separate geosystems with a single governing factor, a spring-summer flood. The annual impact of floods leads to a deviation from the normal state of the productivity of floodplain ecosystems and to structural changes inside them. Each spill interrupts the normal development of biogenic components, causes the deposition of alluvium, and activates the deformation of the riverbed via breakthrough of the necks of meanders and the lateral displacement of the riverbed. These changes demonstrate the great dynamism of the floodplain landscapes.

The scale of the floods depends on the river length and water discharge: the larger the river, the longer the flood period. The rivers of Western Siberia exhibit exclusively long floods often persisting during full summer. This is so called "West Siberian type of hydrological regime."

2.1. The Height and Duration of Inundation of the Floodplain

Periodic flooding with the flood waters is the main feature that distinguishes the modern development of floodplain natural territorial complexes from interfluvial complexes (watershed divides). The role of the flood in the life of the floodplain is hard to overestimate. Flood waters act as organizing component, which determines the main difference of the floodplain complexes from the non-floodplain complexes. The duration and intensity of exposure of the flood waters to all components of the landscape depend on the intensity of the flood. Within the floodplain, the plant communities have to adapt not only to the mineral substrate and micro-climate, but also to the average years of regime of the height and duration of inundation of the floodplain and of "the amplitude of the seasonal moisture" (Annenskaya,

1982). Contrasting relief of the floodplain causes the complexity of the pattern of flooding during periods of rising water levels in the river.

A floodplain "is not a monolithic whole, it consists of many floodplain arrays with a very rough surface separated by the riverbeds, ducts and sleeves" (Petrov, 1979). According to Popov (1969), an array of floodplain is the area of floodplain with a closed hydraulic cycle. Within this territory, there is the equality between the water inflow from the riverbed and the outflow into the channel and accumulation in the floodplain array. Meandering rivers of the floodplain are limited by the main riverbed and ducts in the floodplain. Below, the development of the flood is considered.

With the rise of water level in the river, the water begins to go to the floodplain from below, gradually rising over the common slope of the terrain. The velocity of countercurrents usually does not exceed 0.02 to 0.06 m/s (Chernov, 1983). The river water enters in the mouths of the tributaries, in the lower mouths of the backwaters and straits, and in the sources of lakes and bays. Thus, in the initial phase of flooding, the array of the water flows through the lower hollows of the old riverbed with the its lower downstream side and fills the lowest sites of the floodplain area. At this stage of the flood, there is a noticeable difference between the water level in the main riverbed and in the floodplain (Figure 6).

During further increase in the water level within each segment of the floodplain, the flooded area increases, and, finally, the water from the riverbed begins to flow into the oxbow lakes through their upper banks. At this stage, the levels of the water in the riverbed and within the floodplain in a cross-section are essentially similar, and there is a direct flow in the oxbows. If the upper inputs in the hollows of the old riverbed are blocked with sediments, an overflow of water from the riverbed to the plain through the low sites of the shore can be observed. In this case the slopes of the water surface towards the floodplain are quite significant. This produces concentrated erosion of the bank with the formation of the so-called breakthroughs - erosional hollows on the floodplain.

Figure 6. Start of floods. Riparian floodplain forests already flooded, the water is filling the meadows of the central part of the floodplain.

During the subsequent increase in the water level, a complete inundation of the floodplain and the formation of a single transit stream along the axis of the valley occur. For a single floodplain reservoir, such as an oxbow lake, the dynamics of floods events is as follows: after water passage into the floodplain, an isolated floodplain lake is initially converted into the bay, then into the duct, finally can fully merge with the main river flow.

After the passing of the peak of the flood a reverse order of sequential exemption of the floodplain occurs. In the closed inter-ridge depressions the temporary lakes are formed, where the water level is higher than in the mainstream. The rate of drying of such water reservoirs depends on several factors such as the area and the depth of the lake, the weather conditions of the subsequent period, the hydrogeological characteristics of the site etc. In the case of a combination on the floodplain of segments of different ages, their heights can vary significantly. This often becomes the cause of the emergence of the major secondary lakes, which bring about the waterlogging of the central floodplain. The surface of a near-terrace floodplain is rarely flooded.

The floodplains of the majority of large rivers are characterized by the presence of multiple surfaces of the different elevation. In relation to the river surface they are called "altitude belts of accumulation" (Rodnyanskaya, 1960), or "altitude levels of the floodplain capacity" (Khromykh, 1975). "The differences between these surfaces are defined by the flood frequency and duration: high-level floodplains are rarely submerged, middle-level valleys are frequently submerged and for the more prolonged period. Finally, low-level surface are submerged annually, for longer time, both in spring and duirng the storm flood" (Annenskaya, 1982, p. 50). Consequently, the height and duration of inundation of the floodplain define the vertical differentiation of floodplain landscapes. In the floodplain of the Middle Ob River, four types of inundation are identified: exclusively long-term, long-term, average-term and short-term levels" (Khromykh, 1975).

The characteristics of floods include the frequency, the layer and duration of flooding, the date of the start and the end of the flood. V. P. Bolotnov (1988) suggests also considering the area of flooding, temperature, and water pollution. In each zone of the floodplain with similar tectonic conditions the height of the floods are fluctuated over the years within certain range. The direct observations show that the extreme borders of the floodplain are close to the average value of the maximum water level. The increasing of the thickness of floodplain deposits is limited obviously by the position of maximum water level in the river.

The flow of the spring flood is determined by snow reserves and depends on the density and orientation of the hydrographic network (rivers, gullies, ravines), as well as the melting conditions due to orography and exposition of slopes. Other factors include the accumulation of the flow in closed depressions and its losses on moisture of soils, depending of soil capacity, morphology and mechanical composition" (Petrov, 1976). One can also recognize the importance of soil moisture in late winter, and the amount and mode of precipitations that fall during the period of snowmelt or high water combined with the magnitude of evaporation and the nature of the vegetation cover.

The characteristics of flood depend, among other things, on the structure of the floodplain terrace. As it is known, the broad plots of floodplain, having a flat relief and layered soils, are often replaced with narrow stretches, where the ascent of water in flood is higher, the rate of flow is faster, and the light-textured soils dominate.

The height and duration of inundation of the floodplain exert significant impact on the landscape components. Often high, but short-term flood can bring more significant changes than the low and stretched flood. This feature applies primarily to riverbed deformations. During the flood, this largely depends on the ratio of the velocity and direction of the riverbed to the floodplain flow. During the peak of the flood the flow greatly modifies the formation of the riverbed (Maccaveev, 1955).

The interaction of flood waters with the other components of the landscape is not limited by mere duration of flooding. The chemical composition of river water, DOM concentration, temperature, and turbidity also affect the mechanical and chemical composition of the soils and vegetation of the floodplain zone. The impact of the height and duration of inundation of the floodplain on soil cover of floodplain is manifested in the reduction of the period of active soil formation. Periodic flooding creates specific conditions of soil formation, occurring essentially in aquatic and subaquatic mode.

The height and duration of inundation of the floodplain have a great influence on the ground water regime, as well as on the biodiversity and life conditions of animal kingdom. For example, high and prolonged flood leads to mass death of rodents, mass reproduction of aquatic animals and amphibians.

The result of the height and duration of inundation of the floodplain is different from other landscapes. The structure of the floodplain landscapes is closely dependent on the frequency and duration of the flooding. Numerous factors including seasonal, perennial and global climate change, water availability, riverbed processes, tectonic features, play a pivotal role in these processes.

2.2. The Deposition of Alluvium

The hydrological regime has a tremendous impact on the formation of soil cover and ecological communities of the floodplain, and thus determines the specific shape of the floodplain landscapes. During a flood the floodplain represents the bottom of the stream and thus directly affects the water flux, forming the riverbed. On the other hand, the floodplain is a permanent storage site and a source of solid material, transported by the river. The flooding of the floodplain during high water level determines the formation of accumulative rather than erosive landforms. One of the main results of the activity of the flood is the formation of alluvial floodplain facies (from the Latin: Alluvium = deposits). The layer of deposited silt can reach a few meters over time and may smooth out the bumps of the primary relief. "In general, the silt imposes as cloak, slightly softening the primary relief, and only in some places modifying it" (Chernov, 1983, p. 54).

The intensity of deposits of silt depends on the duration of flooding of the surface, which, in turn, is determined by the water level rising. The deposition of alluvium depends on the turbidity of the river flow, which is defined by the amount of suspended mineral and organic particles. Moving the particles from the bottom to the surface is due to the presence in the flow of eddies or creating turbulence perturbations. The higher the amount of suspended sediments in the river, the greater their deposition on the floodplain during high water, and the faster the accumulation of floodplain alluvial facies.

Of particular importance is the ratio of peak flood and turbidity, as often the course of turbidity does not coincide with the course of the water levels. With a significant entry of suspended particles from catchments (for example, in the event of gully erosion and deflation of soils, or during the flushing of first meltwaters), the peak of the turbidity preceeds the peak of flood. For rivers with forest watersheds and the wetlands, the enrichment of the river water with suspended solids occurs mainly due to riverbed erosion. In this case, the peaks of turbidity and flood occur simultaneously, and the deposition of alluvium is maximal.

The sequence of alluvial sediments deposition within the floodplain segment is described below. The downstream part of the floodplain segment exhibits the lowest height. During certain phases of flooding the water becomes stagnant and deposits significant amount of suspended sediments. As a result, the capacity of the floodplain to form the alluvial deposits increases downstream the segment.

At the top of the segment, the sand sediments accumulate. They were deposited here during overflowing of the floodplain stream with sufficiently high speeds (up to 1.5 m/s). In the floodplain relief, such sediments present as tongue-shaped emissions of sand, penetrating into the floodplain sometimes over hundreds of meters. The velocity of the stream flow in the central part of the segment decreases compared to the places where it enters the floodplain by a factor of 3 to 5. The water speed rarely exceeds 0.3 to 0.5 m/s. This causes precipitation of the fine fractions of suspended sediments. As a result, sandy loams and loams by compaction and drying are being deposited. Towards the downstream part of the segment, the flood water velocity decreases in the near-rear lowering zone where thin silt such and clays are deposited. Thus, floodplain segment is separated into several zones, characterized by different conditions of accumulation of alluvium: riparian, central and rear zone.

According to Chernov (1983), in the floodplain of the middle reaches of the Ob River the average annual turbidity is 98 g/m3, the rate of accumulation of the silt in the riparian zone is 70 mm/year and in the central zone is 1.5 to 6 mm/year.

An important indicator of the flood riverbed is the roughness of the floodplain, which depends on the nature of its vegetation and topography. These parameters affect the speed of the floodplain flow and the intensity of deposition of the sediments. Therefore, "a lot of attention is paid to the morphology of the floodplain, including the degree of contrast of its relief, the horizontal configuration, the orientation of floodplain gullies and ducts and the nature of riparian vegetation" (Chernov, 1983).

Meadow vegetation has the lowest coefficient of roughness. As a result, the floodplain deposits covere relatively evenly the surface of the meadow floodplain, forming quite smoothed terrain compared to forested floodplain. Within the forest floodplain the woody vegetation helps to slow down the water velocity, which leads to massive precipitation of suspended sediments. This creates significant variations in the altitude of the terrain.

The deposited alluvium affects the development and reproduction of plants. Thus, even relatively thin loamy and clayey deposits may form quite dense crust during dry periods, hardly penetrable by shoots of meadow plants. The silt of 2-4 mm thickness may increase the productivity of meadow by 30 to 50%, and does not delay the spring development of grass. A 2-cm alluvium deposits slow down the development of herbs for up to a month. The herbage cannot penetrate the layer thicker than 8 to10 cm. Thus, the deposition of alluvium differentiates forms of plants according to their affinity to mechanical composition of soils and can greatly control the diversity of species. The mechanical composition of the alluvial deposits also affects the distribution of plant communities. Gramineous herbages are characteristic for a coarse-grained alluvium, and the herbs proliferate at the places, where the dust particles dominate in the sediments.

The annual deposition of the floodplain alluvium leads to the siltation of the surface, thus increasing the density of the soils, reducing their porosity and filtration properties. This is essentially determined by the level of flooding and the duration of standing waters and affects the development of vegetation and soil cover. The deposition of alluvium causes continuous growth in the height of the floodplain. The deposition of silt reduces of the height and duration of inundation of the floodplain and decrease the flow of the river water. As a result, the increase of the height of the floodplain slows down and falls to fractions of a millimeter per year. The deposited alluvium significantly controls the groundwater regime of the floodplain which in turn may affect its landscape. In order to change the floodplain landscape significantly, the thickness of the new deposits of loamy texture should exceed 0.5 m. In case of sandy soil, the threshold thickness is 15 to 25 cm, due to the different capillary ability of deposits (Mamai, 1992).

3. Dynamics of Groundwater Level

An important component of the landscape, closely related to the hydrological regime, is groundwater. It plays a major role in the formation of floodplain soils and ecological communities and in their dynamics.

Of great importance is the depth of the water table, because over the horizon of groundwater the capillary fringe is formed. The power of the capillary fringe is determined, first of all, by the texture of the rocks and soil lying above the groundwater table. In the case of shallow groundwater, root system of plants reaches the zone of capillary water.

Groundwater depth experiences significant fluctuations during the year. During the flood the groundwater level rises considerably, can link up with surface waters, thus creating a single water-saturated horizon. As the recession of water, the table of groundwater drops, they are differentiated by not linked perched and actually groundwater. Thus, the dependence of the groundwater level on the hydrological regime of the river is reflected in filtering of the flood waters into aquifers connected hydraulically with the river, during the rising stage of the flood, and their return to the river into flood recession.

Seasonal changes in groundwater levels in various parts of the landscape are closely related to weather conditions. They can be reduced to the following four phases: the spring rise caused by intake of the melted snow water; summer decrease caused by the excess of evaporation over precipitation; autumn lift defined by excess of precipitation over evaporation; winter reduction caused by the absence of moisture infiltration.

Between landscapes there are significant differences in the depth, height and intensity rise in groundwater levels. The amplitude of the fluctuations in the level of groundwater is longer in the landscape units, stacked with heavy mechanical composition of soils, and among them in those where there is a deeper groundwater level. In dry years, the oscillation amplitude of the groundwater is longer in the places with their occurrence close to the surface.

In the floodplain of the Middle Ob, usually in the riparian part, the greatest depth of the groundwater level ranges from 1 to 5 m, in the central floodplain is usually 0.5-3 meters, near-terrace is 0 to 2 m (Khromykh, 1975). Fens are characterized closest to the surface of the groundwater level.

The depth of the groundwater is controlled by low water level of the river and the degree of remoteness of the floodplain area of from the riverbed. In turn, the stable value of the low water is determined by the geological structure and depends on groundwater resources and the conditions of their admission into the riverbeds. Spending groundwater resources occurs gradually over a period of 5-10 years, and replenishment is observed in some years in the long-term spring excess moisture of soils (Petrov, 1976).

Apart from the main aquifer, which lies deep from the surface and associated with level of the river and the groundwater of watershed surface, the floodplain is locally perched by groundwater seeps. This aquifer has different levels in virtually every inter-ridge depression, since its formation is associated with local waterproof layers in the floodplain alluvium. The perched water is not associated with groundwater of watershed surface, it feeds at the expense of the flood water and precipitation, so the dynamics of its level is determined by of flood conditions, thermal regime and atmospheric moisture.

The dynamics of the groundwater level due to the hydrological regime of the river can lead to activation of suffusion processes on the slopes of terraces and the interfluvies, eroded away by lateral erosion. When the flood waters by the end of high water falls, the ground water level decreases much more slowly. Therefore, there is a steep depression curve of groundwater and hydraulic pressure in the direction of the slope to the river creates. The gradient of the pressure of the filtration flow reaches the values, quite sufficient for the removal of solid particles from the aquifer with water. As a result, on the locations of the output of groundwater, the suffusion niches in the form of circus and voids up to 3-5 m in diameter appear. This reduces the stability of slopes and intensify the coastal erosion.

The dynamics of the groundwater level plays an important role in the functioning of the floodplain landscape, its character is often associated with the formation of soil water reserves.

4. The Role of Soil in the Dynamics of Floodplain Landscapes

The soil as a component of the landscape is "the result, the function of total mutual activity of these agents - soil-formers: the climate of the area, its plants and animals, the relief and the age of the country or its absolute height, finally, the substrate (i.e., ground parent rocks). Like any natural body, the soil has its past, its life and genesis" (Dokuchaev, 1899, p. 16).

The soil cover of floodplain has an extremely high level of dynamism. It manifests itself in a significant seasonal and annual variability of water-physical, physical-chemical and chemical properties of floodplain soils, as well as diversity of floodplain soil. "The soil is a non-equilibrium, highly dynamic bio-inert system, rich in free energy" (Perelman, 1977, p. 21).

The floodplain soil is characterized by the special conditions of development, relating with the periodicity of the flood of the floodplain and, therefore, with the break in the forming of soil, as well as with the annual deposition on the floodplain of the alluvial deposit that are contribute to the constant rejuvenation of the soils. The powerful influence of the long river spill and of annual deposition of alluvial sediments has an inhibiting influence on the soil-forming process. Floodplain soils are characterized by intense waterlogging in spring and summer and by more or less significant drying in the summer-autumn period. Under the influence of periodic excess of moisture, the seasonal dynamics of redox processes in soils occurs, with a predominance of the reduction in the period of spring and summer, and the oxidation - during the autumn low water (Rodnyanskaya, 1963).

The deposition of alluvium plays an important role in the development of soil-forming process of the floodplain. Formed by alluvial sediments, the floodplain soils are periodically replenished with elements of plant nutrients that cause an increased and sustainable fertility of floodplain soils. During the flood, the surface of the soil is enriched by the silt sediments, which are contributing to soil nutrients, mainly phosphorus and potassium. The deposition of alluvium has a great influence on the processes of forming of humus, and promotes the formation of soil structure. Due to the long height and duration of inundation of the floodplain and the flood sedimentation, the forming floodplain soils are enriched with calcium carbonate, compounds of phosphorus, potassium, nitrogen, iron, manganese and secondary minerals: kaolinite, hydromicas, montmorillonite, brought from the river basin. During the floods the floodplain is also fed by organic matter, the content of which can be (in terms of humus) from 1 to 5%. There are especially a lot of humus (10-12%) in clay-silty sediments of wetlands, lakes, shallows. All this is the basis for the formation of fertile soils of the floodplain, with high economic value.

After the riparian shore is liberated from the low-water level of the river, the process of soil formation begins on it. It recycles only a thin surface crust of fresh alluvial deposits. On the surface of the low shafts that are covered by the flood water annually, a new deposit is forming so that they grow up. Processes of soil formation do not have time to convert the entire thickness of the sediment, so primitive layered soils are forming. On medium-high and high shafts, not flooded annually, in years of low floods the soil formation can change the entire layer of sediments from previous floods. Sod-layered soils are forming.

The main role in the formation of floodplain soils belongs to the sod soil-forming process. It takes place when there is a strong grass, sod forming, and relatively deep groundwater, and confined mainly to the riparian floodplain and high ridges of the central floodplain. The intensity of the sod soil formation depends on the amount of annually dying off organic matter and on the rate of its mineralization. Soils of high floodplain are filled with the flood waters not every year, and in the event of flooding their period of flood is shorter than those of the soils of a low floodplain. Therefore, soils of high levels experience the greater impact of zonal conditions than soils of low floodplain. It leads to that, on the sod process on high elements of the topography of a floodplain is imposed the zonal podzol process.

In the case where the groundwater table is shallow, along with the sod process a gley soil-forming process also occurs. Finally, in the case of standing water on the surface at a consistently high level of groundwater the marsh soil-forming process occurs, which is typical for the lowlands of central and near-terrace floodplains.

A distinctive feature of the soil generally is in the fact that it has properties and processes of both biotic and abiotic factors. This is link between biotic and abiotic components, substance and energy exchange between them goes through the soil.

The soil where the processes of infiltration and transformation of the precipitation into underground and surface runoff occur, exert great influence on the formation of runoff and the water balance in general. The soil cover, serving the intermediary between the climate and the river, transforms meteorological phenomena into a hydrological one. Hydrological value of soil cover is mainly determined by water permeability (infiltration ability), and water-holding capacity of the soil, on which the size of the surface runoff, evaporation and groundwater recharge depend on. Soils have different permeability and moisture capacity, depending not only on the natural properties, but also on human impacts (plowing and cultivation). Sandy soils having high permeability can usually almost completely absorb the precipitation of any intensity, and therefore, the surface runoff plays a subordinate role in relation to the underground runoff.

Soil acts as a liaison between the plant community and external systems and in relation to phytocenosis (Shepeleva, 1998). The influence of soil conditions on the dynamics of vegetation is manifested in the variation of the reserves of productive soil moisture, their temperature and nutrient content. In different years, optimal for the absorption of moisture and nutrient conditions are formed at different depths. Phytocoenosis responds to the heterogeneity of water-physical and agrochemical properties of horizons of soils on the selection of ecologically and biologically different plant species and their natural placement in space.

Reserves of productive moisture in the soil are an integral characteristic of the hydrological regime of the habitat (function mode flood, weather, and groundwater levels) and acts as the main control factor of floodplain geosystem, defining the change of the biomass and transformation of the structure of the landscape. "For floodplain lands, soil moisture and its change during the warm period and of the whole year, eventually determines the ecological and resource potential of floodplain" (Rusakov, 1991, p. 46).

Floodplain soils, peculiar group of hydromorphic soils, are different from many others soils in that they are constantly growing up. The increased dynamics of soil-formation in the direction from the watershed to the floodplain is also reflected in the increase in the intensity of the biological cycle of substances, reducing the content of many mobile elements in soils, in the strengthening of the morphologically pronounced accumulations in the form of humus, iron-humus and mineralized horizons (Nechayeva, 1974). Also layering, buried humus horizons and mosaicity are inherent for the floodplain soils.

5. Successional Changes of Floodplain Vegetation

Floodplain landscapes are continuously evolving, and their process of development takes place much faster than the development of noninundated landscapes. This is explained by the youth of floodplain systems, and as a result, by poor adaptability to local conditions and by deep reaction to changing of environment. At the same time the best indicator of the development of landscapes is the changing of vegetation cover. Most clearly fluctuations of vegetation, following on the change of habitats, are expressed in floodplains.

Vegetation is the most active component of ecosystems, able to adapt to changing environmental conditions and transform them. A critical role for the vegetation consists in the formation and transformation of soils, as well as topography, water regime and microclimate.

The vegetation is the most accessible to observation component of the landscape. Therefore endoekogenetic (Sukachev, 1950), or autogenous (Rabotnov, 1983) successions of plant communities can serve as a reliable indicator of the development process or of aimed dynamics of the landscapes.

5.1. The Formation of the Forest

After the exit of the riparian shallow from beneath the low level of the river the pioneer vegetation appears on it. For a floodplain of Central Ob it is the marsh horsetail (Equisetum palustre), Rorippa palustris, the russian, white and three-staminate willows (Salix rossica, S. alba, S. triandra) (Khromykh, 1975).

Thus, the pioneers of the settling of floodplains are willows. They are the most resistant of the plants to the annual flooding and deposition of sediment. Good adaptability of willow to the extreme conditions allows it to grow in a wide environmental range (from the riparian floodplain to the near-terrace floodplain). For example, the three-staminate willow (Salix triandra) quickly displays the root system from under the level of flooding and anaerobic conditions due to the abundance of adventitious roots that are develop in two months after the deposition of fresh silt. But still willows are dominant on the youngest riparian areas.

The accumulation of alluvium and growth of riparian shaft up lead to an increase of the high-rise level, to the decrease in moisture (deficit of moisture in the second half of the summer), to an increase in thickness of the humus layer. As a result more favorable conditions are created for the herbal and shrubby vegetation, dense thickets of osier-beds are thinned out, the shrubby underbrush and the rare herbage develop. Within a few decades the conditions for willow become unfavorable, the willow forests on the ridges are giving way to the shrubby formations. In inter-ridge depressions and in the rear parts of the floodplain the willow lasts longer.

The study of the forest forming process in the floodplain revealed certain patterns of distribution of formations of the forest and shrubby vegetation, depending on the course of the riverbed process. "This allows to use the forest cover as an indicator of the speed and direction of the transverse displacement of riverbed" (Vasiliev, 1988, p. 42).

On the alluvial, convex bank of bends and in lower ends of islands the primary riparian zone is formed. Here, as described above, the communities of osier-beds develop. From the concave bank of the bends and on the upper part of islands secondary riparian zone with overlaid alluvium is formed. The character successions occurring in this zone depends on the type of eroded surface, and on the rate of erosion.

When a catastrophic erosion in hundreds of meters per year the plant communities of the eroding territory do not have time to readjust to the new environment of riparian floodplain and to the riverbed are coming almost unchanged (for example, a meadow floodplain).

With intensive erosion in tens of meters per year on a surface of the eroded plot, the riverbed throws significant in volume the portions of alluvium, in the future dockable with the plantings of willow and black poplar. The willow plantings are formed in the case of erosion of meadow floodplain, plantings of black poplar - in the case of erosion of the forest communities of primary riparian floodplain.

With moderate erosion (2-5 m/year), according to S. V. Vasiliev (1988), birch forests are often formed. As a rule, with such intensity, the erosion of peaty plots occurs. At low erosion (less than 1 m/year) in the case of meadow floodplain, the forest vegetation is not formed. The deposition of secondary alluvium is extremely low.

In cases when coastal shaft was formed earlier, and further erosion is suspended - aspen forests representing the final stage of the development of forest vegetation in the secondary riparian floodplain are developing on it. They have a kind of a narrow strip extending along the bank on the shaft (Vasiliev, 1988).

The lifespan of the forest of a three-staminate and Russian willows seldom exceeds age of 25-30 years. Already by 18-20 years many trees have a dry top, are damaged with stem pests and fungal diseases (Bock, 1972). In the absence of renewal under their canopy they are most often replaced by meadow vegetation. Also the increase in terms of flood of the river contributes to it.

Resumption of birch, pine and other tree species is possible only in the areas where the impact of flood waters is affected to a lesser extent, and does not lead the tree species to the death. Therefore, during sufficient exposure to continental (zonal) factors in a mature stage the willow forest or meadow vegetation are replaced by birch forest (Figure 7), there are individual instances of dark coniferous species. The dark coniferous forest belongs to the latest and most stable stage in the development of the floodplain landscapes and is a subclimax of valleys of Western Siberia. Pine forests owe their emergence in the floodplain, mostly to alluvial sandy deposits.

The value of floodplain forests is very high. Floodplain forests, confined to the riparian part, perform the more protective functions, fasten huge masses of bulk material with the root systems and counteract to the development of erosion and aeolian processes. They accumulate within their extensive thickets the floodplain alluvium and thereby protect the lands of the central floodplain from the burial by coarse sediments. Forest slows down the greater part of the surface runoff and transforms it into soil subsurface, which prevents silting and shallowing of rivers and the development of water erosion. Riparian forests of the floodplain have water preserving value.

5.2. The Formation of the Meadows

Grasslands is the main wealth of the floodplain (Figure 8). The productivity of floodplain meadows in Western Siberia is 8-10 times higher than the productivity of the watershed grassland (Chernov, 1983).

With the removal from the operating bed of the river the woodland of floodplain decreases sharply. After several successional stages, the forest in the central floodplain turns into a meadow. The processes of changing of forest landscapes on meadows have not been studied in detail until now, and this issue is debatable. Apparently, this process is defined by the self-development of forest communities and by external factors: changes in the mechanical composition and capacity of alluvial deposits, the depth of the ground water, the amount of mineralized the litter, i.e., the deterioration of forest growing conditions.

There is a view about the origin of the meadows in the floodplains as a result of the increasing human interference: the woods were reduced in floodplains in historical time for household purposes (Kononov, 1971). On the treeless of floodplains as a result of human activities previously indicated V. N. Sukachev, (1934), A. P. Shennikov (1935), V. I. Schrag (1949), I. S. Buddo (1953), L. I. Nomokonov (1959) and other researchers. Some authors hold the opinion about primacy of floodplain meadows (Govorukhin, 1955).

Figure 7. Birch forest in the central floodplain.

A number of researchers (Ilyin, 1930; Baryshnikov, 1933; Lvov, 1963; Afanacyeva, Remezova, 1968 et al.) consider the northern meadows primary, i.e., they emerged without the influence of human activities on them. Many meadows of the southern regions of a floodplain still owe their existence to human society. The anthropogenic factor is manifested here in the felling of tree vegetation and shaping on its place the meadow. Also, human intervention can make the afforestation of young areas of the floodplain impossible.

According to L. F. Shepeleva (1991), the present (of the short and medium flooding) meadows are in origin secondary - they are created at the expense of a burning out and deforestation of floodplain forests on flat surfaces and gentle slopes of the ridges of medium and high levels of the floodplain. The meadows of the long flooding are primary, as they developed in some locations, where the trees and shrubby vegetation due to prolonged flooding by flood waters gives place to the communities of herbaceous plants.

Figure 8. Central floodplain of the Ob. The forb-grass meadow.

Figure 9. Central floodplain. Herb sedge meadow.

Emerging in the floodplain of the Ob meadows are initially polydominant, reminiscent the riparian weeds meadow. Calamagrostis, couch grass (Agropyrum) or brome (Bromus inermis) (so-called period of rhizomatous meadow community) dominate in them depending on local conditions. Later the different herb appears, and finally the dense tussock grasses or sedges (Carex) (Khromykh, 1975) (Figure 9).

Meadow landscapes can exist long enough, as long as the soil accumulates a thick layer of annually decaying organic matter. As a result, the meadow is waterlogged. On marshy or peaty meadow woody plants begin to appear, which each passing year occupy a greater surface of the meadows. First the bushes of osier, then birch and aspen join. Thus the formations of willow thickets with individual instances of birch and aspen arise in the central floodplain, and also the birch-willows grass-sedges bogs.

5.3. The Formation of Bogs

In the floodplain of the Ob, the bogs occupy about 25% of the whole area (Khromykh, 1975). In accordance with the nature of the water supply all the bogs of the floodplain of the Ob belong to the lowland type. On a vegetable cover marshes are divided into three large groups: forest, grassy and mossy. Forest bogs are mainly in the near-terrace floodplain. Grassy bogs are quite often found in the floodplain of the Ob, but they do not form large arrays, occupy only small depressions in central and near-terrace floodplains. Mossy bogs in the floodplain are very rare and are found only in the central parts of large bog massifs. Formed in the floodplain, a bog “passes in its development a number of successive stages from the willows grass-sedges bogs to mossy mesotrophic bogs” (Khromykh, 1975, p. 123) (Figure 10).

Figure 10. Near-terrace floodplain. Willow sedge marsh.

It is known that bogs are formed by draining water bodies or, on the contrary, the moisture of the land areas. In the first case, the lake goes into the bog through a stage of the overgrown pond. Due to the lack of flow and intensive ongoing process of accumulation of allochthonous and autochthonous materials the floodplain lakes become shallow relatively quickly. The underwater, floating and surface vegetation develops richly in them.

The overgrowing of lakes occurs in two ways. If the depth of the lake is significant, the surface carpet from plants impends on the pond. Overgrowing of shallow lakes occurs by gradual filling of the pond with the plants, rooting in mineral bottom. In the oxbows we observe a very slow, going for centuries, siltation with the sediments of clay and loamy particles mixed with organic material, which form a rather uniform thickness. The proportion of organic deposits in the total weight of depositions increases with the increase of the depth. Precipitation acquires the character of sapropels, the lake is replaced by a mire. In the upper part of such deposits often lie layers of peat.

In the second case, a forest or meadow, being a subject of a process of excessive moisture, become wetlands, which turn into bogs, with the accumulation of a peat layer.

Waterlogging can be seen as the first stage in the process of bog forming, is characterized by repetitive flooding of the surface, transformation of the forest litter or meadow sod into peat, subdivision surfaces in micro-depressions and of increase, a peculiar composition of plant component representing a mixture of meadow, forest and bog species and the formation of hydromorphic soils (Bazanov and Lgotin, 1988).

Due to the downturn in the flood the water is delayed for a long time in various negative forms of the floodplain relief, under conditions of constant waterlogging the organo-mineral hummocks develop here a typical form of microrelief of the floodplain depressions (Chernov, 1983) (Figure 11).

The buildup of peat is the second stage in the process of bog formation, which is characterized by the accumulation of peat, composed mainly of water (90%) and partially mineralized plant residues.

Figure 11. Birch tussock bog with overgrown lake.

The peatlands develop on flood plains in a much lesser degree, due to mismatch conditions of their growth (poor ability to drain of the territory in long-term perspectives) and annual flooding and flushing the surface on the flood waters. However, with a small incision of the valley, a small slope and flow rates, high altitude and duration of inundation and infiltration of flood water into the grounds of floodplain, the rivers from drainage systems can become sources of long-term accumulation of water which is favorable for the development of peatlands (Malik, 1977).

Figure 12. Cotton grass-sedge peat bog.

Most often the peatlands capture the floodplain, going down from the party of the terraces, where the conditions for their growth are most favorable. At first the terrace ledge is leveled and then the rear and the central zones of the floodplain are covering with the peat. On the riparian zone of a floodplain, which is drained better than others, peatlands do not penetrate: they remain near the riverbed only at the erosion of the floodplain arrays, and dry quickly here. In some cases, only rear depressions and hollows cover with the peat, where the runoff of flood water for various reasons is difficult.

In contrast to the organo-mineral hummocks, peatlands in the process of their growth completely alter the topography of floodplain – they smooth out all the bumps and increase the floodplain surface (Figure 12). Flooding of their happens less, the rate of the flood flow falls and this prevents the peatlands from erosion. The surface of the peaty floodplain is flat, there are often found rounded lakes on it, connected by slow flowing marsh smoll rivers. The emergence of lakes - the third stage of bog forming - is characterized by cessation of peat formation in discrete foci and the formation in their place the so-called lakelets.

Thus, the structure of vegetation that depend on the whole complex of landscape components reacts very sensitively to changes of the factors and is "the only direct and reliable evaluator of environmental conditions" (Ramensky, 1938, p. 171). The movement of vegetation within the successional series reflects the processes of self-development of the floodplain landscape.

6. The Dynamic States of Floodplain Landscapes

Each floodplain natural complex exhibits significant seasonal and annual variations of its status. Diurnal and seasonal states are determined mainly by the periodic and cyclic factors (diurnal and seasonal variation of climate conditions, hydrological regime, etc.). The main indicator of the seasonal state of the complex is its vertical structure. Its capacity, complexity, and the presence other ecosystem demonstrate the development and attenuation of certain processes. These processes, in turn, determine the changes in the landscape over a specific period of time. Various seasonal states of the floodplain landscapes are characterized by the presence of specific components: flood water, silt, permafrost horizon, phytomass, etc.

In the study of floodplain landscapes it is necessary to consider the development of their structure. It is important to construct a time series of floodplain complexes. These time series will help to determine the location of the object of research and will provide the characteristic of the directions of self-development. The processes of self-development of various floodplain landscapes are not identical.

Dynamic states of facies within serial ranks depend on climatic and hydrological factors, the riverbed processes, natural successions of plant communities, long-term fluctuations of climate, as well as non-recurrent factors (recent tectonics and anthropogenic impacts). Of course, external dynamic factors exert significant impact on the structure development, and the recognition of the leading factor has great practical importance.

The prevailing influence of zonal (continental) factors, such as climate, leads to the formation of the indigenous (climax) landscapes. In case of the dominance of hydrodynamic factors such as erosive-accumulative activity of rivers, the height and duration of inundation of the floodplain and the deposition of alluvium, the short-lived serial facies are subjected to significant tarnsformations. Indigenous to the floodplain landscapes are river terraces. For example, to the north of the mouth of the Tom river the subclimax communities are birch-coniferous forests (Khromykh, 1975). Indigenous facies exhibit certain equilibrium between the constituting components.

The closest to the zonal landscapes in the floodplain are communities occupying the highest rarely inundated floodplain surfaces. These are high ridges of central floodplain that lack soil moisture after the recession of the flood. Areas, located at lower altitude levels, are also "zonal" (Khromykh, 1975).

Facies of the lowest floodplain areas and of the riparian floodplain have short life time without stable structure. Medium-altitude parts of the central floodplain are occupied by semi-serial facies. In serial facies the natural successions cannot be completed because of the dynamism of their location. With the reduction of the duration and intensity of flooding towards indigenous facies, the elementary geosystems tend to stabilize. This finally led to completion of the successions by climax complexes. The near-terrace floodplain is dominated by indigenous facies subjected to significant influence of moisture (waterlogging). Facies of the high central floodplain and near-terrace floodplain are semi-indigenous.

Thus, the study of the dynamics of the floodplain landscapes requires a conjugate analysis of the states at various levels. Formation of a floodplain is fully integrated into formation of landscapes. The overall result of these conditions depends on existing modes (water-flooding, sedimentation, thermal, etc.), imposed on the existing lithological-morphological basis. All the factors shaping the landscapes in a floodplain are extremely variable in time and space and act in their complexity, supplementing and mutually causing each other. In this regard, the hydrological regime of the floodplain is the most important factor, followed by local climate, topography, alluvial deposits, soil, vegetation, and zoocenoses.

The formation of the relief is controlled by erosive-accumulative activity of rivers, the deposition of alluvium, and accumulation of peat. The wind, the ice (during floods), and the waves are of less significance. In the near-terrace floodplain, the changes in the relief are controlled by landslides, cones, gullies and ravines systems. The intensity of erosive-accumulative processes is controlled by geological structures with their inherent tectonic regimes.

The originality of the nature of the floodplain of the Ob is its meridional orientation. It presents the combination of intrazonal properties with zonal influence of landscapes. The impact of zonal climatic factors on the floodplain is reduced from the upper reaches to the lower reaches. In the lower reaches the effect of zonal factors is expected to be minimal. The influence of the mainland factors on floodplain landscapes increases from low levels to high levels, and with the increase of the distance from the river. Mainland factors may include climate and lithology.

Overall, floodplain landscapes are strongly controlled by water and vegetation regime, which makes them from other complexes. They are way more dynamic in comparison with landscapes of lithogenic basis. On the other hand, the presence of biotic component in the floodplain geosystem determines its stability and capacity to adapt to abiotic conditions. The best approach for studying the dynamics of the floodplain landscapes is the comparison of time-series aerospace images of the same area, taken under comparable resolution and environmental conditions.

Conclusion

1. The dynamics of floodplain landscapes is determined by the combination of external factors and internal processes. The impact of zonal and provincial factors on the landscapes of the investigated territory increases with the increase in the elevation level of the floodplain and the distance from the riverbed. A long-term impact of river hydrodynamics on the floodplain is maximal within the riparian zone. The sequence of landscapes evolution involves the processes of formation of forest, meadow and bog.
2. The dynamics of the floodplain landscapes includes irreversible and reversible changes. The irreversible processes include gradual accumulation of elements thus building new landscape elements. They are caused by erosive and accumulative activity of the river, tectonic features of the territory, global climate change, and also self-development of landscapes within serial-dynamic sequence. The reversibility of dynamic changes is linked to the rhythm of external abiotic factors, such as gelio-and geophysical rhythms, cyclic changes of climate and hydrological characteristics. Dynamic changes in the landscapes of the floodplain are best indicated by the vegetation cover.
3. Progressive change in floodplain landscapes occurs with the increase of biological productivity, the complication of their structure and increase of their stability. The landscapes with rich biotic component exhibit the steadiest and easiest adaption for the change of external factors in the floodplain. They also have higher plasticity which is useful for stabilizing them early in the geosystem, with more difficult structural organization and thus they can persist longer periods. These characteristics correspond to the indigenous landscapes, confined to the elevated areas of the central and near-terrace floodplains, and quasi-indigenous (in waterlogged lowlands) floodplain landscapes. The rest of the geosystems are serial, and the major role there is played by reformative dynamics.

The dynamics of floodplain ecosystems have to be taken into account when conducting economic activity in the floodplain. These ecosystems are extremely vulnerable due to their youth and overall instability of the structure. The anthropogenic influence leads to modifications of natural complexes and appearance of secondary natural processes in the landscape.

References

Annenskaya, G. N., 1982. Factors of formation of the morphological structure of the floodplain landscape. Landscape: Theory and Practice. *Questions geography*, 121, 44-55.

Afanasyeva, T. V., Remezova, G. D., 1968. Soils and vegetation of the floodplain of the south taiga part of the Ob River. *Bulletin of Moscow University, VI, Biology, Soil Science*, 2.

Baryshnikov, M. K., 1933. Meadows of Ob and Irtysh of Tobolsk North. 95 pp.

Bazanov, V. A., Lgotin, V. A., 1988. Assessment of intensity of formation of bogs of Middle Ob area on materials of remote researches. *The economic evaluation of the landscape of the Tomsk region*. 150-153.

Bock, E. N., 1972. Osier-beds of a floodplain of Ob. *Biological Resources of the floodplain of the Ob. Proceedings Biol. Institute of the SB RAS.* 19, 325-333.

Bolotnov, V. P., 1988. A method of reservation of a runoff of spring floods on the basis of the analysis of stability of biological components of inundated ecosystems. *The economic evaluation of the landscape of the Tomsk region,* p. 36-39.

Buddo, I. S., 1953. Meadows of Angara area. *Author. diss.*

Chernov, A. V., 1983. Geomorphology of floodplains of lowland rivers. 198 pp.

Dokuchaev, V. V., 1899. To the doctrine about nature zones.

Govorukhin, V. S., 1955. About an origin of primary floodplain meadows of a taiga zone. *Math. Komi branch of UCO*. 3, 27-31.

Ilyin, R. S. 1930. Nature of Narym edge. *Materials on the study of Siberia*. 2.

Khromykh, V. S., 1975. Structure and quality standard of landscapes of a floodplain of Middle Ob (in borders of the Tomsk region). *Diss.* 230 pp.

Kondratiev, N. E., Popov, I. V., 1973. Typification of riverbed and floodplain processes. *Recommendations about the accounting of riverbed process at design of the high voltage line*. 5-22.

Kononov, K. E., 1971. Floodplain meadows of Middle Lena. 127 pp.

Lvov, Y. A., 1963. To the characteristic of vegetation of a floodplain of the Ob River. *Nature of the floodplain of the Ob River and its economic development*. 152, 258-267.

Maccaveev, N. I., 1955. The riverbed and erosion in its basin. 347 pp.

Malik, L. K., 1977. Role of a modern river network in the progressing bogging of the territory. *Scientific background development bogs of Western Siberia*. 104-124.

Mamai, I. I., 1992. The dynamics of the landscape (a technique of studying). 167

Nechayeva, E. G., 1974. Use of landscape-geochemical method for the determination of the dynamic state of geosystems. *VII Meeting on landscape science*. 48-49.

Nomokonov, L. I., 1959. Floodplain meadows of Yenisei.

Perelman, A. I., 1977. Bioinert systems of Earth. 160 pp.

Petrov, G. N., 1976. Hydrological and geographical study of water resources of the Middle Volga area. *Landscape and water. Questions geography.* 102, 13-29.
Petrov, I. B., 1979. Ob-Irtysh floodplain (typing and qualitative assessment of lands). 136 pp.
Popov, I. V., 1968 Types of floodplains and their communication with types of riverbed process. Proceedings of the State. Hydrological. Inst. 155, 39-55.
Popov, I. V., 1969. Deformations of river courses and hydrotechnical construction. 363 pp.
Rabotnov, T. A., 1983. *Phytocenology.* 292 pp.
Ramensky, L. G., 1938. Introduction to the complex soil-geobotanical study of lands. 620 pp.
Rodin, L. E., Bazilevich, N. I., 1965. Dynamics of organic matter and biological cycle in the main types of vegetation.
Rodnyanskaya, E. E., 1960. Typology of floodplain landscapes on the example of the Ob River. *Izvestia VGO.* 92, 1.
Rodnyanskaya, E. E., 1963. Typology of floodplain landscapes on the example of the Ob River within a taiga zone. *Author. diss.* 13.
Rusakov, V. N., 1991. Conditions and methods of complex melioration of floodplains of the rivers of Western Siberia at the regulated river drain. 108 pp.
Schrag, V. I., 1949. To a question of protective forest strips on coast of the rivers. *Forestry management*, 7.
Shennikov, A. P., 1935. The principles of botanical classification of meadows. *Soviet botany*, 5.
Shepeleva, L. F., 1991. Natural haymakings and pastures, their state and ways of improvement. *Natural resources of Tomsk region.* 56-67.
Shepeleva, L. F., 1998. Organization of meadow communities of a floodplain of Middle Ob. *Author. diss.* 34 pp.
Solntsev, N. A., Mamai, I. I., Marcus, J. A., 1976. Landscape researches of river basins for the hydrological purposes. *Landscape and water. Questions geography.* 102, 75-92.
Sukachev, V. N., 1934. Dendrology with fundamentals of forest geobotany.
Sukachev, V. N., 1950. On some basic issues of the phytocenology. *Problems of geobotany.* 1, 449-464.
Surkov, V. V., 1997. Modern evolution of natural territorial complexes of a floodplain of Ob under the influence of natural and anthropogenous factors. *Author. diss.* 24 pp.
Vasiliev, S. V., 1988. Remote assessment of speed of erosion of floodplain coasts of Central Ob. *The economic evaluation of the landscape of the Tomsk region.* 42-44.
Voeikov, A. I., 1884. Climates of the globe.
Voropai, L. I., Kunitsa, N. A., Levitsky, V. I., 1974. Studying of floodplain systems for detection of regularities of development of landscapes in the Holocene. *VII Meeting on landscape science.* 114-115.
Yakovleva, L. V., 1978. Influence of erosive and accumulative processes on development of floodplain soils (on the example of Middle Ob). *Vestn. Mosk. Univ. Ser. geogr.* 5, 74-78.
Yelenevsky, R. A., 1936. Questions of studying and development of floodplains. 100 pp.

About the Editor

Oleg S. Pokrovsky, PhD
Research Director at the CNRS
CNRS, Toulouse, France;
Institute of Ecological Problem of the North, RAS, Arkhangelsk;
and BIO-GEO-CLIM Laboratory, Tomsk State University, Russia
Email: oleg@get.obs-mip.fr

O. S. Pokrovsky, graduated from Geochemistry department, Moscow State University, PhD in geochemistry (1994), entered CNRS (FRANCE) in 1999 and works over past decade on biogeochemistry of arctic and subarctic rivers and lakes. He is now a research director at the CNRS (FRANCE). He possesses both experimental physico-chemical, microbiological and geochemical expertise. Over past decade, O. S. Pokrovsky directed and co directed 12 PhD students, 5 Post-doctoral research associates. Since 2000, O. S. Pokrovsky coordinated and leaded more than 20 various research grants, international cooperation programs and consortia, large-scale national (French) grants and served as important partner of a number of European (FP7) projects. Since 2013, O. S. Pokrovsky, a recipient of a prestigious Mega-grant of Russian Ministry of Science and Education (3 M $ for 3 years), directs BIO-GEO-CLIM Laboratory on Environment, Climate and Permafrost at Tomsk State University.

O. S. Pokrovsky has strong academic records with ~160 papers peer reviewed papers and the same number of conference abstracts published since 1992 on physical chemistry, experimental geochemistry, and aquatic biogeochemistry; his HI factor is equal to 37 and the total citation number is close to 4000.

INDEX

#

A

B

C

D

E

F

M

N

O

P

R

S

T

U